Nonlinear Physical Science

Nonlinear Physical Science

Nonlinear Physical Science focuses on recent advances of fundamental theories and principles, analytical and symbolic approaches, as well as computational techniques in nonlinear physical science and nonlinear mathematics with engineering applications.

Topics of interest in *Nonlinear Physical Science* include but are not limited to:

- New findings and discoveries in nonlinear physics and mathematics
- Nonlinearity, complexity and mathematical structures in nonlinear physics
- Nonlinear phenomena and observations in nature and engineering
- Computational methods and theories in complex systems
- Lie group analysis, new theories and principles in mathematical modeling
- Stability, bifurcation, chaos and fractals in physical science and engineering
- Nonlinear chemical and biological physics
- Discontinuity, synchronization and natural complexity in the physical sciences

Series editors

International Advisory Board

More information about this series at http://www.springer.com/series/8389

Mohamad S. Alwan · Xinzhi Liu

Theory of Hybrid Systems: Deterministic and Stochastic

Higher Education Press

Springer

Mohamad S. Alwan
Department of Applied Mathematics
University of Waterloo
Waterloo, ON, Canada

Xinzhi Liu
Department of Applied Mathematics
University of Waterloo
Waterloo, ON, Canada

ISSN 1867-8440 ISSN 1867-8459 (electronic)
Nonlinear Physical Science
ISBN 978-981-10-8045-6 ISBN 978-981-10-8046-3 (eBook)
https://doi.org/10.1007/978-981-10-8046-3

Jointly published with Higher Education Press, Beijing, China

The print edition is not for sale in China Mainland. Customers from China Mainland please order the print book from: Higher Education Press Limited Company.
ISBN of the China Mainland edition: 978-7-04-050615-0

Library of Congress Control Number: 2018954043

This Springer imprint is published by the registered company Springer Nature Singapore Pte Ltd.
The registered company address is: 152 Beach Road, #21-01/04 Gateway East, Singapore 189721, Singapore

To Our Families

Preface

Hybrid systems have become increasingly popular during the recent decades in various fields of the scientific research and are expected to carry on the potential for further explorations. A hybrid system exhibits a combination or coexistence of continuous and discrete events and has behaviors determined by the interaction between the continuous and discrete components, and/or between them with other environmental factors. From practical perspective, it has been observed that if the interaction, within a single system, is strong, then the above hybridness has to be unified in one model. This unification has paved the path to the study of hybrid systems leading to fascinating outcomes for the following reasons: (i) The hybrid system paradigm has been recognized as a proper tool to represent a wide range of diversified applications in nature or in the human-made world. Among those are systems modeling population growth model, infectious disease models, medical drugs, chemical reaction processes, heating/cooling systems, several control systems, power systems, automated highway systems, air traffic control systems, neural networks, computer synchronization, secure communication networks, just to name a few. (ii) A large class of systems are intrinsically ruled by multimodal dynamics, such as those presented in many control systems, multibody mechanical systems, thermostats in heating/cooling systems, prey–predator systems with finite, different prey sources and epidemic disease models with periodic vaccinations or treatments. (iii) Many systems are asymptotically stabilized by multiple control laws monitored by a high-level supervisory agent, and others are stabilized or state estimated by discrete events. This is the case when the available information is only measured at discrete moments, rather than continuous time period, as in the case of vaccination or drugs administrated by way of injection. On the other hand, systems may undergo impulsive perturbing forces that must be taken into account in the modeling process. (iv) Nowadays, the technology has produced much hierarchically sophisticated machinery that cannot be analyzed as a whole system. Hybrid system representations can also be considered here to minimize the complexity of these systems. Namely, they provide sequential mathematical descriptions of the system that are often manageable for analysis. For these listed reasons, the heterogenous composition in the hybrid systems has become a modeling priority which, as a result, creates an important, fruitful research field applicable to many practical areas.

Mathematically, the typical systematic configuration of hybrid systems can be represented by: (i) a mix of differential equations representing the continuous evolution of a process and a set of difference equations representing the impulsive actions. (ii) Another type of hybrid systems consists of a finite sequence of dynamical subsystems (or modes) combined by a control-based discrete switching signal. The role of the later signal is to organize the switches among the system modes to achieve a coherent performance of the system. The first class of hybrid systems are often called *impulsive systems*, while the second ones are called *switched systems*. (iii) A third class of hybrid systems is referred to as *impulsive switched systems*. These systems arise when the impulsive actions occur as a result of mode switchings. Moreover, hybrid systems become even more complex if time delay and random noise are taken into consideration. The resulting systems are then called *stochastic hybrid system with time delay*.

This monograph aims to give a systematic account on recent developments about deterministic, stochastic hybrid systems with/without time delay. It includes many linear, nonlinear systems, large-scale systems, singularly perturbed systems, systems of differential equations with piecewise constant arguments (EPCA), and systems subject to input disturbance. It is intended to cover the most interesting topics, provide a systematic analysis of system theory and control, and enlighten researchers about further investigations into hybrid systems. The contents of this monograph are largely based on some recent research developments conducted by the authors. Its chapters shed the light on several fundamental, important system properties, such as stability, stabilization, input-to-state stability/stabilization, state estimation, reliable controllers, H_∞-control and variable structure control, also known as a sliding mode control (SMC). The analysis of these properties utilizes a variety of techniques including comparison principle, Lyapunov method, or Lyapunov–Razumikhin technique if time delay is present. Moreover, it has many illustrative examples with numerical simulations.

We are thankful to Dr. T. Sugati and Ms. H. Kiyak for proofreading the book. Our special thanks go to Ms. Liping Wang at Higher Education Press and Springer for her cooperation. The first author wishes to express his profound gratitude to the staff (academic and nonacademic) of the Department of Applied Mathematics for their support during his stay at the University of Waterloo. The second author was supported in part by the Natural Sciences and Engineering Research Council of Canada (NSERC) which is gratefully acknowledged.

Waterloo, Canada

Mohamad S. Alwan
Xinzhi Liu

Contents

Notations and Symbols

$\mathbb{N}$	The set of natural numbers $\{1, 2, 3, \cdots\}$
$\mathbb{R}$	The real number set
$\mathbb{R}_+$	Nonnegative real number set
$\mathbb{R}^n$	n-dimensional vector space
$\mathbb{R}^{m \times n}$	$m \times n$ real matrices
x^T	Transpose of a vector x
$x \cdot y$ or $x^T y$	The dot product of $x, y \subset \mathbb{R}^n$, i.e., $x \cdot y = \sum_{i=1}^n x_i y_i$
$\|x\|$	Norm of a vector x
$S(\rho)$	$\{x \in \mathbb{R}^n \mid \|x\| < \rho$, for some $\rho \geq 0\}$
$\|A\|$	Norm of a matrix A
$\max(\vee), (\min \wedge)$	Maximum, minimum
sup	Supremum, the least upper bound
inf	Infimum, the greatest lower bound
A^T	Transpose of a matrix A
A^{-1}	Inverse of a (nonsingular) matrix A
$\lambda(A)$	Eigenvalue of a matrix A
$\mathrm{Re}\,[\lambda(A)]$	The real part of an eigenvalue of a matrix A
$\lambda_{\max}(A)$	Maximum eigenvalue of a symmetric matrix A
$\lambda_{\min}(A)$	Minimum eigenvalue of a symmetric matrix A
$A > 0$	The real symmetric matrix A is positive definite
I	Identity (unit) matrix
$\mathrm{tr}A$	The trace of a matrix A
$A \backslash B$	The difference between two sets, i.e., $A \backslash B = \{x \mid x \in A$ and $x \notin B\}$
$\dot{x}(t)$	Time derivative of a time-varying vector x
D^+V	Right-hand side (Dini's) derivative of a function V
$\|\phi\|_r$	$\sup_{-r \leq s \leq 0} \|\phi(s)\|$
$\|x_t\|_r$	$\sup_{t-r \leq s \leq t} \|x(s)\|$
$x(t^+)$	$\lim_{t \to t^+} x(t)$
Δx	$x(t^+) - x(t)$

$x \equiv (=)0$ on $[a,b]$	$x(t) = 0$ for all $t \in [a,b]$
$o(\cdot)$	Order of magnitude
$f \circ g$	The composite function of f and g, i.e., $(f \circ g)(x) = f(g(x))$
$\mathscr{C}([a,b];\mathbb{R}^n)$	The space of continuous functions mapping $[a,b]$ into $\mathbb{R}^n$
$\mathscr{C}(\mathbb{R}_+ \times \mathbb{R}^n;\mathbb{R}^n)$	The space of $\mathbb{R}^n$-valued functions continuous on $\mathbb{R}_+ \times \mathbb{R}^n$
$\mathscr{PC}([a,b];\mathbb{R}^n)$	The space of piecewise continuous function mapping $[a,b]$ into $\mathbb{R}^n$
$\mathscr{C}^m(\mathscr{D};\mathbb{R}^n)$	The family of continuously m-times differentiable $\mathbb{R}^n$-valued functions defined on $\mathscr{D}$
$\mathscr{C}^{1,2}(\mathbb{R}_+ \times \mathscr{D};\mathbb{R}_+)$	The family of $\mathbb{R}_+$-valued functions $V(t,x)$ defined on $\mathbb{R}_+ \times \mathscr{D}$, which are continuously once differentiable in $t \in \mathbb{R}_+$ and twice differentiable in $x \in \mathscr{D}$
$\mathscr{C}^{1,1}(\mathbb{R}_+ \times \mathscr{D};\mathbb{R}_+)$	The family of $\mathbb{R}_+$-valued functions $V(t,x)$ defined on $\mathbb{R}_+ \times \mathscr{D}$, which are continuously once differentiable in $t \in \mathbb{R}_+$ and once differentiable in $x \in \mathscr{D}$
$\mathscr{L}_{ad}(\Omega; L^p[a,b])$	The family of $\mathbb{R}^n$-valued $\mathscr{F}_t$-adapted process $f(t)$ such that $\int_a^b \|f(t)\|^p dt < \infty$ (a.s.) for all $t \in [a,b]$
$\mathscr{M}^p([a,b];\mathbb{R}^n)$	The family of processes $f(t)$ in $\mathscr{L}_{ad}(\Omega; L^p[a,b])$ such that $\mathbb{E}\int_a^b \|f(t)\|^p dt < \infty$
$\mathscr{L}^p_{\mathscr{F}_0}([-r,0];\mathbb{R}^n)$	The family of $\mathscr{F}_0$-adapted $\mathscr{PC}([-r,0];\mathbb{R}^n)$-valued random variable ϕ such that $\mathbb{E}[\|\phi\|^p] < \infty$
$L^p_{\mathscr{F}_0}([-r,0];\mathbb{R}^n)$	The family of $\mathscr{F}_0$-adapted $\mathscr{C}([-r,0];\mathbb{R}^n)$-valued random variable ϕ such that $\mathbb{E}[\|\phi\|^p] < \infty$
1_A	Indicator function of a set A, i.e., $1_A(x) = 1$ if $x \in A$ or otherwise 0

List of Figures

Chapter 1
Motivating Examples

To introduce the readers to the notion of hybrid systems, which include switched, impulsive and impulsive-switched systems, we present in this chapter some real and human-made phenomena that are ideally modeled by such systems.

1.1 Switched Systems

By a *switched system*, we mean a dynamical system that consists of multidynamical subsystems (often called modes) and a monitoring device called a *switching signal* also known as *switching rule, switching logic* or *switching law*. The main role of this signal is to orchestrate the switching among system modes to accomplish a desired feature of the system.

1.1.1 Supervisory Switching Control

Many systems are, for instance, asymptotically stabilized or controlled by several feedback control signals (or controllers), rather than one signal, and each of these controllers is set to accomplish a certain desire. The logic-based supervisory control, here, organizes the switching among them to achieve the *overall* system stability or controllability. Figure 1.1 illustrates the supervisory controller [1].

M. S. Alwan and X. Liu, *Theory of Hybrid Systems: Deterministic and Stochastic*, Nonlinear Physical Science, https://doi.org/10.1007/978-981-10-8046-3_1

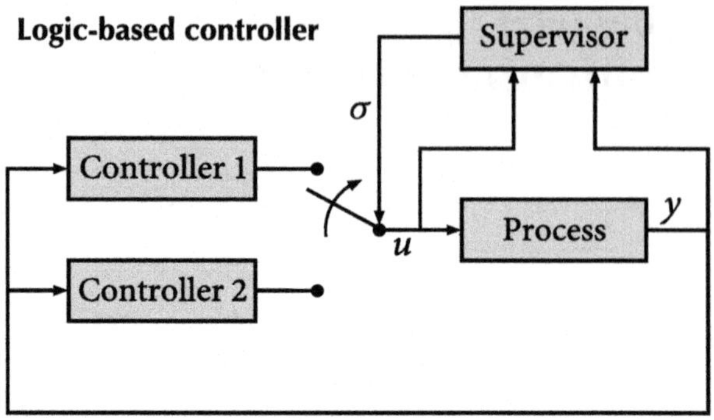

Fig. 1.1 Supervisor controller

1.1.2 Switched Server System

Another class of the discretely controlled continuous system is a *switched flow system*. Consider that the system consists of N *buffers* and one *server*, where the content of the buffer is referred to as *work*. One may think of the buffer as a tank and the work as a fluid. In this system, the server delivers work from any selected buffer at unit rate and the work is removed from buffer i ($i = 1, 2, \ldots, N$) at a fixed rate of $r_i > 0$. If the system is assumed to be closed, then $\sum_{i=1}^{N} r_i = 1$. The switching law for the server can be designed as follows: when the server removes work from a selected buffer for a time period, it instantaneously switches to another buffer that is determined by the switching rule, $\sigma : \mathbb{R}^N \to \{1, 2, \ldots, N\}$. Then, the switching process of the server repeats itself forming a cycle [2].

1.1.3 Singular System with Markov Switching

This type of systems includes an *RLC* electrical circuit in which the position of the switch follows a continuous-time Markov process, $\{r(t),\ t \geq 0\}$, which takes values in the index set having finite states, $\mathscr{S} = \{1, 2, \ldots, N\}$ with following stationary transition probabilities:

$$P[r(t+h) = j | r(t) = i] = \begin{cases} \lambda_{ij} h + o(h), & i \neq j, \\ 1 + \lambda_{ii} h + o(h), & \text{otherwise}, \end{cases}$$

where $h > 0$, $\lim_{h \to 0} \frac{o(h)}{h}$ and $\lambda_{ij} \geq 0$ is the transition probability from mode i to the mode j at time t and $\lambda_{ii} = \sum_{j=1, j \neq i}^{N} \lambda_{ij}$.

For instance, the mathematical model of the circuit shown in Fig. 1.2 with $N = 3$ is given by the following stochastic switched system [3]:

$$E\dot{x}(t) = A(r(t))x(t) + Bu(t)$$

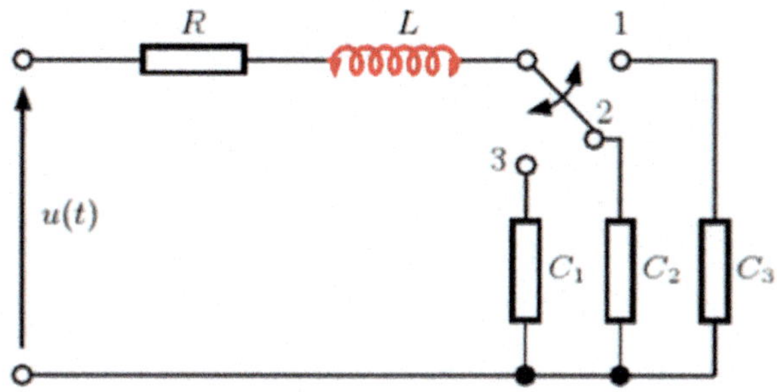

Fig. 1.2 Electrical circuit [3]

where

$$E = \begin{bmatrix} L & 0 & 0 & 0 \\ L & 1 & -1 & 0 \\ 0 & -1 & 1 & 0 \\ L & 0 & 0 & 0 \end{bmatrix}, \quad A(r(t)) = \begin{bmatrix} 0 & 1 & 1 & 0 \\ -a(r(t)) & 1 & 1 & -1 \\ a(r(t)) - R & 0 & 0 & 1 \\ -R & 1 & 0 & 0 \end{bmatrix}, \quad B = \begin{bmatrix} -1 \\ -1 \\ 0 \\ 0 \end{bmatrix}$$

with $a(r(t)) = \begin{cases} \frac{1}{C_1}, & \text{if } r(t) = 1, \\ \frac{1}{C_2}, & \text{if } r(t) = 2, \\ \frac{1}{C_3}, & \text{if } r(t) = 3. \end{cases}$

1.2 Impulsive Systems

Another special class of hybrid systems is *impulsive systems* or *systems of differential (or discrete) equations with impulses* which is a combination of differential (of discrete) equations representing the continuous evolution of the system and a set of difference equation representing jumps or impulsive actions.

1.2.1 SEIRS Epidemic Model with Impulse Vaccinations

An important class of dynamical impulsive systems is the *SEIRS disease models with impulse vaccinations with saturation incidence.* Denote by $S(t)$, $E(t)$, $I(t)$ and $R(t)$ the susceptible, exposed (infected but not infectious), infectious and recovered population at time t, respectively, such that the total population at t is $N(t) = S(t) + E(t) + I(t) + R(t)$. A delayed SEIRS epidemic model with saturation incidence and the effects of pulse vaccination may be given by the following differential equations with time delay and impulsive effects [4]

$$
\begin{aligned}
\dot{S}(t) &= A - bS(t) - \frac{\beta S(t)I(t)}{1+mS(t)} + \eta e^{-b\tau} R(t-\tau),\\
\dot{E}(t) &= \frac{\beta S(t)I(t)}{1+mS(t)} - \frac{\beta e^{-b\omega} S(t-\omega)I(t-\omega)}{1+mS(t-\omega)} - bE(t),\\
\dot{I}(t) &= \frac{\beta e^{-b\omega} S(t-\omega)I(t-\omega)}{1+mS(t-\omega)} - (b+\gamma+\alpha)I(t),\\
\dot{R}(t) &= \gamma(t)I(t) - bR(t) - \eta e^{-b\tau} R(t-\tau), \qquad (k-1)\tau < t \le k\tau, \ \text{for } k \in \mathbb{N},\\
S(t^+) &= (1-\theta)S(t),\\
E(t^+) &= E(t),\\
I(t^+) &= I(t),\\
R(t^+) &= R(t) + \theta S(t), \qquad t = k\tau, \ \text{for } k \in \mathbb{N},
\end{aligned}
$$

where A is the constant recruitment rate of the susceptible population, b is the natural death rate of the population, β is the transmission coefficient, α is extra disease-related death rate of the infectious hosts, γ is the recovery rate of infectious population, η is the rate of losing immunity, ω is the latent period of the disease, τ is the immune period of recovered population, and θ $(0 < \theta < 1)$ is the fraction of susceptible population to whom the vaccination inoculated at times $t = k\tau$.

1.2.2 *Insulin Treatment*

In pharmacokinetics, the process of maintaining the drug level in a body can be adequately modeled by impulsive differential equations especially if the time period in which body responds to the medication is very small that can be reasonably approximated as a time moment. For instance, diabetics aim to maintain the daily sugar level in the body at a certain range, say $[a, b]$. Due to having food, the sugar level continuously increases in the blood approaching the upper bound of the range. As a result, the insulin should be injected so that the sugar level instantly jumps to a lower bound near a. In this example, injection times represent the *impulsive moments*, the insulin injections represent the *impulsive effects or actions*, and the continuous increase in the blood sugar represents the *continuous evolution*.

References

1. Liberzon D (2003) Switching in systems and control. Brikhauser, Boston
2. Li Z, Soh Y, Wen C (eds) (2005) Switched and impulsive systems: analysis, design and applications. Lecture notes in control and information sciences. Springer, Berlin
3. Boukas EK (2006) Stochastic switching systems: analysis and design. Birkhauser
4. Zhang T, Teng Z (2008) An impulsive delayed SEIRS epidemic model with saturation incidence. J Biol Dyn 2(1):64–84

Chapter 2
Mathematical Background

This chapter serves as an introduction to the rest of the book. Particularly, we introduce two types of hybrid systems, impulsive systems and switched systems, with/without time delay and with/without random noise that is represented by Wiener process. Definitions of solutions of these systems and different stability notions in the sense of Lyapunov are given. We also address some comparison principles for these systems.

2.1 Basic Definitions

Denote by $\mathbb{R}_+$ the set of all nonnegative real numbers, $\mathbb{R}^n$ the n-finite-dimensional Euclidean space with the norm $\|\cdot\|$ (i.e., if $x \in \mathbb{R}^n$ then $\|x\| = \sqrt{\sum_{i=1}^{n} x_i^2}$) and $\mathbb{R}^{m\times n}$ the set of all $m \times n$ real matrices. If $A \in \mathbb{R}^{m\times n}$, then we define the induced norm of A by $\|A\| = \sqrt{\mathrm{tr}(A^T A)}$.

Consider the following initial-value problem (IVP)

$$\begin{cases} \dot{x} = f(t, x), \\ x(t_0) = x_0, \end{cases} \tag{2.1}$$

where $x \in \mathbb{R}^n$ is the system state with x_0 being the initial state, $t \geq t_0$ represents the system evolution time with the initial time $t_0 \in \mathbb{R}_+$ and the vector field $f : \mathbb{R}_+ \times \mathscr{D} \to \mathbb{R}^n$ with $\mathscr{D} \subset \mathbb{R}^n$ being the open domain containing the origin $x = 0$. To guarantee that the IVP has a solution $x(t)$ in some interval containing t_0, f is assumed to be continuous in its domain of definition. The solution is unique if f is locally Lipschitz in x; that is, $\forall \bar{x} \in \mathscr{D}$ there exists a ball $\mathscr{B}$ of $\bar{x}$ such that for all $x, y \in \mathscr{B}$ and $t \in \mathbb{R}_+$, there exists an $L \geq 0$ such that $\|f(t,x) - f(t,y)\| \leq L\|x - y\|$. We also assume, without loss of generality, that $f(t, 0) = 0$ for all $t \geq t_0$ so that $x \equiv 0$ is

M. S. Alwan and X. Liu, *Theory of Hybrid Systems: Deterministic and Stochastic*,
Nonlinear Physical Science, https://doi.org/10.1007/978-981-10-8046-3_2

an equilibrium or *trivial solution* of system (2.1). Note that any nonzero equilibrium point can be shifted to the origin by a change of variable. We are now ready to introduce the concepts of (Lyapunov) stability.

Definition 2.1 The trivial solution, $x \equiv 0$, of system (2.1) is said to be

(i) *stable* (in the sense of Lyapunov) if, for any $\varepsilon > 0$ and $t_0 \in \mathbb{R}_+$, there is $\delta = \delta(\varepsilon, t_0) > 0$ such that

$$\|x_0\| < \delta \quad \text{implies} \quad \|x(t)\| < \varepsilon, \quad \forall t \geq t_0, \tag{2.2}$$

where $x(t) = x(t; t_0, x_0)$ is any solution of system (2.1);

(ii) *uniformly stable* if it is stable and δ is independent of t_0;

(iii) *asymptotically stable* if (i) holds and there is a positive constant $c = c(t_0)$ such that, for all $\|x_0\| < c$, $\lim_{t\to\infty} x(t) = 0$;

(iv) *uniformly asymptotically stable* if (ii) holds and there is a positive constant c, independent of t_0, such that, for any $\eta > 0$, there is $T = T(\eta) > 0$ such that, for all $\|x_0\| < c$,

$$\|x(t)\| < \eta, \qquad \forall t \geq t_0 + T(\eta);$$

(v) *exponentially stable* if there are positive constants c, k and λ such that

$$\|x(t)\| \leq k\|x_0\|e^{-\lambda(t-t_0)}, \quad \forall \|x_0\| < c;$$

(vi) *unstable* if (i) fails to hold.

Furthermore, the above stability properties are satisfied *globally* if (i)–(v) hold for any $x_0 \in \mathbb{R}^n$, i.e., $S(\rho)$ (or $\mathscr{D}$) is taken to be the entire space $\mathbb{R}^n$.

Having defined the stability concepts, throughout this book we use the *method of Lyapunov* to determine these qualitative properties. The Lyapunov stability technique requires defining a special class of functions, also known as energy-like functions, which enjoy some positive definiteness features.

Definition 2.2 Let $\mathscr{D} \subset \mathbb{R}^n$ be an open set containing the point $x = 0$. A function $V : \mathscr{D} \to \mathbb{R}$ is said to be *positive semi-definite* if (i) $V(t, 0) = 0$ and (ii) $V(t, x) \geq 0$, for all $t \geq t_0$ and $x \in \mathscr{D} \setminus \{0\}$. It is said to be *positive-definite* if the inequality in (ii) is replaced by $V(t, x) > 0$. Moreover, it is said to be *radially unbounded* (or *proper)* if it is positive definite and, for each fixed t, $\lim_{\|x\|\to\infty} V(t, x) = \infty$.

In the Lyapunov stability theorems, the focus is on the time derivative of V along the trajectories of the dynamical system under consideration. So that, we need the following definition of upper right-hand derivative, which is also known as a *Dini derivative*, of the function V.

Definition 2.3 Let $\mathscr{D}$ be an open subset of $\mathbb{R}^n$. If $V : \mathbb{R}_+ \times \mathscr{D} \to \mathbb{R}_+$, then the *upper right-hand (Dini) derivative* of V with respect to system (2.1) is defined by

$$D^+V(t,x) = \lim_{h\to 0^+} \sup \frac{1}{h}[V\big(t+h, x+hf(t,x)\big) - V(t,x)], \quad \forall (t,x) \in \mathbb{R}_+ \times \mathscr{D},$$

where the limit $h \to 0^+$ means h approaches 0 from right. If, moreover, V has continuous partial derivatives with respect to t and x (i.e., $V \in \mathscr{C}^{1,1}(\mathbb{R}_+ \times \mathscr{D}; \mathbb{R}_+)$), then the Dini derivative becomes the ordinary time derivative

$$D^+V(t,x) = \dot{V}(t,x) = \frac{\partial V(t,x)}{\partial t} + \nabla_x V(t,x) \cdot f(t,x),$$

where $\nabla_x V$ is the gradient vector of V, i.e.,

$$\nabla V_x(t,x) = \left(\frac{\partial V(t,x)}{\partial x_1} \; \frac{\partial V(t,x)}{\partial x_2} \cdots \frac{\partial V(t,x)}{\partial x_n}\right)^T$$

with $x = (x_1\ x_2 \cdots x_n)^T$ and "$\cdot$" refers to the dot product of two vectors.

Analogously, one can define other types of Dini derivatives, such as D_+, D^- and D_-, where for instance

$$D_-V(t,x) = \lim_{h\to 0^-} \inf \frac{1}{h}[V\big(t+h, x+hf(t,x)\big) - V(t,x)], \quad \forall (t,x) \in \mathbb{R}_+ \times \mathscr{D},$$

is called *lower left-hand (Dini) derivative*, where the limit $h \to 0^-$ means h approaches 0 from left. Likewise, if $V \in \mathscr{C}^{1,1}(\mathbb{R}_+ \times \mathscr{D}; \mathbb{R}_+)$, then the Dini derivatives become the ordinary time derivative.

Toward stating the sufficient conditions that guarantee the stability (or stability-like) properties, a special class of functions, known as *comparison functions*, are needed.

Definition 2.4 A function $\alpha \in \mathscr{C}([0,\rho); \mathbb{R}_+)$ (for $\rho > 0$) is said to belong to class $\mathscr{K}$ (i.e., $\alpha \in \mathscr{K}$) if it is strictly increasing and $\alpha(0) = 0$. If, in addition, $\rho = \infty$ and $\alpha(r) \to \infty$ as $r \to \infty$, then α is said to belong to class $\mathscr{K}_\infty$.

Definition 2.5 A function $\beta \in \mathscr{C}([0,\rho) \times \mathbb{R}_+; \mathbb{R}_+)$ is said to belong to class $\mathscr{KL}$ (i.e., $\beta \in \mathscr{KL}$) if, for each fixed s, $\beta(\cdot, s) \in \mathscr{K}$, and, for each fixed r, $\beta(r, \cdot)$ is decreasing and $\beta(r,s) \to 0$ as $s \to \infty$.

The following function classes will be also used in this book.

Definition 2.6 We define the following classes of functions:

$K_1 = \{g \in \mathscr{C}(\mathbb{R}_+; \mathbb{R}_+) \mid g(0) = 0 \text{ and } g(s) > 0 \text{ for } s > 0\};$
$K_2 = \{g \in \mathscr{C}(\mathbb{R}_+; \mathbb{R}_+) \mid g(0) = 0, g(s) > 0 \text{ for } s > 0, \text{ and } \lim\inf_{s\to\infty} g(s) > 0\};$
$K_3 = \{g \in \mathscr{C}(\mathbb{R}_+; \mathbb{R}_+) \mid g(0) = 0, g(s) > 0 \text{ for } s > 0, \text{ and } g \text{ is nondecreasing in } s\};$
$K_4 = \{g \in \mathscr{C}(\mathbb{R}_+; \mathbb{R}_+) \mid g(0) = 0, g(s) > 0 \text{ for } s > 0, \text{ and } \lim_{s\to\infty} g(s) = \infty\}.$

Definition 2.7 A function $\varphi : \mathbb{R} \to \mathbb{R}$ is said to be *convex* if the following holds

$$\varphi(\lambda x + (1-\lambda)y) \le \lambda\varphi(x) + (1-\lambda)\varphi(y), \qquad \lambda \in (0,1).$$

It is said to be *concave* if $\le$ is replaced by $\ge$.

Definition 2.8 A function $a \in \mathscr{C}(\mathbb{R}_+ \times [0,\rho); \mathbb{R}_+)$ is said to be in class $\mathscr{K}_c$ (i.e., $a \in \mathscr{K}_c$) if $a(t,0) = 0$, and $a(t,u)$ is concave and strictly increasing in u for each $t \in \mathbb{R}_+$.

Definition 2.9 A function $g \in \mathscr{C}(\mathbb{R}_+; \mathbb{R}_+)$ is said to be in class $\mathscr{K}_3$ (i.e., $g \in \mathscr{K}_3$) if $g(0) = 0$, and g is concave and nondecreasing.

Definition 2.10 A function $b \in \mathscr{C}([0,\rho); \mathbb{R}_+)$ is said to be in class $\mathscr{K}_v$ (i.e., $b \in \mathscr{K}_v$) if $b(0) = 0$ and, b is convex and strictly increasing.

Theorem 2.1 *Let w_1 and w_2 be positive-definite functions on the domain $\mathscr{D}$ which contains the point $x = 0$. Assume that $V \in \mathscr{C}^{1,1}(\mathbb{R}_+ \times \mathscr{D}; \mathbb{R}_+)$ such that, for all $(t,x) \in \mathbb{R}_+ \times \mathscr{D}$,*

$$w_1(x) \le V(t,x) \le w_2(x), \tag{2.3}$$

$$\frac{\partial V(t,x)}{\partial t} + \nabla V_x(t,x) \cdot f(t,x) \le 0. \tag{2.4}$$

Then, the trivial solution of system (2.1) is uniformly stable. Moreover, if the inequality in (2.4) is strengthened to

$$\frac{\partial V(t,x)}{\partial t} + \nabla V_x(t,x) \cdot f(t,x) \le -w_3(x), \qquad \forall (t,x) \in \mathbb{R}_+ \times \mathscr{D}, \tag{2.5}$$

where the function w_3 is a continuous and positive-definite on $\mathscr{D}$, then $x \equiv 0$ is uniformly asymptotically stable. If, furthermore, there exist positive constants r and c such that $B_r = \{x \in \mathscr{D} \mid \|x\| \le r\}$ and $\min_{\|x\|=r} w_1(x) > c$, then for all x starts in B_r such that $w_2(x) \le c$, we have

$$\|x(t)\| \le \beta(\|x(t_0)\|, t - t_0), \qquad \forall t \ge t_0, \tag{2.6}$$

where $\beta \in \mathscr{K}\mathscr{L}$. If $\mathscr{D} = \mathbb{R}^n$ and w_1 is radially unbounded, then $x \equiv 0$ is globally uniformly asymptotically stable. Particularly, if $\beta(r,s) = re^s \to 0$ as $s \to \infty$, then the asymptotic stability result reduces to the exponential stability.

In reality, systems are often subject to some types of disturbances (or inputs disturbances). The interest will be, then, to investigate how the systems respond to these disturbances. This question motivates the notion of (asymptotic) input-to-state stability (ISS), proposed by Sontag [1, 2]. The importance of ISS is manyfold. It bridges the gap between the input/output stability concept in which a system is being viewed as a black box and the Lyapunov stability (of the equilibrium point); that is, it connects the system equilibrium state (but not the output) to the input. Also, it has many equivalencies or implications to other stability-like concepts, such as integral ISS, global asymptotic stability (for zero input) and finite gain with respect to supremum norms and finite L^2. ISS has found applications in different areas in linear and nonlinear system and control theory, such as coprime factorization, cascade or feedforward systems, small-gain theorems and singularly perturbed systems.

Consider the following nonlinear system with the input

$$\begin{cases} \dot{x} = f(t, x, u), & t \geq t_0, \\ x(t_0) = x_0, \end{cases} \tag{2.7}$$

where $f : \mathbb{R}_+ \times \mathbb{R}^n \times \mathbb{R}^m \to \mathbb{R}^n$ with $t_0 \in \mathbb{R}_+$ and the input $u \in \mathscr{PC}(\mathbb{R}_+; \mathbb{R}^m)$ (i.e., u is a piecewise continuous function) with bounded energy (i.e., $\sup_{t \geq t_0} \|u(t)\| < \infty$). This system can be considered as a perturbation of the unforced system, i.e., $u(t) \equiv 0$,

$$\dot{x} = f(t, x, 0), \tag{2.8}$$

with the same initial state. Assuming that the trivial solution of (2.8) is globally uniformly asymptotically stable and u is bounded, then the state of the corresponding perturbed nonlinear system remains bounded if further sufficient conditions are imposed. In the following, we first define the ISS concept, then state these sufficient conditions.

Definition 2.11 System (2.7) is said to be *input-to-state stable* (ISS) if there exist functions $\beta \in \mathscr{KL}$ and $\gamma \in \mathscr{K}$ such that, for any initial state x_0 and bounded input u, the solution $x(t)$ exists and satisfies

$$\|x(t)\| \leq \beta(\|x_0\|, t - t_0) + \gamma\left(\sup_{t_0 \leq s \leq t} \|u(s)\|\right), \qquad \forall t \geq t_0. \tag{2.9}$$

In fact, this inequality can be written as follows

$$\|x(t)\| \leq \beta(\|x_0\|, t - t_0) + \gamma\left(\sup_{t_0 \leq s \leq t} \|u(s)\|\right), \quad \forall\, t_0 \leq t \leq t_0 + T,$$

$$\|x(t)\| \leq \gamma\left(\sup_{t_0 \leq s \leq t} \|u(s)\|\right), \quad \forall\, t \geq t_0 + T,$$

where $T \geq 0$. Evidently, for large enough T, the $\mathscr{KL}$ function β converges to zero asymptotically and, when $t \geq t_0 + T$, the solution will stay bounded by a class$-\mathscr{K}$

function γ, meaning that the solution of (2.7) has an ultimate bound γ, which is a ball with a radius depending on the input magnitude.

Clearly, from the inequality (2.9), if the input u is set to zero (i.e., $u(t) \equiv 0$), the ISS reduces to the globally uniformly asymptotic stability of the trivial solution of the unforced system (2.8).

The following Lyapunov-type theorem gives sufficient conditions that ensure ISS, which can also prove the asymptotic stability property of $x \equiv 0$ of the unforced system (2.8).

Theorem 2.2 *Assume that there exist class*$-\mathcal{K}_\infty$ *functions a and b, a class* $\mathcal{K}$ *function* ρ *and a positive-definite function c. Let* $V : \mathbb{R}_+ \times \mathbb{R}^n \to \mathbb{R}_+$ *such that the following conditions holds:*

$$
\begin{aligned}
&b(\|x\|) \le V(t,x) \le a(\|x\|), && \forall (t,x) \in \mathbb{R}_+ \times \mathbb{R}^n; \\
&\dot{V}(t,x,u) \le -c(x), \quad \textit{whenever} && \|x\| \ge \rho(\|u\|),
\end{aligned}
$$

for any $(t,x,u) \in \mathbb{R}_+ \times \mathbb{R}^n \times \mathbb{R}^m$. *Then, system (2.7) is ISS with* $\gamma(\cdot) = b^{-1}\big(a(\rho(\cdot))\big)$. *Particularly, if* $u \equiv 0$, *then the trivial solution of the unforced system (2.8) is globally uniformly asymptotically stable.*

2.2 Comparison Method

An important technique often used in the study of differential equations

$$
\dot{x} = f(t,x),
$$

where $x \in \mathbb{R}^n$ and $t \ge t_0$ with $t_0 \in \mathbb{R}_+$, is the so-called *comparison principle*. The advantages of this technique are manyfold including: (i) one can find an upper/lower estimation on $\|x(t)\|$ instead of finding the solution itself. This particularly important if the corresponding IVP does not admit a unique solution. We should also remind the readers that the upper estimation can be, in fact, found by the well-known Gronwall–Bellman inequality and by using Bihari's Lemma, as will be illustrated later. (ii) It connects the vector differential equation

$$
\dot{x} = f(t,x)
$$

to an auxiliary scalar differential equation

$$
\dot{u} = g(t,u)
$$

through the scalar differential inequality

$$\dot{V}(t,x) \leq g(t, V(t,x))$$

for some a scalar-valued function V; this process will be seen throughout many chapters of the book. (iii) The comparison principle can also be used if V is not differentiable, but has a Dini derivative. That is, for instance if D^+V exists then V satisfies the differential inequality

$$D^+V(t,x) \leq g(t, V(t,x)).$$

The following theorems summarize the comparison principle.

Theorem 2.3 (Comparison theorem) *Consider the vector IVP*

$$\dot{x} = f(t,x), \tag{2.10a}$$

$$x(t_0) = x_0. \tag{2.10b}$$

Let $V : \mathbb{R}_+ \times \mathbb{R}^n \to \mathbb{R}_+$ be continuous on $\mathbb{R}_+ \times \mathbb{R}^n$ and locally Lipschitz in x. Assume that V satisfies

$$D^+V(t,x) \leq g(t, V(t,x)), \tag{2.11}$$

where $g : \mathbb{R}_+ \times \mathbb{R}_+ \to \mathbb{R}$ is continuous on $\mathbb{R}_+ \times \mathbb{R}_+$. Let $r(t) = r(t; t_0, u_0)$ be the maximal solution of the auxiliary scalar system

$$\dot{u} = g(t,u), \tag{2.12a}$$

$$u(t_0) = u_0 \geq 0, \quad t_0 \in \mathbb{R}_+. \tag{2.12b}$$

Then, $V(t_0, x_0) \leq u_0$ implies that

$$V(t, x(t)) \leq r(t), \qquad \textit{for all } t \geq t_0 \tag{2.13}$$

where $x(t) = x(t; t_0, x_0)$ is any solution of (2.10) defined on $[t_0, \infty)$.

The following theorem states the sufficient conditions regarding the implementation of the comparison principle in establishing the stability properties of system (2.10).

Theorem 2.4 (Stability theorem) *Suppose that there exist class$-\mathcal{K}$ functions a and b. Assume that $V \in \mathscr{C}(\mathbb{R}_+ \times S(\rho); \mathbb{R}_+)$, V is locally Lipschitz in x and the following conditions are satisfied:*

(i) $b(\|x\|) \leq V(t,x) \leq a(\|x\|), \quad \forall (t,x) \in \mathbb{R}_+ \times S(\rho)$; *and*
(ii) $D^+V(t,x) \leq g(t, V(t,x)), \quad \forall (t,x) \in \mathbb{R}_+ \times S(\rho)$,

where $g \in \mathscr{C}(\mathbb{R}_+ \times \mathbb{R}; \mathbb{R})$ and $g(t,0) = 0$ for all $t \in \mathbb{R}_+$. Then, the stability properties of the trivial solution, $u \equiv 0$, of (2.12) imply the corresponding stability properties of the trivial solution, $x \equiv 0$, of (2.10).

2.3 Delay Systems

In contrast to the ordinary differential equations, where the system initial state (or initial condition) is given at a certain initial time, in the *delay-type differential equations*, where the derivative of the unknown function, say $x(t)$, at a specific time, t, also depends on the values of the function, x, at previous times (i.e., a part of the history of x). So that, the initial data in this case are generally continuous functions defined on a time interval, but not only the initial time. To define the initial-value problem of a delay-type system, we need some definitions and notations.

Let $\mathscr{C}_r = \mathscr{C}([-r, 0]; \mathbb{R}^n)$ be the set of all continuous functions from $[-r, 0]$ to $\mathbb{R}^n$ where $r > 0$ represents a time delay. If $\phi \in \mathscr{C}_r$, the r-norm of this function is defined by $\|\phi\|_r = \sup_{-r\leq s\leq 0} \|\phi(s)\|$, where $\|\cdot\|$ is the Euclidean norm on $\mathbb{R}^n$.

Definition 2.12 Let $t^* \in \mathbb{R}$ and $a > 0$. If x is a function mapping $[t^* - r, t^* + a]$ into $\mathbb{R}^n$, then, for each $t \in [t^*, t^* + a]$, we define a new function x_t which maps $[-r, 0]$ into $\mathbb{R}^n$ by $x_t(s) = x(t + s)$, for all $s \in [-r, 0]$ (i.e., $x_t : [-r, 0] \to \mathbb{R}^n$) and its norm is defined by $\|x_t\|_r = \sup_{t-r\leq\theta\leq t} \|x(\theta)\|$.

Here, for each $t \in [t^* - r, t^*]$, $x_t(s)$ (or simply x_t) is the segment of the function x from $t^* - r$ to t^* that has been shifted to the interval $[-r, 0]$. A general nonlinear delay-type differential equation may have the form

$$\dot{x}(t) = f(t, x_t) \tag{2.14a}$$

and is called *functional differential equation*, where f is called a *functional operator* mapping $\mathbb{R}_+ \times \mathscr{C}_r$ to $\mathbb{R}^n$. In fact, the functional f in (2.14a) may also depend on the system state, that is $f = f(t, x(t), x_t)$. If $t = t_0$, then an initial state function is simply given by

$$x_{t_0} = \phi(s), \quad s \in [-r, 0]. \tag{2.14b}$$

Thus, the *initial-value problem of a delay-type system* is defined by (2.14). Here, we assume that f is completely continuous and smooth enough to guarantee that the IVP in (2.14) admits a unique solution. A special class of (2.14) is called a *delay system* in which $s = -r$, i.e., $x_t(s) = x(t - r)$, and the corresponding differential equation and initial function are defined accordingly.

In the following, we define some stability concepts of $x \equiv 0$ for the delay system in (2.14), where it is assumed that $f(t, 0) = 0$ for all $t \in \mathbb{R}_+$.

Definition 2.13 The trivial solution $x \equiv 0$ of (2.14a) is said to be

(i) stable if, for each $\varepsilon > 0$ and $t_0 \in \mathbb{R}_+$, there exists a $\delta = \delta(t_0, \varepsilon) > 0$ such that, if $\phi \in \mathscr{C}_r$ with $\|\phi\|_r \leq \delta$, then $\|x(t)\| \leq \varepsilon$, where, for all $t \geq t_0$, $x(t) = x(t; t_0, \phi)$ is any solution of (2.14);

(ii) uniformly stable if δ in (i) is independent of t_0;

(iii) asymptotically stable if (i) holds and for each $t_0 \in \mathbb{R}_+$, there exists an $\eta_0 = \eta(t_0) > 0$ such that, if $\phi \in \mathscr{C}_r$ with $\|\phi\|_r \leq \eta_0$, then $\lim_{t\to\infty} x(t) = 0$;
(iv) uniformly asymptotically stable if (iii) holds and there exists an $\eta_0 > 0$ such that, for each $\gamma > 0$, there exists some $T = T(\eta, \gamma) > 0$ such that if $\phi \in \mathscr{C}_r$ with $\|\phi\|_r \leq \eta_0$, then $\|x(t)\| \leq \gamma$ for all $t \geq t_0 + T$; and
(ii) unstable if (i) fails to hold.

One can similarly define the Dini derivatives with respect to the system of functional differential equations in (2.14).

Definition 2.14 Let $J \subseteq \mathbb{R}_+$ and $\mathscr{D}$ be an open subset of $\mathbb{R}^n$. If $V : J \times \mathscr{D} \to \mathbb{R}_+$, then the *upper right-hand (Dini) derivative* of V with respect to system (2.14) is defined by

$$D^+V(t, \psi(0)) = \lim_{h\to 0^+} \sup \frac{1}{h}\big[V\big(t + h, \psi(0) + hf(t, \psi)\big) - V(t, \psi(0))\big],$$

for all $(t, \psi) \in J \times \mathscr{C}([-r, 0]; \mathscr{D})$.

If, moreover, V has continuous partial derivatives with respect to its variables, then we have

$$D^+V(t, \psi(0)) = \dot{V}(t, \psi(0)) = \frac{\partial V(t, \psi(0))}{\partial t} + \nabla_{\psi(0)} V(t, \psi(0)) \cdot f(t, \psi).$$

In the following theorem, we state the sufficient conditions that guarantee some stability properties for (2.14) by using the Razumikhin–Lyapunov technique in which the time derivative of a Lyapunov function, but not functional, is investigated.

Theorem 2.5 *Suppose that f maps $\mathbb{R} \times \mathscr{D}$ into $\mathbb{R}^n$ where $\mathscr{D} \subset \mathscr{C}_r$ with $x = 0 \in \mathscr{D}$. Assume there exist functions $u \in \mathscr{K}$, $v \in \mathscr{K}_\infty$ and $w \in \mathscr{C}(\mathbb{R}_+; \mathbb{R}_+)$ that is nondecreasing. If there is a continuous function $V \in \mathscr{C}(\mathbb{R} \times \mathscr{D}; \mathbb{R})$ satisfying the following conditions:*

(i) $u(\|x\|) \leq V(t, x) \leq v(\|x\|)$ for all $(t, x) \in \mathbb{R} \times \mathscr{D}$; and
(ii) $\dot{V}(t, \psi(0)) \leq -w(\|\psi(0)\|)$ whenever $V(t + \theta, \psi(\theta)) \leq V(t, \psi(0))$,

for $\theta \in [-r, 0]$, then the trivial solution $x \equiv 0$ of (2.95) is uniformly stable. If, moreover, the condition in (ii) is replaced by

(iii) $\dot{V}(t, \psi(0)) \leq -w(\|\psi(0)\|)$ whenever $V(t + \theta, \psi(\theta)) \leq p(V(t, \psi(0)))$,

where w is strengthened to $w(s) > 0$ for $s > 0$ and the function p is continuous nondecreasing and $p(s) > s$ for $s > 0$, then the trivial solution $x \equiv 0$ of (2.14a) is uniformly asymptotically stable. If $\lim_{s\to\infty} u(s) = \infty$ (i.e., $u \in \mathscr{K}_\infty$), then the trivial solution $x \equiv 0$ of (2.14a) is globally uniformly asymptotically stable.

An important special class of the delay differential equations, which often appear in applications, has the following linear differential equation with state delay

$$\dot{x}(t) = Ax(t) + Bx(t - r), \tag{2.15}$$

where $x \in \mathbb{R}^n$, A and B are $n \times n$ constant matrices, and $r > 0$ represents the time delay. When analyzing the stability properties of this system by using the Lyapunov method, one encounters the following scalar differential inequality

$$\dot{v}(t) \leq -\alpha v(t) + \beta \sup_{t-r \leq s \leq t} v(s), \tag{2.16}$$

where α and β are positive constants. The interest here is to calculate an upper estimate on $v(t)$ for all $t \geq t_0$ with $t_0 \in \mathbb{R}_+$. The following lemmas are concerned with different estimations depending on the values of α and β.

Lemma 2.1 *Suppose that the scalar differential inequality in (2.16) is satisfied where $\alpha > \beta > 0$, then there exist $\gamma > 0$ and $k > 0$ such that*

$$v(t) \leq k e^{-\gamma(t-t_0)}, \qquad \forall t \geq t_0, \tag{2.17}$$

where the decay rate, γ, is the unique positive solution of the nonlinear equation $-\gamma = -\alpha + \beta e^{\gamma r}$ and $k = \inf_{t_0 - r \leq s \leq t_0} y(s)$.

On the other hand, if the scalar differential inequality has the form

$$\dot{v}(t) \leq \alpha v(t) + \beta \sup_{t-r \leq s \leq t} v(s), \tag{2.18}$$

then the upper estimate along with the growth rate is provided by the following lemma.

Lemma 2.2 *Suppose that the scalar differential inequality in (2.18) is satisfied where $\alpha > 0$ and $\beta > 0$, then there exist $\gamma > 0$ and $k > 0$ such that*

$$v(t) \leq k e^{\gamma(t-t_0)}, \qquad \forall t \geq t_0,$$

where $\gamma = \alpha + \beta$ and $k = \sup_{t_0 - r \leq s \leq t_0} v(s)$.

Generalization of the last two lemmas to an n-dimensional vector differential inequality is stated in the following lemmas.

Lemma 2.3 *For all $t \geq t_0$ with $t_0 \in \mathbb{R}_+$, let $A(t)$ and $B(t)$ be $n \times n$ matrices of continuous functions such that $A(t)$, $B(t)$ and $\dot{A}(t)$ are bounded, and $A(t)$ is Hurwitz. Furthermore, assume that, for all t, the following conditions hold*

(i) $\lambda(A^T(t) + A(t)) \leq -\alpha(t) < 0,$ *with* $\alpha(t) > 0$;
(ii) $-\alpha(t) + 2\|B(t)\| \leq -\beta < 0,$ *with β being a positive constant; and*
(iii) the differential inequality

$$\dot{y}(t) \leq A(t)y(t) + B(t) \sup_{t-\tau \leq \theta \leq t} y(\theta),$$

where $y(t) = (y_1(t)\ y_2(t)\ \cdots\ y_n(t))^T \geq 0$ *and*
$\sup_{t-\tau\leq\theta\leq t} y(\theta) = (\sup_{t-\tau\leq\theta\leq t} y_1(\theta)\ \sup_{t-\tau\leq\theta\leq t} y_2(\theta)\ \cdots\ \sup_{t-\tau\leq\theta\leq t} y_n(\theta))^T$.

Then, for all $t \geq t_0$, $y(t)$ *satisfies*

$$\|y(t)\| \leq \|y_{t_0}\|_\tau e^{-\zeta(t-t_0)},$$

where ζ *is the unique positive solution of the nonlinear equation*

$$\zeta - \alpha(t) + \|B(t)\| + \|B(t)\|e^{\zeta\tau} = 0.$$

Lemma 2.4 *For all* $t \in [t_0, t_0 + a)$ *with* $t_0 \in \mathbb{R}_+$ *and* $a > 0$, *let* $A(t)$ *and* $B(t)$ *be* $n \times n$ *matrices of continuous functions,* $\alpha(t) = \lambda\Big(A(t) + A^T(t)\Big)$, $\|B(t)\| \leq \beta_1$ *and* $\alpha(t) + \|B(t)\| \leq \beta_2$. *Assume that the vector differential inequality*

$$\dot{y}(t) \leq A(t)y(t) + B(t) \sup_{t-r\leq s\leq t} y(s)$$

holds where $y(t) = \big(y_1(t)\ y_2(t)\ \cdots y_n(t)\big)^T$ *with* $y_i(t) \geq 0$ *for all* $t \in [t_0, t_0 + a)$ *and*

$$\sup_{t-r\leq s\leq t} y(s) = \Big(\sup_{t-r\leq s\leq t} y_1(s)\ \sup_{t-r\leq s\leq t} y_2(s)\ \cdots\ \sup_{t-r\leq s\leq t} y_n(s)\Big)^T.$$

Then, there exists a $\xi > 0$, *defined by* $\xi = \frac{1}{2}(\beta_1 + \beta_2)$, *such that*

$$\|y(t)\| \leq \|y_{t_0}\|_r e^{\xi(t-t_0)}, \qquad t \in [t_0, t_0 + a).$$

2.4 Impulsive Systems

A general class of impulsive systems or systems with impulsive differential equations may have the form

$$\dot{x}(t) = f(t, x(t)), \qquad \kappa(t, x) \neq 0, \tag{2.19a}$$

$$\Delta x(t) = \mathscr{I}(t, x(t)), \qquad \kappa(t, x) = 0, \tag{2.19b}$$

$$x(t_0^+) = x_0, \tag{2.19c}$$

where $x \in \mathbb{R}^n$ is system state vector, $\Delta x(t) = x(t^+) - x(t)$ for some $t \in \mathbb{R}_+$ with $x(t^+) = \lim_{\epsilon\to 0^+} x(t + \epsilon)$ and $x(t) = x(t^-)$, i.e., x is assumed to be left-continuous. Also, in the difference equation (2.19b), the function $\mathscr{I}(t, x(t))$ is state-dependent representing impulsive amount. In this system, the impulses occur if a spatio-temporal relation $\kappa(t, x) = 0$ is satisfied. Moreover, if we assume that there is no impulsive action at the initial time t_0, then the initial condition in (2.19c) has the form $x(t_0) = x_0$.

The solution of this system evolves as follows: the system state starts when $\kappa(t_0, x_0) \neq 0$. Then, whenever $\kappa(t, x) \neq 0$, the system process is governed by the ordinary differential equation (2.19a) until $t = \tau_1$ such that $\kappa(\tau_1, x(\tau_1)) = 0$ is satisfied. At this moment, the process is subject to an impulse and instantly changes by some amount $\mathscr{I}(t, x(t))$, given by the difference equation in (2.19b), causing a jump discontinuity in the system state. For $t > \tau_1$, if the relation $\kappa(t, x) \neq 0$ holds, the process continues according to the differential equation in (2.19a) until an impulsive action occurs again. This continues in the same manner as long as the solution exists. Consequently, the resulting solution is either continuous or piecewise continuous with simple jump discontinuities at the moments of impulse t for which $\mathscr{I}(t, x(t)) \neq 0$.

Due to the difficulty in dealing with relations of the type $\kappa(t, x) = 0$, the interest deflects to a particular type of relation, where the set of points $(t, x) \in \mathbb{R}_+ \times \mathbb{R}^n$ for which $\kappa(t, x) = 0$ are assumed to be represented by a sequence of hypersurfaces of the form $t = \tau_k(x)$, where generally $\tau_k \in \mathscr{C}(\mathbb{R}^n; \mathbb{R}_+)$ for $k = 0, 1, 2, \ldots$ and $0 = \tau_0(x) < \tau_1(x) < \tau_2(x) < \cdots$ with $\lim_{k\to\infty} \tau_k(x) = \infty$ for each $x \in \mathbb{R}^n$. Therefore, the particular system can be written as

$$\dot{x}(t) = f(t, x(t)), \qquad t \neq \tau_k(x), \tag{2.20a}$$

$$\Delta x(t) = \mathscr{I}(t, x(t)), \qquad t = \tau_k(x), \tag{2.20b}$$

$$x(t_0) = x_0. \tag{2.20c}$$

In this case, the system is said to have impulses at variable times. Indicative features of this system are that solutions start at different points will be subject to impulses (or jump discontinuities) at different times. This problem breaks down the classical continuous dependence or stability since neighbouring solutions tend to undergo impulses at different times. Also, a solution may hit the same hypersurface several times or not at all, or intersect it more than once after intersecting other hypersurfaces. The frequent interception of the same hypersurface is called *pulse or beating phenomenon*. To avoid this circumstance, further restrictions have to be made on the impulsive hypersurface, as will be seen in the following chapter.

If the functions τ_k's are constants (i.e., $\tau_k(x) = \tau_k$ for all k and x), system (2.20) is said to have impulses at fixed times and all solutions undergo impulses at the same times.

Another challenging issue arising in impulsive systems, which makes the theory of ordinary differential equation not directly applicable, is known as *confluence* (or solution merging), which happens when, for instance, two solutions start at different points merge after a certain impulse. The reason is that, for specific impulse amount represented by the function $\mathscr{I}$, the mapping $x + \mathscr{I}(\tau_k, x)$ is not one-to-one in x. On the other hand, if the mapping is not onto, the backward continuation of solutions would be impossible.

So far, we have assumed that the solutions of impulsive systems are left-continuous, instead, one may consider solutions to be right-continuous. Accordingly, system (2.20) is written as

$$\dot{x}(t) = f(t, x(t)), \qquad t \neq \tau_k(x(t^-)), \tag{2.21a}$$

$$\Delta x(t) = \mathscr{I}(t, x(t^-)), \qquad t = \tau_k(x(t^-)), \tag{2.21b}$$

$$x(t_0) = x_0. \tag{2.21c}$$

The choice of right-continuous is advantageous when time delay is involved in impulsive systems [4].

Due to its importance in this book, the solution of the general impulsive system presented in (2.19) is given in the following definition, where the definitions of solutions of special classes, such as the one in (2.21), can be directly extracted.

Definition 2.15 A function $x : (t_0, \beta) \to \mathbb{R}^n$, for $0 \leq t_0 < \beta \leq \infty$, is said to be a solution of system (2.19) if the following conditions are satisfied:

(i) for $t \in (t_0, \beta)$, $(t, x(t)) \in \mathbb{R}_+ \times \mathscr{D}$;
(ii) the right-hand limit $x(t_0^+) = \lim_{t \to t_0^+} x(t)$ exists and $(t_0, x(t_0^+)) \in \mathbb{R}_+ \times \mathscr{D}$;
(iii) $\forall t \in (t_0, \beta)$, if $\kappa(t, x(t)) \neq 0$, then x is continuously differentiable at t and satisfies the differential equation $\dot{x}(t) = f(t, x(t))$;
(iv) the set of impulsive moments $T = \{t \in (t_0, \beta) \mid \kappa(t, x(t)) = 0\}$ is finite or consists of countable increasing sequence of points with limit β; and
(v) if the moment of impulse $t \in T$, then the left-hand limit $x(t^-) = \lim_{t \to t^-} x(t)$ exists and $x(t^-) = x(t)$ for $t \neq t_0$, meaning that the solution is left-continuous, and $x(t^+)$ exists with $x(t^+) = x(t) + \mathscr{I}(t, x(t))$ for $t \neq \beta$.

Generally, a solution $x(t) = x(t; t_0, x_0)$ of (2.19) defined on an interval (t_0, β) and experiencing impulses at points $T = \{t_k\}_{k=1}^{\infty}$ with $t_k < t_{k+1}$ can be described as follows:

$$x(t; t_0, x_0) = \begin{cases} x(t; t_0, x_0), & t_0 < t \leq t_1, \\ x(t; t_1, x(t_1^+)), & t_1 < t \leq t_2, \\ \vdots & \\ x(t; t_k, x(t_k^+)), & t_k < t \leq t_{k+1}, \\ \vdots & \end{cases} \tag{2.22}$$

where $x(t_k^+) = x(t_k) + \mathscr{I}(t_k, x(t_k))$.

Having defined the solution of an impulsive system, we turn our interest to address the stability concepts of a solution of the impulsive systems. As stated earlier, the stability property of impulsive systems with time-dependent impulsive moments is more challenging. As a result, the stability concepts of a nontrivial solution cannot be shifted by a change of variables to the stability of the trivial solution. This situation urges a modification to the stability of an ordinary system.

This complicated situation has limited the study of stability for systems undergoing impulses at state-independent fixed times, i.e., $\tau_k(x) = \tau_k$ for all $x \in \mathbb{R}^n$,

$$\dot{x}(t) = f(t, x(t)), \qquad t \neq \tau_k, \tag{2.23a}$$
$$\Delta x(t) = \mathscr{I}(t, x(t)), \qquad t = \tau_k, \tag{2.23b}$$
$$x(t_0) = x_0, \tag{2.23c}$$

where the impulsive times τ_k (for $k = 0, 1, \ldots$) satisfy $\tau_k < \tau_{k+1}$ and $\lim_{k\to\infty} \tau_k = \infty$. So that, if moreover $f(t, 0) = 0$ and $\mathscr{I}(t, 0) = 0$, that is system (2.23) possesses the trivial solution, then the stability notions are identical to those of ordinary systems. Throughout this book, the stability results are developed for systems with impulses occurring at fixed times. In the following definition, we define different stability notions of the trivial solution, $x \equiv 0$, of (2.23).

Definition 2.16 The trivial solution, $x \equiv 0$, of (2.23) is said to be

(i) stable if, for each $\varepsilon > 0$ and $t_0 \in \mathbb{R}_+$, there exists $\delta = \delta(t_0, \varepsilon) > 0$ such that

$$\|x_0\| < \delta \quad \text{implies} \quad \|x(t)\| < \varepsilon, \quad \forall t > t_0,$$

where $x(t) = x(t; t_0, x_0)$ is any solution of (2.23);

(ii) uniformly stable if δ is independent of t_0;

(iii) asymptotically stable if (i) is satisfied and for every $t_0 \in \mathbb{R}_+$, there exists $\eta = \eta(t_0) > 0$ such that

$$\|x_0\| \le \eta \quad \text{implies} \quad \lim_{t\to\infty} x(t) = 0;$$

(iv) uniformly asymptotically stable if (ii) is satisfied and there exists $\eta > 0$ such that for every $\gamma > 0$, there exists some time $T = T(\eta, \gamma)$ such that

$$\|x_0\| \le \eta \quad \text{implies} \quad \|x(t)\| \le \gamma, \qquad \forall\, t \ge t_0 + T;$$

(v) unstable if (i) fails to hold.

In the following theorem and as a warm-up, we state and prove the exponential stability (in the sense of Lyapunov) for a linear impulsive system given by

$$\begin{cases} \dot{x}(t) = Ax(t), & t \neq t_k, \quad k \in \mathbb{N} \\ \Delta x(t) = B_k x(t), \quad \text{or} \quad x(t^+) = [I + B_k]x(t), & t = t_k, \\ x(t_0^+) = x_0, \end{cases} \tag{2.24}$$

where $t \ge t_0$ with $t_0 \in \mathbb{R}_+$, $x \in \mathbb{R}^n$, $A \in \mathbb{R}^{n\times n}$, $B_k \in \mathbb{R}^{n\times n}$ and $\Delta x(t) = x(t^+) - x(t)$.

When investigating the stability properties of an impulsive system, we consider the following assumptions.

Assumption A1 There exist $0 \le \varrho_1 \le \varrho$ such that, for all $\tau_k \in \mathbb{R}_+$ (with $k \in \mathbb{N}$) and x defined on $\mathscr{PC}(\mathbb{R}_+; \mathscr{D})$, for some open set $\mathscr{D} \in \mathbb{R}^n$, if

$$\|x(\tau_k^-)\| < \varrho_1, \quad \text{then} \quad \|x(\tau_k)\| < \varrho.$$

Assumption A2 For all $k \in \mathbb{N}$, we have

$$\tau_{\sup} = \sup\{\tau_k - \tau_{k-1}\} < \infty \quad \text{and} \quad \tau_{\inf} = \inf\{\tau_k - \tau_{k-1}\} > 0.$$

Remark 2.1 Assumption A1 is made to ensure that the solution be bounded just after any impulsive effect (i.e., at $t = \tau_k$) so long as it is bounded just before the impulsive effects (i.e., at $t = \tau_k^-$). While Assumption A2 is made to guarantee that the interval between any two consecutive impulses be neither infinity nor zero, respectively. The reason behind imposing Assumption A2 is to avoid the trivialness.

Theorem 2.6 *Assume that A is Hurwitz. Then, the trivial solution of (2.24) is globally exponentially stable if the following condition holds:*

$$\ln \alpha_k - \nu(t_k - t_{k-1}) \le 0, \qquad k \in \mathbb{N}, \tag{2.25}$$

where $\alpha_k = \frac{\lambda_{\max}([I+B_k]^T P[I+B_k])}{\lambda_{\min}(P)}$ *with P being a positive-definite matrix satisfying the Lyapunov matrix equation:*

$$A^T P + PA = -Q$$

for any positive-definite matrix Q and $0 < \nu < \xi$ *with* $\xi = \lambda_{\min}(Q)/\lambda_{\min}(P)$.

Proof For all $t \ge t_0$ with $t_0 \in \mathbb{R}_+$, let $x(t) = x(t; t_0, x_0)$ (or simply x) be the solution of (2.24) and $V(x) = x^T Px$ for $P \in \mathbb{R}^{n\times n}$. Define the time-varying function $v(t) = V(x(t))$. Then, the derivative of v along the trajectory of (2.24) is given by

$$\dot{v}(t) \le -\xi v(t), \qquad t \in (t_{k-1}, t_k]$$

where $\xi = \lambda_{\min}(Q)/\lambda_{\min}(P)$, and

$$v(t) \le v(t_{k-1}^+)e^{-\xi(t-t_{k-1})}, \qquad t \in (t_{k-1}, t_k]$$

while at $t = t_k^+$, we have

$$\begin{aligned} v(t_k^+) &= x(t_k^+)^T Px(t_k^+) \\ &= x(t_k)^T [I + B_k]^T P[I + B_k]x(t_k) \\ &\le \lambda_{\max}([I + B_k]^T P[I + B_k])x(t_k)^T x(t_k) \\ &= \alpha_k v(t_k). \end{aligned}$$

Namely, we have

$$v(t_k^+) \le \alpha_k v(t_k), \tag{2.26}$$

where $\alpha_k = \frac{\lambda_{\max}([I+B_k]^T P[I+B_k])}{\lambda_{\min}(P)}$. Now, for instance when $t \in (t_0, t_1]$, we have

$$v(t) \leq v(t_0^+)e^{-\xi(t-t_0)}$$

and

$$v(t_1^+) \leq \alpha_1 v(t_1) \leq \alpha_1 v(t_0^+)e^{-\xi(t_1-t_0)}.$$

Similarly, for $t \in (t_1, t_2]$, we have

$$v(t) \leq v(t_0^+)\alpha_1 e^{-\xi(t_1-t_0)}e^{-\xi(t-t_1)} = v(t_0^+)\alpha_1 e^{-\xi(t-t_0)}.$$

That is, for all $t \in (t_0, t_2]$, we have obtained

$$v(t) \leq v(t_0^+)\alpha_1 e^{-\xi(t-t_0)}.$$

Generally, we have for $t \in (t_k, t_{k+1}]$

$$\begin{aligned} v(t) &\leq v(t_0^+)\alpha_1\alpha_2 \cdots \alpha_k e^{-\xi(t-t_0)} \\ &= v(t_0^+)\alpha_1\alpha_2 \cdots \alpha_k e^{-\nu(t-t_0)}e^{-(\xi-\nu)(t-t_0)} \\ &= v(t_0^+)\alpha_1 e^{-\nu(t_1-t_0)}\alpha_2 e^{-\nu(t_2-t_1)} \cdots \alpha_k e^{-\nu(t_k-t_{k-1})}e^{-(\xi-\nu)(t-t_0)}. \end{aligned}$$

Provoking the assumption in (2.25), we get

$$v(t) \leq v(t_0^+)e^{-(\xi-\nu)(t-t_0)}, \qquad t \geq t_0$$

which implies that

$$\|x(t)\| \leq K\|x(t_0^+)\|e^{-(\xi-\nu)(t-t_0)/2}, \qquad t \geq t_0$$

where $K = \sqrt{\mu}$; this shows that the trivial solution, $x \equiv 0$, of (2.24) is globally exponentially stable.

2.5 Comparison Method for Impulsive Systems

In this section, the comparison principle presented in Sect. 2.2 is used to analyse the stability properties of systems with impulses occurring at fixed times, i.e.,

$$\dot{x}(t) = f(t, x), \qquad t \neq t_k, \tag{2.27a}$$

$$\Delta x(t) = \mathscr{I}_k(x), \qquad t = t_k, \tag{2.27b}$$

$$x(t_0^+) = x_0, \qquad t_0 \geq 0 \tag{2.27c}$$

where $0 < t_1 < t_2 < \cdots < t_k < \cdots$ with $\lim_{k\to\infty} t_k \to \infty$, $f : \mathbb{R}_+ \times \mathbb{R}^n \to \mathbb{R}^n$ is continuous in $(t_{k-1}, t_k] \times \mathbb{R}^n$, for all $x \in \mathbb{R}^n$ and $k \in \mathbb{N}$, $\lim_{(t,y)\to(t_k^+,x)} f(t, y) = f(t_k^+, x)$ exists and $\mathscr{I}_k : \mathbb{R}^n \to \mathbb{R}^n$ for any $k \in \mathbb{N}$. Moreover, $f(t, 0) = 0$ and $\mathscr{I}_k(0) = 0$ for all $k \in \mathbb{N}$; that is, the system has the trivial solution.

As done before, we start with following comparison theorem.

Theorem 2.7 (Comparison theorem) *Let* $V : \mathbb{R}_+ \times \mathbb{R}^n \to \mathbb{R}_+$ *be continuous in* $(t_{k-1}, t_k] \times \mathbb{R}^n$ *for each* $x \in \mathbb{R}^n$ *and* $k \in \mathbb{N}$, $\lim_{(t,y)\to(t_k^+,x)} V(t, y) = V(t_k^+, x)$ *exists and locally Lipschitz in* x. *Assume that* V *satisfies*

$$\begin{aligned} D^+V(t, x) &\leq g(t, V(t, x)), && t \neq t_k \\ V(t, x + \mathscr{I}(t, x)) &\leq \psi_k(V(t, x)), && t = t_k, \end{aligned} \tag{2.28}$$

where $g : \mathbb{R}_+ \times \mathbb{R}_+ \to \mathbb{R}$ *is continuous in* $(t_{k-1}, t_k] \times \mathbb{R}_+$ *and, for all* $x, y \in \mathbb{R}_+$ *and* $k \in \mathbb{N}$, $\lim_{(t,y)\to(t_k^+,x)} g(t, y) = g(t_k^+, x)$ *exists and* ψ_k *is nondecreasing. Let* $r(t) = r(t; t_0, u_0)$ *be the maximal solution of the scalar auxiliary comparison system*

$$\dot{u}(t) = g(t, u), \qquad t \neq t_k, \tag{2.29a}$$

$$\Delta u(t) = \psi_k(u), \qquad t = t_k, \tag{2.29b}$$

$$u(t_0^+) = u_0 \geq 0, \qquad t_0 \geq 0. \tag{2.29c}$$

Then, $V(t_0^+, x_0) \leq u_0$ *implies that*

$$V(t, x(t)) \leq r(t), \qquad \textit{for all } t \geq t_0, \tag{2.30}$$

where $x(t) = x(t; t_0, x_0)$ *is any solution of (2.27) defined on* $[t_0, \infty)$.

Theorem 2.8 (Stability theorem) *Suppose that there exist class*$-\mathscr{K}$ *functions* a *and* b. *Assume that the Lyapunov function* $V : \mathbb{R}_+ \times S(\rho) \to \mathbb{R}_+$ *is continuous in* $(t_{k-1}, t_k] \times S(\rho)$ *for all* $x \in S(\rho)$ *and* $k \in \mathbb{N}$, $\lim_{(t,y)\to(t_k^+,x)} V(t, y) = V(t_k^+, x)$ *exists. Moreover,* V *is assumed to be locally Lipschitz in* x *and the following conditions are satisfied:*

(i) $b(\|x\|) \leq V(t, x) \leq a(\|x\|)$, $\forall (t, x) \in \mathbb{R}_+ \times S(\rho)$;
(ii) $D^+V(t, x) \leq g(t, V(t, x))$, $\forall t \neq t_k$ *and* $x \in S(\rho)$;
(iii) $V(t, x + \mathscr{I}(t, x)) \leq \psi_k(V(t, x))$, *for all* $t = t_k$ *and* $x \in S(\rho)$, $g : \mathbb{R}_+ \times \mathbb{R}_+ \to \mathbb{R}$ *is continuous in* $(t_{k-1}, t_k] \times \mathbb{R}_+$, $g(t, 0) = 0$ *and, for all* $x, y \in \mathbb{R}_+$ *and* $k \in \mathbb{N}$, $\lim_{(t,y)\to(t_k^+,x)} g(t, y) = g(t_k^+, x)$ *exists; and*
(iv) there exists a $\rho_0 > 0$ *such that* $x \in S(\rho_0)$ *implies that* $x + \mathscr{I}_k(x) \in \mathscr{S}()$ *for all* k *and* $V(t, x + \mathscr{I}_k(x)) \leq \psi_k(V(t, x))$, *for all* $t = t_k$ *and* $x \in S(\rho_0)$ *where* ψ_k *is nondecreasing.*

Then, the stability properties of the trivial solution, $u \equiv 0$, *of (2.12) imply the corresponding stability properties of the trivial solution,* $x \equiv 0$, *of (2.27).*

2.6 Impulsive Systems with Time Delay

Incorporating impulsive effects of the variable time type in the delay system (2.14) leads to *impulsive system with time delay* (ISD)

$$\dot{x}(t) = f(t, x_t), \qquad t \neq \tau_k(x(t^-)), \tag{2.31a}$$

$$\Delta x(t) = \mathscr{I}(t, x_{t^-}), \qquad t = \tau_k(x(t^-), \tag{2.31b}$$

$$x_{t_0} = \phi(s), \qquad s \in [-r, 0]. \tag{2.31c}$$

The extended theory of this system of impulsive functional differential equations was initially developed in [4]. Studying the fundamental properties of this system, compared to the impulse-free (or continuous) delay system, can be very challenging unless further restrictions are imposed on the functional f. In the continuous case on one hand, if the system state $x(t)$ is continuous for all $t \in [t_0 - r, t_0 + a]$ (for some $a > 0$), then x_t is a continuous function of t with respect $\|\cdot\|_r$ for all $t \in [t_0, t_0 + a]$ and, hence, f is continuous. On the other hand, if $x(t)$ is discontinuous at a point, say $t^* \in [t_0 - r, t_0 + a]$, then x_t may be discontinuous at some or all $t \in [t_0, t_0 + a]$ and, hence, we cannot draw any conclusion about the continuity of $f(\cdot, \cdot)$ even if it is continuous in its two arguments. To support this argument, the function

$$x(t) = \begin{cases} 0, & t \in [1, 0) \\ 1, & t \in [0, 1] \end{cases}$$

is discontinuous at $t = 0$, while x_t is discontinuous at any $t \in [0, 1]$ with respect to the norm $\|\cdot\|_r$ [4]. In fact, this problem was ruled out by defining a new functional space called *composite piecewise continuous* [4]. In this book, it suffices for us to state some definitions and theorems regarding the stability notions via Lyapunov–Razumikhin technique for system (2.31). Later in Chap. 3, we fully address the fundamental properties of the ISD with random noise.

Definition 2.17 For any $a, b \in \mathbb{R}$ with $a < b$ and for some set $\mathscr{D} \in \mathbb{R}^n$, define

$$\begin{aligned} \mathscr{PC}\big([a,b];\mathscr{D}\big) = \Big\{ &\psi : [a,b] \to \mathscr{D} \mid \psi(t^+) = \psi(t),\ \forall t \in [a,b),\ \psi(t^-) \text{ exists in } \mathscr{D}, \\ &\forall t \in (a,b], \text{and } \psi(t^-) = \psi(t) \ \text{ for all except at most a} \\ &\text{finite number of points } t \in (a,b] \Big\}, \end{aligned}$$

$$\begin{aligned} \mathscr{PC}\big([a,b);\mathscr{D}\big) = \Big\{ &\psi : [a,b) \to \mathscr{D} \mid \psi(t^+) = \psi(t),\ \forall t \in [a,b),\ \psi(t^-) \text{ exists in } \mathscr{D}, \\ &\forall t \in (a,b),\ \text{and } \psi(t^-) = \psi(t) \ \text{ for all except at most a finite} \\ &\text{number of points } t \in (a,b) \Big\}, \end{aligned}$$

$$\mathscr{PC}\big([a,\infty);\mathscr{D}\big)=\Big\{\psi:[a,\infty)\to\mathscr{D}\mid\forall c>a,\psi|_{[a,c]}\in\mathscr{PC}([a,c];\mathscr{D})\Big\}.$$

In these spaces, the functions are right-continuous on their domains and left-continuous except at simple jump discontinuities where the left-hand limits exist. The number of discontinuities is finite if the functions are defined on finite intervals; otherwise (i.e., on infinite interval) the number of discontinuities is countably infinite, which form an increasing sequence of points tending to infinity.

Let $\mathscr{PC}_r\big([-r,0];\mathbb{R}^n\big)=\{\phi\mid\phi\in\mathscr{PC}([-r,0];\mathbb{R}^n)\}$ and define the r-norm of $\phi\in\mathscr{PC}_r$ by $\|\phi\|_r=\sup_{-r\le s\le 0}\|\phi(s)\|$. If $x\in\mathscr{PC}([t_0-r,\infty);\mathbb{R}^n)$ with $t_0\in\mathbb{R}_+$, we define a function $x_t\in\mathscr{PC}([-r,0];\mathbb{R}^n)$ by $x_t(s)=x(t+s)$ for all $s\in[-r,0]$. Let $J\subseteq\mathbb{R}_+$ and $\mathscr{D}\subset\mathbb{R}^n$ be an open set containing $x\equiv 0$. So that, in (2.31), we have $f:J\times\mathscr{PC}([-r,0];\mathscr{D})\to\mathbb{R}^n$ and $\phi\in\mathscr{PC}([-r,0];\mathscr{D})$. If the impulses in (2.31) occur at fixed times, i.e., when $t=\tau_k$ for all $k\in\mathbb{N}$, then we have

$$\dot{x}(t)=f(t,x_t),\qquad t\neq\tau_k,\tag{2.32a}$$

$$\Delta x(t)=\mathscr{I}(t,x_{t^-}),\qquad t=\tau_k,\tag{2.32b}$$

$$x_{t_0}=\phi(s),\qquad s\in[-r,0].\tag{2.32c}$$

We also assume that $f(t,0)=0$ for all $t\in\mathbb{R}_+$, $\mathscr{I}(\tau_k,0)=0$ for all $\tau_k\in\mathbb{R}_+$, and due to the local nature of stability analysis, we assume that impulsive system (2.32) has a local solution, say $\|x\|\le\rho$ for sufficiently small $\rho>0$, and the Lyapunov function, V, is defined on $\mathbb{R}_+\times\mathscr{PC}([-r,0];S(\rho))$ with $S(\rho)\subset\mathscr{D}$.

Before giving the conditions that guarantee the stability results, we state the definitions of different stability notions.

Definition 2.18 The trivial solution, $x\equiv 0$, of (2.32) is said to be

(i) stable if, for each $\varepsilon>0$ and $t_0\in\mathbb{R}_+$, there exists a $\delta=\delta(t_0,\varepsilon)>0$ such that, if $\phi\in\mathscr{PC}([-r,0];S(\rho))$ with $\|\phi\|_r\le\delta$, then $\|x(t)\|\le\varepsilon$, where, for all $t\ge t_0$, $x(t)=x(t;t_0,\phi)$ is any solution of (2.32);
(ii) uniformly stable if δ in (i) is independent of t_0;
(iii) asymptotically stable if (i) holds and for each $t_0\in\mathbb{R}_+$, there exists an $\eta=\eta(t_0)>0$ such that, if $\phi\in\mathscr{PC}([-r,0];S(\rho))$ with $\|\phi\|_r\le\eta$, then $\lim_{t\to\infty}x(t)=0$;
(iv) uniformly asymptotically stable if (ii) holds and there exists an $\eta>0$ such that, for each $\gamma>0$, there exists some $T=T(\eta,\gamma)>0$ such that if $\phi\in\mathscr{PC}([-r,0];S(\rho))$ with $\|\phi\|_r\le\eta$, then $\|x(t)\|\le\gamma$ for all $t\ge t_0$; and
(v) unstable if (i) fails to hold.

Theorem 2.9 *Suppose that there exist functions $a,b,c\in K_1$, $p\in\mathscr{PC}(\mathbb{R}_+;\mathbb{R}_+)$ and $g\in K_3$. Assume that the function $V:[-r,\infty)\times S(\rho)\to\mathbb{R}_+$ is continuous in $[-\tau,\tau_0]\times S(\rho)$ and in $[\tau_{k-1},\tau_k)\times S(\rho)$ for $k\in\mathbb{N}$ and that, for each $x\in S(\rho)$ and $k\in\mathbb{N}$, $\lim_{(t,y)\to(\tau_k^-,x)}V(t,y)=V(\tau_k^-,x)$ exists. Moreover, V is assumed to be locally Lipschitz in x and the following conditions are satisfied:*

(i) $b(\|x\|) \le V(t,x) \le a(\|x\|)$ *for all* $(t,x) \in [-r,\infty) \times S(\rho)$;
(ii) $D^+V(t,\psi(0)) \le p(t)V(t,\psi(0))$ *for all* $t \ne \tau_k \in \mathbb{R}_+$ *and* $\psi \in \mathscr{PC}([-r,0]; S(\rho))$ *whenever* $V(t,\psi(0)) \ge g(V(t+s,\psi(s)))$ *for* $s \in [-r,0]$;
(iii) $V(\tau_k, \psi(0) + \mathscr{I}(\tau_k,\psi)) \le g(V(\tau_k^-, \psi(0)))$ *for all* $(\tau_k,\psi) \in \mathbb{R}_+ \times \mathscr{PC}([-r,0]; S(\rho))$ *for which* $\psi(0^-) = \psi(0)$; *and*
(iv) $\tau = \sup_{k\in\mathbb{N}}\{\tau_k - \tau_{k-1}\} < \infty$, $M_1 = \sup_{t\ge 0}\int_t^{t+\tau} p(s)ds < \infty$ *and* $M_2 = \inf_{q>0}\int_{g(s)}^{q} \frac{ds}{c(s)} > M_1$.

Then, the trivial solution $x \equiv 0$ *of (2.32) is uniformly asymptotically stable.*

Theorem 2.10 *Suppose that there exist functions* $a, b, c \in K_1$, $p \in \mathscr{PC}(\mathbb{R}_+; \mathbb{R}_+)$ *and* $g, \hat{g} \in K_3$ *such that* $s \le \hat{g}(s) < g(s)$ *for* $s > 0$. *Assume that the function* $V : [-r,\infty) \times S(\rho) \to \mathbb{R}_+$ *is continuous on* $[-\tau,\tau_0] \times S(\rho)$ *and on* $[\tau_{k-1},\tau_k) \times S(\rho)$ *for* $k \in \mathbb{N}$ *and that for each* $x \in S(\rho)$ *and* $k \in \mathbb{N}$, $\lim_{(t,y)\to(\tau_k^-,x)} V(t,y) = V(\tau_k^-,x)$ *exists. Moreover,* V *is assumed to be locally Lipschitz in* x *and the following conditions are satisfied:*

(i) $b(\|x\|) \le V(t,x) \le a(\|x\|)$ *for all* $(t,x) \in [-r,\infty) \times S(\rho)$;
(ii) $D^+V(t,\psi(0)) \le -p(t)V(t,\psi(0))$ *for all* $t \ne \tau_k \in \mathbb{R}_+$ *and* $\psi \in \mathscr{PC}([-r,0]; S(\rho))$ *whenever* $g(V(t,\psi(0))) \ge V(t+s,\psi(s))$ *for* $s \in [-r,0]$;
(iii) $V(\tau_k, \psi(0) + \mathscr{I}(\tau_k,\psi)) \le \hat{g}(V(\tau_k^-, \psi(0)))$ *for all* $(\tau_k,\psi) \in \mathbb{R}_+ \times \mathscr{PC}([-r,0]; S(\rho))$ *for which* $\psi(0^-) = \psi(0)$; *and*
(iv) $\mu = \inf_{k\in\mathbb{N}}\{\tau_k - \tau_{k-1}\} > 0$, $M_2 = \sup_{q\ge 0}\int_q^{g(s)} \frac{ds}{c(s)}$, *and* $M_1 = \inf_{t\ge 0}\int_t^{t+\mu} p(s)ds > M_2$.

Then, the trivial solution $x \equiv 0$ *of (2.32) is uniformly asymptotically stable.*

2.7 Stochastic Differential Equations

In this section, some basic concepts that will be used throughout this book are presented. We start with introducing some notations and definitions from the probability theory. Then, we give the definition of stochastic processes and particularly the so-called Wiener (or Brownian motion) process. After that, we define a particularly important class of stochastic integrals, namely Itô integrals, followed by stochastic differential equations.

2.7.1 *Notations and Basic Definitions*

Denote by ω the outcome of an experience and Ω the probability sample space. If the event ω is a possible outcome of a certain random experiment, we suitably write $\omega \in \Omega$. Denote by $\mathscr{F}$ the family of all interesting events of Ω. For further purposes, $\mathscr{F}$ is required to be a σ-algebra (or σ-field), which is defined below.

Definition 2.19 A collection of subsets (or events) $\mathscr{F}$ of Ω is said to be a *σ-algebra on Ω* if the following conditions hold:

(i) the empty subset $\emptyset \in \mathscr{F}$;
(ii) if $A \in \mathscr{F}$, then $A^c \in \mathscr{F}$ where A^c stands for the complement of A; and
(iii) if $\{A_i\}_{i\geq 1} \in \mathscr{F}$, then $\cup_{i\geq 1} A_i \in \mathscr{F}$.

A *measure space* can then be defined by the pair $(\Omega, \mathscr{F})$ and the elements of $\mathscr{F}$, in this case, are called *$\mathscr{F}$-measurable sets*. If $\mathscr{S}$ is a class of subsets of Ω, then one can find a smallest σ-algebra $\sigma(\mathscr{S})$ on Ω that contains $\mathscr{S}$. Particularly, if $\Omega = \mathbb{R}^d$ and $\mathscr{S}$ is the smallest class of all open set in $\mathbb{R}^d$, then $\mathscr{B}^d = \sigma(\mathscr{S})$ is called the *Borel σ-algebra* and its elements are called *Borel sets*. We can now give the definitions of a random variable and probability measure.

Definition 2.20 A real-valued function $X : \Omega \to \mathbb{R}$ is said to be a *random variable* or *$\mathscr{F}$-measurable* if $\{\omega \mid X(\omega) \leq x\} \in \mathscr{F}$ for all $x \in \mathbb{R}$. Also, an $\mathbb{R}^d$-valued function $X(\omega) = (X_1(\omega)\ X_2(\omega)\ \cdots\ X_d(\omega))^T$ is said to be $\mathscr{F}$-measurable if all the elements X_i are $\mathscr{F}$-measurable. Analogously, an $\mathbb{R}^{d\times m}$-valued function $X(\omega) = [X_{ij}(\omega)]_{d\times m}$ is said to be $\mathscr{F}$-measurable if all the elements X_{ij} are $\mathscr{F}$-measurable.

Definition 2.21 A function $\mathbb{P} : \mathscr{F} \to [0, 1]$ is said to be a *probability measure* on the measurable space $(\Omega, \mathscr{F})$ if the following conditions hold:

(i) $\mathbb{P}(\emptyset) = 0$ and $\mathbb{P}(\Omega) = 1$; and
(ii) for any pairwise disjoint sequence or collection of subsets $\{A_i\}_{i\geq 1} \subset \mathscr{F}$ (i.e., $A_i \cap A_j = \emptyset$ for all $i \neq j$),

$$\mathbb{P}\big(\cup_{i\geq 1} A_i\big) = \sum_{i=1}^{\infty} \mathbb{P}(A_i).$$

Moreover, the triplet $(\Omega, \mathscr{F}, \mathbb{P})$ is called a *probability space*. Also, the probability space is said to be *complete* if the σ-algebra is complete, i.e., $\mathscr{F} = \bar{\mathscr{F}}$, where $\bar{\mathscr{F}}$ is the completion of $\mathscr{F}$. In this book, we will always assume that the probability space is complete.

It is well known that the probabilistic behavior of a random variable is completely and uniquely described by its *distribution function* $F(x)$, which is defined by

$$F(x) = \mathbb{P}\{\omega \mid X(\omega) \leq x\}, \qquad \text{for all } x \in \mathbb{R}.$$

Assume that X is a continuous random variable, then there exists a nonnegative and integrable function $f(x)$ such that, for every x,

$$F(x) = \int_{-\infty}^{x} f(s)ds,$$

which implies that $f(x) = \frac{dF(x)}{dx}$, which is called the (probability) *density function* of X.

Let $(\Omega, \mathscr{F}, \mathbb{P})$ be a probability space and X be a random variable that is integrable with respect to the probability measure $\mathbb{P}$, then the *mathematical expectation*, also known as *mean* or *average value* of $x = X(\omega)$ with respect to $\mathbb{P}$, is a real number defined by

$$\mathbb{E}[X] = \int_\Omega X(\omega)\, d\,\mathbb{P}(\omega) = \int_{-\infty}^{\infty} x\, dF(x),$$

the pth moment of X is defined by

$$\mathbb{E}[X^p] = \int_\Omega X^p(\omega)\, d\,\mathbb{P}(\omega) = \int_{-\infty}^{\infty} x^p\, dF(x),$$

where $p > 0$. Particularly, if $p = 2$, $\mathbb{E}[X^2]$ is the *mean square* (m.s.) of X. Also, the *variance* of X is defined by

$$V(X) = \mathbb{E}\big[X - \mathbb{E}[X]\big]^2$$

and, if Y is another random variable, the *covariance* of X and Y is defined by

$$Cov(X, Y) = \mathbb{E}\big[(X - \mathbb{E}[X])(Y - \mathbb{E}[Y])\big],$$

where all involved integrals exist.

Consider the probability space $(\Omega, \mathscr{F}, \mathbb{P})$, and let $X_1(\omega), X_2(\omega), \cdots$, be a sequence of random variables, and $X(\omega)$ be defined on the given probability space. Then, the sequence $\{X_k(\omega)\}_{k\geq 1}$ is said to converge to $X(\omega)$ *with probability one* (w.p.1) or *almost surely* (a.s.) if

$$\mathbb{P}\big\{\omega \mid \lim_{k\to\infty} X_k(\omega) = X(\omega)\big\} = 1;$$

it is said to converge to $X(\omega)$ *in probability* or *stochastically* if, for every $\varepsilon > 0$,

$$\lim_{k\to\infty} \mathbb{P}\big\{\omega \mid |X_k(\omega) - X(\omega)| > \varepsilon\big\} = 0;$$

it is said to converge to $X(\omega)$ *in the pth moment* if

$$\lim_{k\to\infty} \mathbb{E}\big[|X_k(\omega) - X(\omega)|^p\big] = 0,$$

where all involved integrals exist, and it is said to converge to $X(\omega)$ *in the m.s.* if $p = 2$. Furthermore, if $\{X_k(\omega)\}_{k\geq 1}$ and $X(\omega)$ have distribution functions $F_k(x)$ and $F(x)$, respectively, then the sequence of the random variables is said to converge to $X(\omega)$ *in distribution* if

$$\lim_{k\to\infty} F_k(x) = F(x)$$

in every continuity point of $F(x)$.

2.7.2 Stochastic Processes

Let $(\Omega, \mathscr{F}, \mathbb{P})$ be a complete probability space. A *filtration* is a family (or a sequence) of increasing sub$-\sigma$-algebra $\{\mathscr{F}_t\}_{t\geq 0}$ of $\mathscr{F}$ (i.e., $\mathscr{F}_t \subset \mathscr{F}_s \subset \mathscr{F}$ for all $t \in I = [0, s)$ with $s < \infty$. The filtration $\{\mathscr{F}_t\}_{t\geq 0}$ is said to be *right continuous* if $\mathscr{F}_t = \cap_{s>t}\mathscr{F}_s$, and it is said to satisfy the *usual conditions* if it is right-continuous and $\mathscr{F}_0$ contains all $\mathbb{P}$-null sets (i.e., any random event $A \in \mathscr{F}_0$ with $\mathbb{P}(A) = 0$). From now on, the complete probability space under consideration satisfies the usual conditions and, in this case, we use the quadruple $(\Omega, \mathscr{F}, \{\mathscr{F}_t\}_{t\geq 0}, \mathbb{P})$.

Definition 2.22 A *stochastic process* $X(t)$ is a family of random variables

$$\{X_t(\omega) \mid \forall t \in I \text{ and } \omega \in \Omega\},$$

which is also denoted by $X(t, \omega)$ (or for simplicity by $X(t)$) for the same t and ω.

Throughout this book, we restrict ourselves to a parameter or (index) set $I \subseteq \mathbb{R}_+$ and state space Ω that is $\mathbb{R}$ or $\mathbb{R}^n$, unless stated otherwise. Apparently, a stochastic process is a function of two variables; for each fixed $t \in I$, $X_t(\omega)$ is a scalar real-valued random variable (or $\mathbb{R}^n$-valued), while, for each fixed $\omega \in \Omega$, $X_t(\omega)$ is real-valued (or $\mathbb{R}^n$-valued) function defined on I. The latter is called a *sample path* or *realization* of the stochastic process.

Let $X(t)$ be an $\mathbb{R}^d$-valued stochastic process. It is said to be *continuous* (respectively, *right continuous, left continuous*) if, for almost all $\omega \in \Omega$, $X_t(\omega)$ is continuous (respectively, right continuous, left continuous) for all $t \in \mathbb{R}_+$. It is said to be *cadlag* if it is right-continuous and, for almost all $\omega \in \Omega$, the left limit $\lim_{s\to t} X_s(\omega)$ exists and is finite for all $t > 0$. It is said to be *integrable* if, for all $t \in \mathbb{R}_+$, $X_t(\omega)$ is an integrable random variable. It is said to be *$\mathscr{F}_t$−adapted* (or *nonanticipated*) if, for all $t \in \mathbb{R}_+$, it is $\mathscr{F}_t$-measurable. If $Y_t(\omega)$ is another stochastic process, then the two processes are said to be *indistinguishable* if

$$\mathbb{P}\big\{\omega \mid X_t(\omega) = Y_t(\omega),\ \forall t \in \mathbb{R}_+\big\} = 1.$$

Let $X(t)$ be an $\mathbb{R}^d$-valued cadlag $\mathscr{F}_t$-adapted process and $\mathscr{D}$ be an open subset of $\mathbb{R}^d$. Then, the *first exit time of the process $X(t)$ from $\mathscr{D}$* is defined by

$$\tau = \inf\{t \in \mathbb{R}_+ \mid X(t) \notin \mathscr{D}\},$$

where $\inf \emptyset = \infty$.

Like random variables, stochastic processes can be characterized by their moments, variance and autocorrelation.

Definition 2.23 Let $X(t)$ be a continuous stochastic process. Then, the *mathematical expectation* (or *mean* or the *first moment*) of $X(t)$ is defined by

$$m(t) = \mathbb{E}[X(t)] = \int_{-\infty}^{\infty} x f(x,t)\, dx,$$

where f (or $f(x,t)$) is the probability density function of $x = X(t)$; the *second moment* (or the *mean square*) is defined by

$$m_2(t) = \mathbb{E}[X^2(t)] = \int_{-\infty}^{\infty} x^2 f(x,t)\, dx;$$

the *variance* is defined by

$$Var[X(t)] = \mathbb{E}[(X(t) - m(t))^2] = m_2(t) - m^2(t);$$

and the *autocorrelation* is defined by

$$R(t_1, t_2) = \mathbb{E}[X(t_1)X(t_2)] = \int_{-\infty}^{\infty}\int_{-\infty}^{\infty} x_1 x_2 f(x_1, t_1; x_2, t_2)\, dx_1 dx_2.$$

Definition 2.24 Let $(\Omega, \mathscr{F}, \mathbb{P})$ be a complete probability space with a filtration $\{\mathscr{F}_t\}_{t\geq 0}$. A stochastic process $W(t)$ for all $t \in \mathbb{R}_+$ that is continuous (a.s.) and $\mathscr{F}_t$-adapted is said to be *Wiener (or Brownian motion) process* if

(i) $\mathbb{P}\{\omega \mid W(0) = 0\} = 1$;
(ii) for any $0 \leq s < t < \infty$, the increment $W(t) - W(s)$ is independent of $\mathscr{F}_s$; and
(iii) for any $t \in \mathbb{R}_+$ and $h > 0$, the increment $W(t+h) - W(t)$ is Gaussian (or normally) distributed with

$$\mathbb{E}[W(t+h) - W(t)] = \mu h; \text{ and}$$
$$\mathbb{E}[(W(t+h) - W(t))^2] = \sigma^2 h,$$

where the mean $\mu \in \mathbb{R}$ and the variance σ^2 is a positive constant.

If $\mu = 0$ and $\sigma^2 = 1$, W is said to be a *standard Wiener process*.

Following the definition of distribution function F, the *jointly distribution function* of $X(t_1), \ldots, X(t_n)$ is defined by

$$F_{X(t_1),\ldots,X(t_n)}(x_1, \ldots, x_n) = \mathbb{P}\{X(t_1) \leq x_1, \ldots, X(t_n) \leq x_n\}$$

and, if F has partial derivatives at $x_1, \ldots, x_n$, then the corresponding probability density function of $(x_1, \ldots, x_n)$ is given by

$$f(x_1, \ldots, x_n) = \frac{\partial^n}{\partial x_1 \cdots \partial x_n} F_{X(t_1),\ldots,X(t_n)}(x_1, \ldots, x_n).$$

A stochastic process $X(t)$ is said to be *stationary* if and only if, for all time instants $t_1, \ldots, t_n$ and any time difference τ,

$$f_{X(t_1),\ldots,X(t_n)}(x_1, \ldots, x_n) = f_{X(t_1+\tau),\ldots,X(t_n+\tau)}(x_1, \ldots, x_n).$$

We conclude this subsection with a mathematically useful stochastic process called Gaussian white noise process.

Definition 2.25 A stochastic process $\mathscr{N}$ is said to be a *Gaussian white noise process* if and only if it is a stationary Gaussian process with mean zero and autocorrelation given by

$$R(\tau) = C\delta(\tau),$$

where C is a constant and δ is a Dirac delta or impulse function.

Clearly, the variance of the Gaussian white noise is $Var[\mathscr{N}(t)] = \infty$.

2.7.3 Stochastic Differential Equations

Suppose that a physical process is described by the following ordinary differential equation

$$\frac{dx}{dt} = f(t, x). \tag{2.33}$$

If it is perturbed by some disturbance having a stochastic behavior, say $\xi = \xi(t)$ for all $t \geq t_0$ with $t_0 \in \mathbb{R}_+$, then (2.33) may be written as

$$\frac{dX}{dt} = F(t, X, \xi). \tag{2.34}$$

Due to the random part, this differential equation cannot be interpreted as its ordinary counterpart in (2.33). To better understand the new situation, we consider the following special form of (2.34)

$$\frac{dX}{dt} = f(t, X) + g(t, X)\mathscr{N}(t), \tag{2.35}$$

with a deterministic *drift* coefficient $f(t, X)$ perturbed by a noise term $g(t, X)\mathscr{N}(t)$ with $\mathscr{N}$ being a Gaussian white noise process and the *diffusion* coefficient $g(t, X)$ is the noise intensity. Integrating (2.35) over $[t_0, t]$ yields

$$X(t) = X(t_0) + \int_{t_0}^{t} f(s, X(s))ds + \int_{t_0}^{t} g(s, X(s))\mathscr{N}(s)ds, \tag{2.36}$$

where the first integral is deterministic for almost every $\omega \in \Omega$, while the second one cannot be defined in any meaningful manner.

To cope with this difficulty, we replace the aforementioned second integral by an integral of the form

$$\int_{t_0}^{t} g(s, X)dW(s), \tag{2.37}$$

where W is a Wiener process with the *formal* relationship with the Gaussian white noise process being given by $\dot{W}(t) = \mathscr{N}(t)$ and so $dW(t) = \mathscr{N}(t)dt$. The resulting integral in (2.37) cannot be defined as a Riemann–Stieltjes integral because, for almost all $\omega \in \Omega$, the Wiener sample path $W(\omega)$ is nowhere differentiable and has unbounded variation over every time interval.

However, one can define this integral on a larger class of stochastic processes depending on the properties of Wiener process. This definition was first proposed by K. Itô, and the integral is now known as *Itô stochastic integral.*

Consider the integral of the form

$$\int_{a}^{b} g(s, \omega)dW(s, \omega), \tag{2.38}$$

where g is a stochastic process with appropriate conditions and W is a Wiener process, where we generally assume that the two processes are not mutually independent and $g(t, \omega)$ is not absolutely continuous for almost all $\omega \in \Omega$.

The core feature of the Itô integral is that the random function g is nonanticipative or adapted to the filtration $\{\mathscr{F}_t\}_{t\geq 0}$; that is, $g(t, \omega)$ can at most depend on the present and past, and not on the future, values of the Wiener process $W(t, \omega)$. More precisely, let $(\Omega, \mathscr{F}, \{\mathscr{F}_t\}_{t\geq 0}, \mathbb{P})$ be a complete probability space on which the Wiener process $W(t, \omega)$ is defined for all $t \in \mathbb{R}_+$ and

(i) for every $t_1, t_2 \in \mathbb{R}_+$, $t_1 < t_2$ implies that $\mathscr{F}_{t_1} \subset \mathscr{F}_{t_2}$;
(ii) for all $t \in \mathbb{R}_+$, the random variable $W(t, \omega)$ is $\mathscr{F}_t$-measurable; and
(iii) for $t_{i+1} > t_i \geq t$, the increment $W(t_{i+1}, \omega) - W(t_i, \omega)$ is independent of $\mathscr{F}_t$.

For $a, b \in \mathbb{R}_+$ with $a \leq b$, denote by $L^2[a, b]$ the class of all real-valued random processes (functions) $g(t)$ defined on $[a, b]$ and satisfying the following conditions:

(iv) for all $t \in [a, b]$, $g(t, \omega)$ is $\mathscr{F}_t$-measurable; and
(v) the integral

$$\int_a^b g^2(t,\omega)dt \tag{2.39}$$

is finite w.p.1.

To define the Itô (stochastic) integral, consider the partition $a = t_1 < t_2 < \cdots < t_{k+1} = b$, and let $g(t,\omega)$ be a step or simple function, i.e., $g(t,\omega) = g(t_i,\omega)$ for all $t \in [t_i, t_{i+1}]$, which is assumed to be $\mathscr{F}_{t_i}$-measurable, bounded random variable. Then, the Itô integral is defined by

$$\int_a^b g(t,\omega)dW(t) = \sum_{i=1}^{k} g(t_i,\omega)[W(t_{i+1}) - W(t_i)]. \tag{2.40}$$

Another way to define Itô integral is as a limit of a m.s. convergent sequence of simple processes. Let $g_n(t,\omega) \in L^2[a,b]$ be an arbitrary sequence of simple processes. Then, the Itô integral is defined by

$$\int_a^b g(t,\omega)dW(t) = \lim_{n\to\infty} \int_a^b g_n(t,\omega)dW(t) \tag{2.41}$$

in $L^2[a,b]$, i.e.,

$$\lim_{n\to\infty} \mathbb{E} \int_a^b |g(t,\omega) - g_n(t,\omega)|^2 dt = 0.$$

The Itô integral in (2.41) has some nice properties. Assuming that $g \in \mathscr{L}_{ad}([a,b];\mathbb{R}^d)$, i.e., g is an $\mathbb{R}^d$-valued $\mathscr{F}_t$-adapted process such that $\int_a^b \mathbb{E}\|g(t)\|^2 dt < \infty$, some of these properties are

(i) $\mathbb{E}\left[\int_a^b g(t)dW(t)\right] = 0$; and

(ii) $\mathbb{E}\left[\left\|\int_a^b g(t)dW(t)\right\|^2\right] = \int_a^b \mathbb{E}\|g(t)\|^2\, dt.$

Replacing the stochastic integral in (2.36) by the Itô integral in (2.38) results in the following *stochastic integral equation*

$$X(t) = X(t_0) + \int_{t_0}^t f(s,X(s))ds + \int_{t_0}^t g(s,X(s))dW(s), \tag{2.42}$$

which is equivalent to the symbolic *stochastic differential equation (SDE) of Itô type*

$$dX(t) = f(t,X(t))dt + g(t,X(t))dW(t), \tag{2.43}$$

with the initial state $X(t_0) = X_0$. Before presenting the solution of this equation, we need to define the following class of random processes (functions).

Definition 2.26 Let $(\Omega, \mathscr{F}, \{\mathscr{F}_t\}_{t\geq 0}, \mathbb{P})$ be a complete probability space. For any $\omega \in \Omega, a, b \in \mathbb{R}_+$, with $a < b$ and $p \geq 1$, a random process $f(t, \omega)$ is said to belong to class $\mathscr{L}_{ad}(\Omega; L^p[a, b])$ if it is $\mathscr{F}_t$-adapted and almost all its sample paths are pth integrable in the Riemann sense.

Definition 2.27 For any $t_0, T \in \mathbb{R}_+$, the $\mathbb{R}^n$-valued stochastic process $x(t) = x(t; t_0, x_0)$ is said to be a *solution* of n-dimensional initial-value problem

$$dx(t) = f(t, x(t))dt + g(t, x(t))dW(t), \qquad t \in [t_0, T], \tag{2.44a}$$

$$x(t_0) = x_0, \tag{2.44b}$$

where $W(t) = (W_1(t)\ W_2(t)\ \cdots\ W_m(t))^T \in \mathbb{R}^m$ and x_0 is an $\mathscr{F}_{t_0}$-measurable $\mathbb{R}^n$-valued random variable such that $\mathbb{E}[\|x_0\|^2] < \infty$, if the the following properties hold:

(i) $x(t)$ is continuous and $\mathscr{F}_t$-adapted;
(ii) the $\mathbb{R}^n$-valued $f \in \mathscr{L}_{ad}(\Omega; L^1[a, b])$ and the $\mathbb{R}^{n\times m}$-valued $g \in \mathscr{L}_{ad}(\Omega; L^2[a, b])$;
(iii) for all $t \in [t_0, T]$, $x(t)$ satisfies the SDE in (2.44a) w.p.1; and
(iv) at $t = t_0$, x satisfies the initial condition in (2.44b) w.p.1.

Furthermore, a solution x is said to be *unique* if any other solution y is indistinguishable from x, i.e.,

$$\mathbb{P}\big\{x(t) = y(t),\ \forall t \in [t_0, T]\big\} = 1.$$

When working on Itô SDEs, there arise some peculiarities, and among them is that if x is a solution of an Itô equation and $V(t, x(t))$ is a sufficiently smooth function, we cannot use the chain rule of the classical calculus to set up the SDE governing $V(t, x(t))$. Instead, we use the stochastic version of the chain rule, which is called *Itô formula*. Before stating the definition of Itô formula, we define $\mathscr{C}^{1,2}(\mathbb{R}_+ \times \mathbb{R}^n; \mathbb{R}_+)$ to be the space of all real-valued functions $V(t, x)$ defined on $\mathbb{R}_+ \times \mathbb{R}^n$ such that they are continuously differentiable once in t and twice in x. For instance, if $V(t, x) \in \mathscr{C}^{1,2}(\mathbb{R}_+ \times \mathbb{R}^n; \mathbb{R}_+)$, then we have

$$V_t(t, x) = \frac{\partial V(t, x)}{\partial t},\ V_x(t, x) = \Big(\frac{\partial V(t, x)}{\partial x_1}\ \cdots\ \frac{\partial V(t, x)}{\partial x_n}\Big)_{1\times n},\ V_{xx}(t, x) = \Big(\frac{\partial^2 V(t, x)}{\partial x_i \partial x_j}\Big)_{n\times n}.$$

Definition 2.28 (*Itô formula*) For all $t \geq t_0$ with $t_0 \in \mathbb{R}_+$, let $x(t) = x(t; t_0, x_0)$ be an $\mathbb{R}^n$-dimensional $\mathscr{F}_t$-adapted stochastic process satisfying

$$x(t) = x(t_0) + \int_{t_0}^{t} f(s, x(s))ds + \int_{t_0}^{t} \sigma(s, x(s))dW(s), \qquad \text{(a.s.)}$$

where f and g are as defined in the last definition and W is the Wiener process. Suppose that $V \in \mathscr{C}^{1,2}(\mathbb{R}_+ \times \mathbb{R}^n; \mathbb{R})$. Then, for all $t \geq t_0$, $V(t, x(t))$ is a stochastic process satisfying

$$\begin{aligned} V(t, x(t)) = \; & V(t_0, x(t_0)) + \int_{t_0}^{t} \mathscr{L}V(s, x(s))ds \\ & + \int_{t_0}^{t} V_x(s, x(s))\sigma(s, x(s))dW(s), \qquad \text{(a.s.)} \end{aligned}$$

where

$$\mathscr{L}V(t, x(t)) = V_t(t, x(t)) + V_x(t, x(t))f(t, x(t)) + \frac{1}{2}\text{tr}[\sigma^T(t, x(t))V_{xx}(t, x(t))\sigma(t, x(t))]$$

is the infinitesimal operator acting on the process $V(t, x(t))$ with $V_t(t, x(t))$, $V_x(t, x(t))$ and $V_{xx}(t, x(t))$ being the partial differentials of the process $V(t, x(t))$ as defined above.

In fact, the Itô formula may be stated in an equivalent form in which the two integral equations take the differential forms. That is, if x is a stochastic process satisfying

$$dx(t) = f(t, x(t))dt + \sigma(t, x(t))dW(t), \qquad \text{(a.s.)}$$

and $V \in \mathscr{C}^{1,2}(\mathbb{R}_+ \times \mathbb{R}^n; \mathbb{R})$. Then, for all $t \geq t_0$, $V(t, x(t))$ is a stochastic process satisfying

$$dV(t, x(t)) = \mathscr{L}V(t, x(t))dt + V_x(t, x(t))\sigma(t, x(t))dW(s), \qquad \text{(a.s.)}$$

where $\mathscr{L}V(t, x(t))$ is defined above.

The operator $\mathscr{L}$ (or $\mathscr{L}V$ as a single notation) is also called the *averaged derivative* (or *infinitesimal diffusion operator*) at a point (t, x) and can be generally defined as

$$\mathscr{L}V(t, x) = \lim_{h \to 0^+} \frac{1}{h}\big[\mathbb{E}[V(t + h, x(t + h))] - V(t, x)\big].$$

As mentioned earlier, a more general system than (2.44) is when the system states are subject to time lag. This leads to *stochastic systems with time delay* or *systems with stochastic functional differential equations*, which are typically defined by

$$\begin{cases} dx(t) = f(t, x_t)dt + g(t, x_t)dW(t), & t \in [t_0, T], \\ x_{t_0}(s) = \phi(s), & s \in [-r, 0], \end{cases} \tag{2.45}$$

for all $T > t_0$ with $t_0 \in \mathbb{R}_+$.

We have stated clearly that one of the main discrepancies between ordinary and delay systems is the amount of the initial data, which, in the latter case, must be given over a certain period of time rather than at a specific time instance. Moreover, due to the randomness that drives the system states, the given initial condition function is generally defined as a stochastic process. Consequently, to define a solution of the initial-value problem given in (2.45), it is natural to consider the initial function ϕ to be $\mathscr{F}_{t_0}$-measurable, continuous random variable mapping $[-r, 0]$ into $\mathbb{R}^n$ such that $\mathbb{E}[\|\phi\|_r^p] < \infty$ for some $p > 0$. The solution of (2.45) can then be defined similarly to that of (2.44) except, of course, $x(t)$ is defined over the interval $[t_0 - r, T]$ for all $T \in \mathbb{R}_+$ (or $[t_0 - r, t_0 + \alpha]$ for $\alpha > 0$).

Having defined the solution x of (2.45) and the Itô formula, we can present the definition of some stochastic properties of the trivial solution of (2.45).

Definition 2.29 The trivial solution $x \equiv 0$ of (2.45) is said to be

(i) *almost-surely stable (or stable w.p.1)* if, for any given $\varepsilon, \varepsilon' > 0$ and $t_0 \in \mathbb{R}_+$, there exists $\delta = \delta(\varepsilon, \varepsilon', t_0)$ such that

$$\|\phi\|_r < \delta \quad \text{implies} \quad \mathbb{P}\{\omega \mid \sup_{t \geq t_0} \|x(t)\| > \varepsilon'\} < \varepsilon;$$

where $x(t) = x(t; t_0, \phi)$ is any solution of system (2.45);

(ii) *pth moment stable* if, for any $\varepsilon > 0$ and $t_0 \in \mathbb{R}_+$, there exists $\delta = \delta(\varepsilon, t_0)$ such that, for $p > 0$,

$$\|\phi\|_r^p < \delta \quad \text{implies} \quad \mathbb{E}[\sup_{t \geq t_0} \|x(t)\|^p] < \varepsilon;$$

(iii) *asymptotically stable* if, for any $\varepsilon \in (0, 1)$, there exists $\delta = \delta(\varepsilon, t_0)$ such that

$$\|\phi\|_r < \delta \quad \text{implies} \quad \mathbb{P}\{\omega \mid \lim_{t \to \infty} \sup \|x(t)\| = 0\} < 1 - \varepsilon;$$

(iv) *almost-surely asymptotically stable* if it is almost-surely stable and

$$\mathbb{P}\{\omega \mid \lim_{t \to \infty} \sup \|x(t)\| = 0\} = 1;$$

(v) *pth moment asymptotically stable* if it is stable in the pth moment and

$$\lim_{t \to \infty} \mathbb{E}[\sup \|x(t)\|^p] = 0;$$

(vi) *pth moment exponentially stable* if there exist positive constants p, K and λ such that, for any $t_0 \in \mathbb{R}_+$,

$$\|\phi\|_r^p < \delta \quad \text{implies} \quad \mathbb{E}[\|x(t)\|^p] \leq K \|\phi\|_r^p e^{-\lambda(t-t_0)}.$$

Moreover, the above stability properties are said to be satisfied *globally* if they hold for arbitrarily large δ. Also, they are said to hold *uniformly* if δ is chosen to be independent of t_0.

2.7.4 Comparison Method for Stochastic Systems

In this subsection, we state the comparison results for the initial-value problem with the stochastic differential equations of Itô type

$$dx(t) = f(t, x(t))dt + g(t, x(t))dW(t), \tag{2.46a}$$
$$x(t_0) = x_0, \tag{2.46b}$$

where $f \in \mathscr{C}(\mathbb{R}_+ \times \mathbb{R}^n; \mathbb{R}^n)$, $g \in \mathscr{C}\left(\mathbb{R}_+ \times \mathbb{R}^n; \mathbb{R}^{n\times m}\right)$ and W is a Wiener process defined on the complete probability space $(\Omega, \mathscr{F}, \{\mathscr{F}_t\}_{t\geq 0}, \mathbb{P})$ for all $t \in \mathbb{R}_+$.

Theorem 2.11 (Comparison theorem) *Let $V \in \mathscr{C}^{1,2}(\mathbb{R}_+ \times \mathbb{R}^n; \mathbb{R})$ such that the differential inequality*

$$D^+V(t, x) \leq h(t, V(t, x)) \tag{2.47}$$

holds (a.s.) for all $(t, x) \in \mathbb{R}_+ \times \mathbb{R}^n$, where $h \in \mathscr{C}(\mathbb{R}_+ \times \mathbb{R}^n; \mathbb{R})$, and $h(t, z)$ is concave and quasi-monotone nondecreasing in z for all fixed $t \in \mathbb{R}_+$. Let $r(t) = r(t; t_0, u_0)$ be the maximal solution of the auxiliary scalar differential system

$$\dot{u}(t) = h(t, u), \qquad t \in \mathbb{R}_+ \tag{2.48a}$$
$$u(t_0) = u_0. \tag{2.48b}$$

Then, for all $t \in \mathbb{R}_+$, $\mathbb{E}[V(t_0, x_0)] \leq u_0$ implies that $\mathbb{E}[V(t, x(t))] \leq r(t; t_0, u_0)$ with $x(t)$ being the solution process of (2.46).

Theorem 2.12 (Stability theorem) *Suppose that there exist functions $a \in \mathscr{K}_c$ and $b \in \mathscr{K}_v$. Let $V \in \mathscr{C}^{1,2}(\mathbb{R}_+ \times S(\rho); \mathbb{R})$ with $S(\rho) \subset \mathbb{R}^n$ for $\rho > 0$ such that, for all $(t, x) \in \mathbb{R}_+ \times S(\rho)$, the following conditions are satisfied*

(i) $b(\|x\|^p) \leq V(t, x) \leq a(\|x\|^p)$, *(a.s.)*
(ii) $\mathscr{L}V(t, x) \leq h(t, V(t, x))$, *(a.s.)*

where $p \geq 1$, $h \in \mathscr{C}(\mathbb{R}_+ \times \mathbb{R}_+; \mathbb{R})$, $h(t, 0) \equiv 0$, and $h(t, z)$ is concave and quasi-monotone nondecreasing in z for all fixed $t \in \mathbb{R}_+$. Then, the stability properties of the trivial solution, $u \equiv 0$, (2.48) imply the corresponding pth moment stability of the trivial solution, $x \equiv 0$, of (2.46).

On the other hand, if the system states of (2.46) experience impulsive effects at fixed times, we are led to *stochastic impulsive systems* or *systems of stochastic impulsive differential equations*, which are generally given by

$$\begin{cases} dx(t) = f(t, x(t))dt + g(t, x(t))dW(t), & t \neq \tau_k, \\ \Delta x(t) = \mathscr{I}(t, x(t^-)), & t = \tau_k, \\ x(t_0) = x_0. \end{cases} \tag{2.49}$$

2.8 Stochastic Impulsive System with Time Delay

We have previously described stochastic systems with time delay and systems of stochastic impulsive differential equations. In this section, these systems are combined to lead us to consider *stochastic impulsive system with time delay* (SISD). Before formulating and for convenient reading, we restate some of the notations that have been presented in previous sections.

Let $(\Omega, \mathscr{F}, \{\mathscr{F}_t\}_{t\geq 0}, \mathbb{P})$ be a complete probability space with filtration $\{\mathscr{F}_t\}_{t\geq 0}$ satisfying the usual conditions (i.e., it is right-continuous and $\mathscr{F}_0$ contains all $\mathbb{P}$-null sets). Let $W(t) = (W_1(t)\ W_2(t)\ \cdots\ W_m(t))^T$ be an m-dimensional Wiener process defined on the above probability space. Let $r > 0$ represent time delay and denote by $\mathscr{C}([-r, 0]; \mathbb{R}^n)$ (and $\mathscr{PC}([-r, 0]; \mathbb{R}^n)$) the space of continuous (piecewise continuous) functions ϕ mapping $[-r, 0]$ into $\mathbb{R}^n$. Moreover, if $x : [t - r, \infty) \to \mathbb{R}^n$, we define x_t by $x_t = x(t + s)$ for $s \in [-r, 0]$ and the corresponding r-norm is $\|x_t\|_r = \sup_{t-r\leq s\leq t} \|x(s)\|$. We also define $x_{t^-} \in \mathscr{PC}([-r, 0]; \mathbb{R}^n)$ by $x_{t^-}(s) = x(t + s)$ for $-r \leq s < 0$ and $x_{t^-}(s) = x(t^-)$ for $s = 0$. We should mention that this does not mean $x_{t^-} = \lim_{s\to t^-} x_s$ because, if $x \in \mathscr{PC}([-r, 0]; \mathbb{R}^n)$, the limit $\lim_{s\to t^-} x_s$ does not generally exist. For $p > 0$, let $\mathscr{L}^p_{\mathscr{F}_0}([-r, 0]; \mathbb{R}^n)$ be the set of all $\mathscr{F}_0$-measurable $\mathscr{PC}([-r, 0]; \mathbb{R}^n)$-valued random variables $\phi = \{\phi(s) \mid -r \leq s \leq 0\}$ such that $\mathbb{E}[\|\phi\|^p_r] \leq c$, for some $c \geq 0$. We also assume that ϕ is independent of $W(t, \omega)$. For a given Wiener process $W(t, \omega)$ and filtration $\{\mathscr{F}_t \,|\, a \leq t \leq b\}$, we assume that $W(t, \omega)$ is $\mathscr{F}_t$-adapted (i.e., for each $t \in [a, b]$, $W(t, \omega)$ is $\mathscr{F}_t$-measurable) and for any $s \leq t$, the random variable $W(t, \omega) - W(s, \omega)$ is independent of the σ-algebra $\mathscr{F}_s$.

Since the solution of a stochastic initial-value problem is a random process, rather than merely a deterministic function, we need to define the piecewise continuous function.

Definition 2.30 For $a, b \in \mathbb{R}$ with $a < b$ and $\mathscr{D} \subset \mathbb{R}^n$, a random process $\psi : [a, b] \times \Omega \to \mathscr{D}$ is said to be an element of the space $\mathscr{PC}([a, b] \times \Omega; \mathscr{D})$ (or $\mathscr{D}$-cadlag) if, for almost all $\omega \in \Omega$, $\psi(t^+, \omega) = \psi(t, \omega)\ \forall t \in [a, b)$ and $\psi(t^-, \omega)$ exists in $\mathscr{D}\ \forall t \in (a, b]$ and $\psi(t^-, \omega) = \psi(t, \omega)$ for all but at most a finite number of points $t \in (a, b]$. Furthermore, a random process $\psi : [a, \infty) \times \Omega \to \mathscr{D}$ is said to be an element of $\mathscr{PC}([a, \infty) \times \Omega; \mathscr{D})$ if, for almost all $\omega \in \Omega$, $c > a$, where $t \in [a, c]$, $\psi(t, \omega) \in \mathscr{PC}([a, c] \times \Omega; \mathscr{D})$.

Consider now the following nonlinear SDE with time delay

$$dx(t) = f(t, x_t)dt + g(t, x_t)\,dW(t), \quad t \in [a, b], \tag{2.50a}$$

where $x \in \mathbb{R}^n$ is the system state random process, $f \in \mathbb{R}^n$ and $g \in \mathbb{R}^{n\times m}$. The initial condition is given by

$$x_{t_0} = \phi(s), \quad s \in [-r, 0], \tag{2.50b}$$

where $\phi \in \mathscr{L}^2_{\mathscr{F}_0}([-r, 0]; \mathbb{R}^n)$ (i.e., the initial state is assumed to be $\mathscr{F}_0$-adapted, piecewise continuous with finite pth moment); thus, the corresponding stochastic integral equation takes the form

$$x(t) = \phi(0) + \int_{t_0}^{t} f(s, x_s)ds + \int_{t_0}^{t} g(s, x_s)\, dW(s), \quad \text{(a.s.)} \tag{2.51}$$

for all $t \geq t_0$. The first integral is a Riemann integral almost surely (a.s.) and the second one is an Itô integral satisfying

$$\mathbb{E}\Big[\int_{t_0}^{t} g(s, x_s)\, dW(s)\Big] = 0, \text{ and } \mathbb{E}\Big\|\int_{t_0}^{t} g(s, x_s)\, dW(s)\Big\|^2 = \int_{t_0}^{t} \mathbb{E}\|g(s, x_s)\|^2\, ds.$$

Considering impulse effects (of variable times) in (2.50a) leads to the following SISD

$$dx(t) = f(t, x_t)dt + g(t, x_t)\, dW(t), \quad t \neq \tau_k(x(t^-)), \tag{2.52a}$$

$$\Delta x(t) = \mathscr{I}(t, x_{t^-}), \quad t = \tau_k(x(t^-)), \tag{2.52b}$$

where $\tau_k \in \mathscr{C}^2(\mathbb{R}^n; \mathbb{R}_+)$ represents an impulsive hypersurface, for $k \in \mathbb{N}$, and satisfies $0 = \tau_0(x) < \tau_1(x) < \tau_2(x) < \cdots$ and $\lim_{k\to\infty} \tau_k(x) = \infty$ for $x \in \mathbb{R}^n$. The initial condition is given by

$$x_{t_0} = \phi(s), \quad s \in [-r, 0]. \tag{2.52c}$$

We also assume that the solution of (2.52) is right-continuous (i.e., $x(t^+) = x(t)$). In difference equation (2.52b), $\Delta x = x(t) - x(t^-)$ and the functional $\mathscr{I}(\cdot)$ is the impulse amount, which is assumed to be $\mathscr{F}_{t_k}$-adapted.

In the following, we define the solution of the initial-value problem (2.52).

Definition 2.31 For any $t_0 \in \mathbb{R}_+$ and $\alpha > 0$, an $\mathbb{R}^n$-valued random process $x \in \mathscr{PC}([t_0 - r, t_0 + \alpha]; \mathbb{R}^n)$ is said to be a solution of (2.52) if it satisfies the following conditions:

(i) the set of impulses $\mathbb{T} = \{t \in (t_0, t_0 + \alpha] \,\big|\, t = \tau_k(x(t^-)) \text{ for some } k\}$ is finite;
(ii) $x(t)$ is continuous for all $t \in (t_0, t_0 + \alpha]\backslash\mathbb{T}$ and $\mathscr{F}_t$-adapted;
(iii) the functionals $f \in \mathscr{L}_{ad}(\Omega; L[t_0, t_0 + \alpha])$ and $g \in \mathscr{L}_{ad}(\Omega; L^2[t_0, t_0 + \alpha])$;
(iv) for any $t \in (t_0, t_0 + \alpha]$, $\phi \in \mathscr{L}^2_{\mathscr{F}_0}([-r, 0]; \mathbb{R}^n)$, and $\mathscr{I}(t_k, x_{t_k^-})$ that is $\mathscr{F}_{t_k}$-adapted, the following equation

$$x(t) = \begin{cases} \phi(t - t_0), & t \in [t_0 - r, t_0] \\ \phi(0) + \int_{t_0}^{t} f(s, x_s)ds + \int_{t_0}^{t} g(s, x_s)\, dW(s) & \\ \quad + \sum_{\{k: t_k \in (t_0, t]\}} \mathscr{I}(t_k, x_{t_k^-}), & t \in (t_0, t_0 + \alpha] \end{cases} \tag{2.53}$$

holds w.p.1;
(v) for any $t \in \mathbb{T}$, $x(t)$ satisfies the difference equation in (2.52b) w.p.1; and
(vi) x satisfies the initial condition in (2.52c) w.p.1.

We should also mention that, in this definition, we have restricted ourselves to the case where solutions undergo a finite number of impulses over any finite interval. However, letting $t \in (t_0, \infty)$, there would be a countably infinite number of impulses, which represent the simple jump discontinuities of x.

A special class of the SISD (2.52) is when the impulsive instances occur at fixed times, i.e.,

$$dx(t) = f(t, x_t)dt + g(t, x_t)\, dW(t), \quad t \neq \tau_k, \tag{2.54a}$$

$$\Delta x(t) = \mathscr{I}(t, x_{t^-}), \quad t = \tau_k, \tag{2.54b}$$

$$x_{t_0} = \phi(s), \quad s \in [-r, 0]. \tag{2.54c}$$

This system will be studied in later chapters for the stability-like properties.

2.9 Switched Systems

As described in the introductory chapter of this book, a switched system is a combination of a finite number of subsystems (or modes) and a control-based switching logic to organize the switching among the subsystems. In this section, we focus on a mathematical formulation of such a system, including defining what is meant to be a switching signal or law. Then, we state and prove a stability property of the switched system. This result is considered as a warm-up for further stability theorems of the system that will be presented in this book.

2.9.1 System Formulation

Consider the following controlled system

$$\dot{x} = f(t, x) + u(t), \tag{2.55}$$

with the initial state $x(t_0) = x_0 \in \mathbb{R}^n$, where $x : \mathbb{R}_+ \to \mathbb{R}^n$ is the system state, $f : \mathbb{R}_+ \times \mathbb{R}^n \to \mathbb{R}^n$ is the system vector field, and $u \in \mathbb{R}^n$ is the system input having the form

$$u(t) = \sum_{k=1}^{\infty} C_k x(t) l_k(t), \tag{2.56}$$

with C_k being a control gain matrix with appropriate dimensions and $l_k(\cdot)$ being the ladder function defined by

$$l_k = \begin{cases} 1, & t_{k-1} \leq t < t_k, \\ 0, & \text{otherwise.} \end{cases} \tag{2.57}$$

Controller (2.56) can be written as

$$u(t) = C_k x(t), \qquad t \in [t_{k-1}, t_k), \quad k \in \mathbb{N},$$

meaning that the controller $u(t)$ switches its values at every time instant $t = t_k$, i.e., u is a *switching controller*. Accordingly, closed-loop system (2.55) becomes

$$\begin{cases} \dot{x} = f(t, x) + C_k x, \; t \in [t_{k-1}, t_k), \quad k \in \mathbb{N} \\ x(t_0) = x_0. \end{cases} \tag{2.58}$$

This system is called *switched system*. Typically, a general nonlinear switched system takes the form

$$\begin{cases} \dot{x} = f_{\sigma(t)}(t, x), \qquad t \geq t_0, \\ x(t_0) = x_0, \end{cases} \tag{2.59}$$

where $\sigma : [t_0, \infty) \to \mathscr{S} = \{1, 2, \ldots, N\}$, for some $N \in \mathbb{N}$ representing the number of subsystems in the entire switched system, is a piecewise constant function called *switching signal*, also known as a *switching law* or *switching rule*, and takes values in the compact set $\mathscr{S}$, which is also named by the finite state space. The role of σ is to switch among the vector fields on the right-hand side of (2.59), i.e., f_i for all $i \in \mathscr{S}$, so as to accomplish a certain desired task. The solution of (2.59) is generally equipped with a proper switching signal, i.e., it is represented by the pair (x, σ) to emphasize the switching signal in use.

As in systems and control theory, one of the most important problems in switched systems is the search for conditions assuring stability. The basic problems in stability of switched systems are introduced in [5] and classified into the following three categories.

Problem A (*Stability under arbitrary switching*) Finding sufficient conditions to guarantee asymptotic stability of a switched system for an arbitrary switching signal.

Problem B (*Stability by a constrained switching*) Identifying the switching signals for which a switched system is asymptotically stable.

Problem C (*Stabilizability*) Constructing a switching signal that makes a switched system asymptotically stable.

Problems A and B are usually considered under the hypotheses that the individual subsystems are asymptotically stable, while Problem C is considered under the assumption that the individual subsystems are unstable. In this book, we are mainly concerned with Problems B and C.

We have mentioned earlier that switched systems inherit the stability properties of the fundamental theory of single mode systems. However, a possible strange behavior is that switching among all asymptotically stable subsystems does not necessarily guarantee the stability of switched system. The remedy to this undesirable situation is to design a logic-based switching law in order to control the transition among the involved subsystems. It has been shown in [5–7] that, if the running time of each single mode is sufficiently large to allow the switching effect to diminish, then it ensures that the entire switched system preserves the same stability property. This type of switching is often named by *slow* or *constrained switching* and the running time between any two successive switching moments, say t_k for any $k \in \mathbb{N}$, is called *dwell time* and is denoted by τ. This type of switching signals can be represented by

$$\mathscr{S}_{\text{inf}}(\tau) = \{\tau \mid \inf t_k - t_{k-1} \geq \tau,\ \forall k \in \mathbb{N}\}, \tag{2.60}$$

for some $\tau > 0$.

In fact, the above switching signals (or dwell-time conditions) are particularly useful for linear switched or restricted nonlinear systems. Along this line of dwell-time-type conditions that is applicable for general nonlinear systems are *state-dependent dwell-time* [8, 9] or *initial-state-dependent dwell-time condition*, denoted by τ_{isd}, [10]. For time being, it suffices to state the latter one for a stochastic switched system as it will be used later in this monograph. For $k \in \mathbb{N}$ and $i \in \mathscr{S}$, the τ_{isd} condition is defined as follows:

$$\tau_{\text{isd}} = \left\{ t_k - t_{k-1} \geq \ln \frac{\theta_{2_1}(a_{k-1}\mathbb{E}[\|x_0\|^p])}{\theta_{1_i}(a_k\mathbb{E}[\|x_0\|^p])} \right\},$$

where a_k are positive real constants with $a_0 = 1$, $a_k < a_{k-1}$ and $\lim_{k\to\infty} a_k = 0$, and θ_{1_i} and θ_{2_i} are some nonlinear class$-\mathscr{K}_\infty$ functions. Clearly, if the switched system is linear and, hence, θ's are identity functions, i.e., $\theta(s) = s$, then the τ_{isd} reduces to the dwell-time condition (2.60).

From a practical perspective, it may not be suitable to activate every individual subsystem over a time period τ to accomplish the asymptotic stability property. Instead, to achieve the same qualitative property, as proposed in [7], the *average dwell time*, denoted by τ_{ave}, can be taken sufficiently large. This type of switching signals, denoted by $\mathscr{S}_{\text{ave}}(\tau, N_0)$, is defined as follows: for any $T \geq t \geq t_0$,

$$N_\sigma(T,t) \leq N_0 + \frac{T-t}{\tau_{\text{ave}}}, \tag{2.61}$$

where $N_\sigma(T,t)$ represents the number of switching moments of σ in the interval (t, T) and N_0 is the chatter bound.

A more general class of switching signal than $\mathscr{S}_{\text{inf}}(\tau)$ is called *Markovian switching*, in which the signal σ is a right-continuous Markov chain (or process), which takes values in a finite state space $\mathscr{S}$ with generator $\Gamma = (\gamma_{ij})_{N\times N}$; that is, the jumps among the system modes follow a probabilistic rule defined by

$$\mathbb{P}\{r(t+h) = j|r(t) = i\} = \begin{cases} \gamma_{ij}h + o(h), & \text{if } i \neq j, \\ 1 + \gamma_{ii}h + o(h), & \text{if } i = j, \end{cases} \tag{2.62}$$

where $h > 0$. Here, $\gamma_{ij} > 0$ is the transition rate from i to j if $i \neq j$, and $\gamma_{ii} = -\sum_{j=1, j\neq i}^{N} \gamma_{ij}$ and $o(h)$ is such that $\lim_{h\to 0} \frac{o(h)}{h} = 0$.

Conventionally, if the switching signal is represented by a Markov process, the corresponding switched system (2.59) has the form

$$\begin{cases} \dot{x}(t) = f(t, x(t), \sigma(t)), & t \geq t_0, \\ x(t_0) = x_0, \quad \sigma(t_0) = \sigma_0, \end{cases} \tag{2.63}$$

for some initial state $\sigma_0 \in \mathscr{S}$.

2.9.2 Systems with Stable Subsystems

In the following theorem, we state sufficient conditions that guarantee exponential stability of the linear system

$$\dot{x} = A_i x, \qquad t \in [t_{k-1}, t_k), \tag{2.64a}$$

$$x(0) = x_0, \tag{2.64b}$$

where $k \in \mathbb{N}$, $t_{k-1} < t_k$ with $\lim_{k\to\infty} = \infty$ and $A_i \in \mathbb{R}^{n\times n}$ for each $i \in \mathscr{S}$.

Theorem 2.13 *Consider the switched system (2.64). Let A_i be a Hurwitz matrix for each $i \in \mathscr{S}$. Then, trivial solution, $x \equiv 0$, of (2.64) is exponentially stable if the following inequality holds:*

$$\ln \mu - \nu(t_k - t_{k-1}) \leq 0, \qquad k \in \mathbb{N} \tag{2.65}$$

where $\mu = \frac{\lambda_M}{\lambda_m}$ with $\lambda_M = \max\{\lambda_{\max}(P_i)$ for all $i \in \mathscr{S}\}$, $\lambda_m = \min\{\lambda_{\min}(P_i)$ for all $i \in \mathscr{S}\}$, P_i is a positive-definite matrix satisfying Lyapunov matrix equation

$$A_i^T P_i + P_i A_i = -Q_i, \tag{2.66}$$

for any positive-definite matrix Q_i, and ν is such that $0 < \nu < \lambda_i$ where $\lambda_i = c_i/\lambda_M$ with c_i being a positive constant such that

$$\frac{\partial V_i}{\partial x} A_i x \leq -c_i \|x\|^2. \tag{2.67}$$

Proof For all $t \geq t_0$ with $t_0 \in \mathbb{R}_+$, let $(x(t), \sigma(t))$ (or for simplicity $x(t) = x(t; t_0, x_0)$) with σ taking values in $\mathcal{S}$ be the solution of (2.64). For any $i \in \mathcal{S}$, define the Lyapunov function by

$$V_i(x) = x^T P_i x. \tag{2.68}$$

Then, it is guaranteed that

$$\lambda_m \|x\|^2 \leq V_i(x) \leq \lambda_M \|x\|^2 \tag{2.69}$$

and along the trajectories of (2.64) the time derivative of V_i satisfies

$$\frac{\partial V_i}{\partial x} A_i x \leq -c_i \|x\|^2. \tag{2.70}$$

Combining (2.69) and (2.70) leads to

$$\dot{V}_i(x) \leq -\lambda_i V_i(x)$$

where $\lambda_i = c_i/\lambda_M$. The solution of this differential inequality is

$$V_i(x(t)) \leq V_i(x(t_{k-1}))e^{-\lambda_i(t-t_{k-1})} \tag{2.71}$$

From (2.69), we have, for any $i, j \in \mathcal{S}$

$$V_j(x(t)) \leq \mu V_i(x(t)), \qquad \text{where } \mu = \frac{\lambda_M}{\lambda_m}. \tag{2.72}$$

For instance, activating subsystems 1 and 2 on the first and second intervals, respectively, gives

$$\begin{aligned} V_1(x(t)) &\leq e^{-\lambda_1(t-t_0)} V_1(x_0), \qquad t \in [t_0, t_1) \\ V_2(x(t)) &\leq e^{-\lambda_2(t-t_1)} V_2(x(t_1)), \qquad t \in [t_1, t_2) \\ &\leq e^{-\lambda_2(t-t_1)} \mu V_1(x(t_1)) \\ &\leq \mu e^{-\lambda_2(t-t_1)} e^{-\lambda_1(t_1-t_0)} V_1(x_0) \end{aligned}$$

Generally, for any $i \in \mathcal{S}$ and $t \in [t_{k-1}, t_k)$, one may get

$$V_i(x(t)) \leq \mu^{i-1} e^{-\lambda_i(t-t_{k-1})} e^{-\lambda_{i-1}(t_{k-1}-t_{k-2})} \cdots e^{-\lambda_1(t_1-t_0)} V_1(x_0). \tag{2.73}$$

Let $\lambda = \min\{\lambda_i \text{ for all } i \in \mathcal{S}\}$. Then

$$\begin{aligned}
V_i(x(t)) &\le \mu^{i-1} e^{-\lambda(t-t_0)} V_1(x_0) \\
&= \mu^{i-1} e^{-\nu(t-t_0)} e^{-(\lambda-\nu)(t-t_0)} V_1(x_0) \\
&\le \mu^{i-1} e^{-\nu(t_k-t_0)} e^{-(\lambda-\nu)(t-t_0)} V_1(x_0) \\
&= \mu^{i-1} e^{-\nu(t_1-t_0)} e^{-\nu(t_2-t_1)} \cdots e^{-\nu(t_k-t_{k-1})} V_1(x_0) e^{-(\lambda-\nu)(t-t_0)}
\end{aligned} \tag{2.74}$$

or, for $t \in [t_{k-1}, t_k)$

$$V_i(x(t)) \le \mu e^{-\nu(t_1-t_0)} \mu e^{-\nu(t_2-t_1)} \cdots \mu e^{-\nu(t_k-t_{k-1})} V_1(x_0) e^{-(\lambda-\nu)(t-t_0)} \tag{2.75}$$

Provoking the switching signal in (2.65), we obtain

$$V_i(x(t)) \le V_1(x_0) e^{-(\lambda-\nu)(t-t_0)}, \qquad \forall t \ge t_0.$$

By (2.69), we have

$$\|x(t)\| \le K \|x_0\| e^{-(\lambda-\nu)(t-t_0)/2}, \qquad \forall t \ge t_0$$

where $K = \sqrt{\mu}$. This shows that the trivial solution $x \equiv 0$ of the switched system is exponentially stable.

Remark 2.2 In Theorem 2.13, one can write condition (2.65) as follows:

$$t_k - t_{k-1} \ge \frac{\ln \mu}{\nu} =: \tau \qquad k \ge 1. \tag{2.76}$$

The fixed positive constant τ is called *dwell time*. Also, Theorem 2.13 says that if the switched system has exponentially stable subsystems and the interval between any two consecutive discontinuities is larger than τ, then the trivial solution of system (2.64) is exponentially stable. Hespanha and Morse in [7] showed that a similar result still holds if the dwell time condition is not satisfied, but the average interval between consecutive discontinuities in no smaller than τ. In the latter case, τ (or often denoted by τ_{ave}) is called the *average dwell time* [7, 11]. To consider τ_{ave} in the proof, let the number of switchings in the time interval (t_0, t), $N(t_0, t)$, satisfies

$$N(t_0, t) \le N_0 + \frac{t - t_0}{\tau_{\text{ave}}}, \tag{2.77}$$

where N_0 is defined as the chatter bound. Then, rewrite the inequality in (2.73) as follows:

$$V_i(x) \le e^{(i-1)\ln \mu - \lambda(t-t_0)} V_1(x_0).$$

Let $N_0 = \eta / \ln \mu$ (with $\mu \neq 1$) where η is an arbitrary constant and $\tau_{\text{ave}} = \ln \mu / (\lambda - \lambda^*)$ where $(\lambda^* < \lambda)$. Then, applying the average dwell-time in (2.77) leads to

$$V_i(x) \leq e^{\eta - \lambda^*(t-t_0)} V_1(x_0).$$

It is worth mentioning that the last inequality can be found as follows:

$$\begin{aligned}(i-1)\ln \mu - \lambda(t-t_0) &\leq \left(N_0 + \frac{t-t_0}{\tau_{\text{ave}}}\right) \ln \mu - \lambda(t-t_0) \\ &= \left(\frac{\eta}{\ln \mu} + \frac{(\lambda - \lambda^*)(t-t_0)}{\ln \mu}\right) \ln \mu - \lambda(t-t_0) \\ &= \eta - \lambda^*(t-t_0).\end{aligned}$$

2.10 Stochastic Switched Systems with Time Delay

In the nonlinear switched system (2.59), if we consider time delay and random noise, we are led to the following nonlinear *stochastic switched systems with time delay* (SSSD)

$$\begin{cases} dx(t) = f_{\sigma(t)}(t, x_t)dt + g_{\sigma(t)}(t, x_t)dW(t), & t \geq t_0, \\ x_{t_0}(s) = \phi(s), \quad s \in [-r, 0], \end{cases} \tag{2.78}$$

where $f_\sigma : \mathbb{R}_+ \times \mathscr{C}([-r, 0]; \mathbb{R}^n) \to \mathbb{R}^n$ is assumed to belong to the function class $\mathscr{L}_{ad}(\Omega; L[a, b])$ for some $a, b \in \mathbb{R}_+$ with $a < b$, $g_\sigma : \mathbb{R}_+ \times \mathscr{C}([-r, 0]; \mathbb{R}^n) \to \mathbb{R}^{n\times m}$ represents the noise intensity, which belongs to the function class $\mathscr{L}_{ad}(\Omega; L^2[a, b])$, $W : \mathbb{R}_+ \times \Omega \to \mathbb{R}^m$ is m-dimensional Wiener process defined on the complete probability space $(\Omega, \mathscr{F}_t, \{\mathscr{F}_t\}_{t\geq t_0}, \mathbb{P})$ and $\phi : \mathbb{R}_+ \to \mathbb{R}^n$ is the initial function, which belongs to a class of $\mathscr{F}_t$-measurable $\mathscr{C}([-r, 0]; \mathbb{R}^n)$ random variable ϕ with $\mathbb{E}[\|\phi\|_r^p] < \infty$. The latter function class is denoted by $L^p_{\mathscr{F}_0}([-r, 0]; \mathbb{R}^n)$ for some $p > 0$.

In the following, we define the solution of SSSD.

Definition 2.32 For all $t \in [t_0, T]$ with $t_0, T \in \mathbb{R}_+$ and $t_0 < T$, and $\mathbb{R}^n$-valued random process $x(t) = x(t; t_0, \phi)$, the pair $(x(t), \sigma(t))$ is said to be a *solution* of SSSD in (2.78) if it has the following properties:

(i) $x(t)$ is continuous and adapted with respect to the filtration $\{\mathscr{F}_t\}_{t\geq t_0}$;
(ii) $f_{\sigma(t)}(t, x_t) \in \mathscr{L}_{ad}(\Omega; L[t_0, T])$ and $g_{\sigma(t)}(t, x_t) \in \mathscr{L}_{ad}(\Omega; L^2[t_0, T])$; and
(iii) the stochastic integral equation

$$x(t) = \phi + \int_{t_0}^{t} f_{\sigma(s)}(s, x_s)ds + \int_{t_0}^{t} g_{\sigma(s)}(s, x_s)dW(s) \tag{2.79}$$

holds w.p.1, where $x(t) = \phi(t)$ for all $t \in [-r, 0]$.

For simplicity of notation, we denote the solution of (2.78) by the process x, after dropping out the switching signal σ. Also, to avoid any confusion between the domains of the solution x and switching signal σ, we state it clearly that x is defined for all $t \geq -r$, while σ is defined over $\mathbb{R}_+$.

A solution $x(t)$ of a stochastic differential equation is said to be unique if any other solution $y(t)$ is indistinguishable form $x(t)$ for all $t \geq -r$.

Classical hypotheses that ensure the existence of a unique solution of SSSD are that the vector fields satisfy a linear growth condition and Lipschitz condition in the second variable. The following theorem summarizes these conditions.

Theorem 2.14 *Let $\sigma : \mathbb{R}_+ \to \mathscr{S}$ be a switching signal. Assume that there exist a positive constant C such that functionals f_σ and g_σ satisfy the following conditions:*

$$\|f_{\sigma(t)}(t, \psi)\|^2 + \|g_{\sigma(t)}(t, \psi)\|^2 \leq C(1 + \|\psi\|_r^2), \qquad (a.s.) \tag{2.80}$$

for all $t \in \mathbb{R}_+$ and $\psi \in \mathscr{C}([-r, 0]; \mathbb{R}^n)$, and

$$\begin{aligned}\|f_{\sigma(t)}(t, \psi_1) &- f_{\sigma(t)}(t, \psi_2)\|^2 \\ &+ \|g_{\sigma(t)}(t, \psi_1) - g_{\sigma(t)}(t, \psi_2)\|^2 \leq C\|\psi_1 - \psi_2\|_r^2, \qquad (a.s.)\end{aligned} \tag{2.81}$$

for all $t \in \mathbb{R}_+$ and $\psi_1, \psi_2 \in \mathscr{C}([-r, 0]; \mathbb{R}^n)$. Then, there exists a unique solution x defined for all $t \geq -r$ with the initial function $\phi \in L^p_{\mathscr{F}_0}([-r, 0]; \mathbb{R}^n)$. Furthermore, the solution x satisfies

$$\mathbb{E}\Big[\sup_{-r \leq t \leq T} \|x(t)\|^2\Big] < \infty, \qquad \textit{for all } T > 0. \tag{2.82}$$

Once again, if the switching signal σ is a Markov process, which is assumed to be independent of the Wiener process, the corresponding SSSD is conventionally written as

$$\begin{cases} dx(t) = f(t, x_t, \sigma(t))dt + g(t, x_t, \sigma(t))dW(t), & t \geq t_0, \\ x_{t_0}(s) = \phi(s), \quad s \in [-r, 0], \\ \sigma(t_0) = \sigma_0, \end{cases} \tag{2.83}$$

where $f : \mathbb{R}_+ \times \mathscr{C}([-r, 0]; \mathbb{R}^n) \times \mathscr{S} \to \mathbb{R}^n$ and $g : \mathbb{R}_+ \times \mathscr{C}([-r, 0]; \mathbb{R}^n) \times \mathscr{S} \to \mathbb{R}^{n\times m}$, for some $\sigma_0 \in \mathscr{S}$, and $W : \mathbb{R} \to \mathbb{R}^m$ is a Weiner process. The solution x of SSSD in (2.83) can be similarly defined as the solution of (2.78) except that the stochastic integral is slightly modified as follows:

$$x(t) = \phi + \int_{t_0}^t f(s, x_s, \sigma(s))ds + \int_{t_0}^t g(s, x_s, \sigma(s))dW(s), \tag{2.84}$$

which is required to hold w.p.1. Common assumptions guaranteeing the existence of a unique solution are stated in the following theorem.

Theorem 2.15 *Let $\sigma : \mathbb{R}_+ \to \mathscr{S}$ be a switching signal that is represented by a Markov process. Assume that there exist a positive constant C such that the functionals f and g satisfy the following conditions:*

$$\|f(t, \psi, \sigma(t))\|^2 + \|g(t, \psi, \sigma(t))\|^2 \le C(1 + \|\psi\|_r^2), \qquad \text{(a.s.)} \tag{2.85}$$

for all $t \in \mathbb{R}_+$ and $\psi \in \mathscr{C}([-r, 0]; \mathbb{R}^n)$, and

$$\begin{aligned}\|f(t, \psi_1, \sigma(t)) - f(t, \psi_2, \sigma(t))\|^2 \\ + \|g(t, \psi_1, \sigma(t)) - g(t, \psi_2, \sigma(t))\|^2 \le C\|\psi_1 - \psi_2\|_r^2, \qquad \text{(a.s.)}\end{aligned} \tag{2.86}$$

for all $t \in \mathbb{R}_+$ and $\psi_1, \psi_2 \in \mathscr{C}([-r, 0]; \mathbb{R}^n)$. Then, there exists a unique solution x defined for all $t \ge -r$ with the initial function $\phi \in L^p_{\mathscr{F}_0}([-r, 0]; \mathbb{R}^n)$. Furthermore, the solution x satisfies

$$\mathbb{E}\Big[\sup_{-r \le t \le T} \|x(t)\|^2\Big] < \infty, \qquad \textit{for all } T > 0. \tag{2.87}$$

In previous section, we introduced an important diffusion operator ($\mathscr{L}$, or $\mathscr{L}V$ as a single notation) associated with the underlying stochastic differential equation and then examined its estimated upper bound along the trajectories of the system solutions. In SSSD, we continue to present such an operator. However, due to the deterministic or probabilistic nature of the switching signal σ, the operator can be defined accordingly. Particularly, if σ is of a deterministic type, then we define $\mathscr{L}_i$ (or $\mathscr{L}V_i$) as before, where i is such that $\sigma = i \in \mathscr{S}$; that is, $\mathscr{L}_i$ (or $\mathscr{L}V_i$) is the operator of the solution process of the subsystem associated with the $\mathscr{C}^{1,2}$-function V_i, which is designated to the same subsystem. If σ, on the other hand, is a Markov process, one has to take into account the transition rates of this jump process when writing this operator. In the following definition, we state the generalized Itô formula [12].

Definition 2.33 (*Generalized Itô Formula*) If $x : [-r, \infty) \to \mathbb{R}^n$ is an Itô process governed by (2.83) and $V(t, x, i) \in \mathscr{C}^{1,2}(\mathbb{R}_+ \times \mathbb{R}^n \times \mathscr{S}; \mathbb{R}_+)$ with $\sigma = i \in \mathscr{S}$, then $V(t, x, i)$ is an Itô process with its differential equation given by

$$dV(t, x_t, i) = \mathscr{L}V(t, x_t, i)dt + V_x(t, x, i)g(t, x_t), i)dW(t), \quad \text{(a.s.)} \tag{2.88}$$

where

$$
\begin{aligned}
\mathscr{L}V(t, x_t, i) = & V_t(t, x(t), i) + V_x(t, x(t), i) f(t, x_t, i) \\
& + \frac{1}{2}\mathrm{tr}[g^T(t, x_t, i) V_{xx}(t, x(t), i) g(t, x_t, i)] \\
& + \sum_{j=1}^{N} \gamma_{ij} V(t, x(t), j), \qquad \text{(a.s.).}
\end{aligned} \tag{2.89}
$$

Remark 2.3 For simplicity of notation, we wrote the differential dV and functional operator $\mathscr{L}V$ in terms of x_t only while they also depend on state $x(t)$.

In analyzing a certain switched system, it may be convenient to specify the switching signal σ in $\mathscr{S}$ to indicate the system mode in action and the subinterval on which the selected mode is being activated. If, for instance, we have chosen a switching law, say Θ, then generally, we use i_k to refer to the ith mode, for any $i \in \mathscr{S}$, and kth subinterval $[t_{k-1}, t_k)$, for any $k \in \mathbb{N}$. Also, we denote by $\{t_k\}_{k\in\mathbb{N}}$ the switching sequence or signal, which is generated by the switching law Θ. Furthermore, whenever investigating a system property, we always assume that the switching sequence is strictly increasing and that $\lim_{k\to\infty} t_k = \infty$, so long as $t \in \mathbb{R}_+$, to avoid a problem trivialness. The second issue of importance is that any mode cannot be activated on any two successive subintervals $[t_{k-1}, t_k)$ and $[t_k, t_{k+1})$, and the switching sequence in this case is usually called *minimal*. Consequently, following the above particular notation, SSSD in (2.83) is simply written as follows:

$$
\begin{cases}
dx(t) = f(t, x_t, i)dt + g(t, x_t, i)dW(t), & t \in [t_{k-1}, t_k), \\
x_{t_0}(s) = \phi(s), \quad s \in [-r, 0], \\
\sigma(t_0) = \sigma_0.
\end{cases} \tag{2.90}
$$

One more issue about switched systems is the stability definition. In fact, it can be formulated parallel to that of a single-mode system except that, in switched systems, we should highlight the switching law under consideration. In the following, we state some stochastic stability properties of the trivial solution of SSSD in (2.83), which of course imply the corresponding definitions of the other special systems.

Definition 2.34 For any $t_0 \in \mathbb{R}_+$ and a given switching law σ with an initial state $\sigma_0 \in \mathscr{S}$, the trivial solution, $x \equiv 0$, of (2.83) is said to be

(i) *stable in the pth moment* if, for any given $\varepsilon > 0$, there exists a $\delta = \delta(t_0, \varepsilon) > 0$ such that

$$
\mathbb{E}[\|\phi\|_r^p] < \delta \qquad \text{implies} \qquad \mathbb{E}[\|x(t)\|^p] < \varepsilon, \quad \forall t \geq t_0,
$$

where $(x(t), \sigma(t))$ or simply $x(t) = x(t; t_0, \phi) \in \mathscr{C}([t_0 - r, t_0 + \alpha]; \mathbb{R}^n)$, for some $\alpha > 0$, is any solution of (2.83) with $\phi \in L^p_{\mathscr{F}_0}\mathscr{C}([-r, 0]; \mathbb{R}^n)$;

(ii) *uniformly stable in the pth moment* if it is stable in the pth moment and $\delta = \delta(\varepsilon)$;

(iii) *asymptotically stable in the pth moment* if it is stable in the pth moment and there exists an $\eta = \eta(t_0) > 0$ such that

$$\mathbb{E}[\|\phi\|_r^p] < \eta \quad \text{implies} \quad \lim_{t\to\infty} \mathbb{E}[\|x(t)\|^p] = 0;$$

(iv) *uniformly asymptotically stable in the pth moment* if it is uniformly stable in the pth moment and there exists $\eta > 0$ such that, for a given $\gamma > 0$, there exists $T = T(\eta, \gamma) > 0$ such that

$$\mathbb{E}[\|\phi\|_r^p] < \eta \quad \text{implies} \quad \mathbb{E}[\|x(t)\|^p] < \gamma, \quad \forall t \geq t_0 + T;$$

(v) *exponentially stable in the pth moment* if there exist positive constants K and λ such that

$$\mathbb{E}[\|x(t)\|^p] \leq K\mathbb{E}[\|\phi\|_r^p]e^{-\lambda(t-t_0)}, \qquad \text{whenever} \quad \mathbb{E}[\|\phi\|_r^p] < \eta.$$

Moreover, the above stability properties are said to hold *globally* if δ and η are chosen arbitrarily large.

Having familiarized ourselves with impulsive and switched systems, we are in a position to define another type of hybrid systems, namely *impulsive-switched systems*, also known as *switched systems with impulsive effects*. The impulses arise when a switched system transits from one mode to another. Such systems have applications in biology, pulse vaccination and engineering. An early study that formulated this system and developed some of its qualitative results was in [13]. Later, this type of systems was appeared in some other works including papers [14, 15] and a book [16].

A nonlinear *deterministic impulsive-switched system* can have the following form

$$\dot{x}(t) = f_{\sigma(t)}(t, x(t)), \qquad t \neq t_k, \tag{2.91a}$$
$$\Delta x(t) = \mathscr{I}(t, x(t^-)), \qquad t = t_k, \tag{2.91b}$$
$$x(t_0) = x_0, \tag{2.91c}$$

where $\sigma : [t_0, \infty) \to \mathscr{S}$ for any $t_0 \in \mathbb{R}_+$ is the switching signal that is a piecewise constant function. The discontinuities of σ, which represent both the impulsive moments and at the same time switching moments, form a strictly increasing sequence $\mathbb{T} = \{t_k\}_{k\in\mathbb{N}}$ with $\lim_{k\to\infty} t_k = \infty$. As elaborated above, if one is interested in labeling a system mode which is operating on the kth subinterval, we will write $\sigma = i_k$ for any $i_k \in \mathscr{S}$. It follows that the differential equation (2.91a) is written as follows:

$$\dot{x}(t) = f_{i_k}(t, x(t)), \qquad t \in [t_{k-1}, t_k).$$

We next define a solution of the initial-value problem in (2.91).

Definition 2.35 For any $t \geq t_0$ with $t_0 \in \mathbb{R}_+$, $x \in \mathscr{PC}([t_0 - r, t_0 + \alpha]; \mathbb{R}^n)$, for some $\alpha > 0$, and a given switching signal σ, the pair $(x(t), \sigma(t))$ is said to be a *solution* of the impulsive-switched system in (2.91) if

(i) $x(t)$ is continuous for all $t \in \mathbb{R}_+$ except at the switching (or impulsive) moments $\mathbb{T} = \{t_k\}_{k \in \mathbb{N}}$ (i.e., $\forall t \in \mathbb{R}_+ \setminus \mathbb{T}$);
(ii) the derivative of x exists and continuous for all $t \neq t_k$ and, at $t = t_k$, the right-hand derivative exists;
(iii) the right-hand derivative of x satisfies the differential equation in (2.91a) for all $t \in \mathbb{R}_+ \setminus \mathbb{T}$;
(iv) x satisfies the difference equation (2.91b) for all $t \in \mathbb{T}$; and
(v) x satisfies the initial condition in (2.91c).

Finally, it could be of special interest to write the general form of the above solution, which is, after using the so-called method of steps,

$$x(t) = x_0 + \int_{t_0}^{t} f_{i_k}(s, x(s))ds + \sum_{\{k: t_0 < t_k \le t\}} \mathscr{I}(t_k, x(t_k^-)), \tag{2.92}$$

for all $t \geq t_0$.

2.11 Singularly Perturbed Systems

In networks or in models of large-scale interconnected systems such as power systems, large economies, control systems, biochemical or nuclear reactor models, one encounters dynamics with different speeds or multiple timescales. The corresponding dynamical systems are often known as *singularly perturbed systems* or *multiscale systems*. Mathematically, a singularly perturbed system is a dynamical system in which a small parasitic parameter multiplies time derivatives of some of the system states.

Assume that the dynamics in the aforementioned systems have the the following form:

$$\begin{aligned} \dot{x} &= f(t, x, z) \\ \dot{z} &= G(t, x, z) \end{aligned} \tag{2.93}$$

where $x \in \mathbb{R}^m$ is the slow variable and $z \in \mathbb{R}^n$ is the fast variable. Here, we assume that during the fast transients the slow dynamics remain approximately constant and that, over longer time, they become noticeable, while the fast dynamics have already reached their quasi-steady states. Therefore, as we shall see in later chapters of this book, in a short period of time, slow variables are considered constant, and fast variables eventually reach their quasi-steady state. Over long period of time, the system variables are represented by slow variables and the quasi-steady state of the fast variables, as shown in the following system:

$$\begin{aligned} \dot{x}_s &= f(t, x_s, z_s) \\ 0 &= G(t, x_s, z_s) \end{aligned} \tag{2.94}$$

where x_s and z_s are referred to as quasi-steady states. Clearly, the second equation has degenerated into an algebraic (or transcendental) equation, meaning that the time-varying variable is treated as constant ($\dot{z} = 0$). To remove this mathematical inconsistency, system (2.93) is treated as a two-time-scale singular perturbation problem with a perturbation parameter, say ε. Re-scaling the timescale of system (2.93) yields the so-called singularly perturbed system or fast–slow system:

$$\begin{aligned} \dot{x} &= f(t, x, z) \\ \varepsilon\dot{z} &= g(t, x, z) \end{aligned} \tag{2.95}$$

where $g = \varepsilon G$ with $0 < \varepsilon \ll 1$.

Setting $\varepsilon = 0$ reduces the dimension of the full state from $m + n$ to m. Then, system (2.95) becomes

$$\begin{aligned} \dot{x} &= f(t, x, h(t, x)) \\ 0 &= g(t, x, z), \end{aligned}$$

where $h(t, x)$ is the solution of the algebraic equation $0 = g(t, x, z)$ (i.e., $z = h(t, x)$). The result is the same as that of (2.95), but the derivation is now different.

In reality, the perturbation parameter ε has different meaning in different systems. For instance, in some power systems it indicates machine reactance, in a biochemical model ε might represent a small quantity of an enzyme, and in nuclear reactors model ε is due to the fast neutrons.

In fact, to study the stability notion of this system, it is very convenient to treat this system as a large-scale system or an interconnection of lower order subsystems. As will be seen later in this book, a proper way to deal with such complex systems is to decompose interconnected systems into small isolated subsystems and study the stability of each individual subsystem.

That is, after initially ignoring the interconnection between the subsystems, we study the stability property of each isolated subsystem. In the next step, we combine our results from the first step with the connection among these subsystems, which are viewed as a perturbation, to draw a conclusion about the stability of the interconnected system. The following analysis explains these two steps:

Let the nth-order interconnected system has the form

$$\dot{x}_i = f_i(t, x_i) + g_i(t, x), \quad i = 1, 2, \ldots, m \tag{2.96}$$

where $x_i \in \mathbb{R}^{n_i}$ with $\sum_{i=1}^{m} n_i = n$ and $x = \left(x_1^T \; x_2^T \; \cdots \; x_m^T\right)^T$. Assume that for every i and all $t \geq t_0$ with $t_0 \in \mathbb{R}_+$,

$$f_i(t, 0) = 0 \quad \text{and} \quad g_i(t, 0) = 0.$$

That is, system (2.96) admits a trivial solution $x \equiv 0 \in \mathbb{R}^n$. Ignoring the interconnection between the subsystems, $g_i(t, x)$, results in the following m isolated subsystems

$$\dot{x}_i = f_i(t, x_i). \tag{2.97}$$

Assume that the trivial solution $x_i \equiv 0$ of (2.97) (for every i) is uniformly asymptotically stable. Define the scalar composite Lyapunov function candidate

$$V(t, x) = \sum_{i=1}^{m} d_i V_i(t, x_i), \tag{2.98}$$

where $V_i(t, x_i)$ is the Lyapunov function related to the ith subsystem and d_i are positive constant. Then, the time derivative of V along the trajectories of (2.96) is

$$\dot{V}(t, x) = \sum_{i=1}^{m} d_i \Big[\frac{\partial V_i}{\partial t} + \frac{\partial V_i}{\partial x_i} f_i(t, x_i) \Big] + \sum_{i=1}^{m} d_i \frac{\partial V_i}{\partial x_i} g_i(t, x). \tag{2.99}$$

The first term on the right-hand side is negative definite since $V_i s$ are Lyapunov functions for the m asymptotical stable subsystems, while the second term is, generally, indefinite; so that, we assume that $[\partial V_i / \partial x_i] g_i$ is bounded by a nonnegative upper bound. To pursue the analysis mathematically, assume that $V_i(t, x_i)$ satisfies

$$\frac{\partial V_i}{\partial t} + \frac{\partial V_i}{\partial x_i} f_i(t, x_i) \le -\alpha_i \phi_i^2(x_i), \quad t \ge t_0 \tag{2.100}$$

$$\Big\| \frac{\partial V_i}{\partial x_i} \Big\| \le \beta_i \phi_i(x_i) \tag{2.101}$$

where α_i and β_i are positive constants and ϕ_i is a positive-definite function. Suppose that the $g_i(t, x)$ satisfies

$$\| g_i(t, x) \| \le \sum_{j=1}^{m} \gamma_{ij} \phi_j(x_j), \quad i = 1, 2, \ldots, m \tag{2.102}$$

where γ_{ij} are nonnegative constants. Then, the equality in (2.99) leads to

$$\dot{V}(t, x) \le \sum_{i=1}^{m} d_i \Big[-\alpha_i \phi_i^2(x_i) + \sum_{j=1}^{m} \beta_i \gamma_{ij} \phi_i(x_i) \phi_j(x_j) \Big].$$

The right-hand side is a quadratic in $\phi_1, \phi_2, \ldots, \phi_m$; that is,

$$\dot{V}(t,x) \le -\frac{1}{2}\phi^T(DS + S^T D)\phi$$

where $\phi = (\phi_1\ \phi_2\ \cdots\ \phi_m)^T$, $D = \text{diag}(d_1, d_2, \ldots, d_m)$ and S is an $n \times n$ matrix whose elements are given by

$$s_{ij} = \begin{cases} \alpha_i - \beta_i\gamma_{ij}, & i = j \\ -\beta_i\gamma_{ij}, & i \neq j. \end{cases} \tag{2.103}$$

Clearly, the asymptotic stability of the interconnected system is guaranteed if the diagonal matrix D is chosen such that the matrix

$$DS + S^T D > 0. \tag{2.104}$$

The existence of such a diagonal matrix D is ensured if S is an M-matrix as stated in the following definition.

Definition 2.36 An $n \times n$ matrix S is said to be an M-matrix if its leading (successive) principal minors are positive, i.e.,

$$det \begin{pmatrix} s_{11} & s_{12} & \cdots & s_{1k} \\ s_{21} & s_{22} & \cdots & s_{2k} \\ \cdots & \cdots & \cdots & \cdots \\ s_{k1} & s_{k2} & \cdots & s_{kk} \end{pmatrix} > 0, \quad k = 1, 2, \ldots, n.$$

The following lemma provides the sufficient condition that guarantees the existence of D.

Lemma 2.5 *There exists a positive diagonal matrix D that satisfies (2.104) if and only if S is an M-matrix.*

The next theorem summarizes the above results.

Theorem 2.16 *Consider the interconnected system (2.96). Assume that, for* $i = 1, 2, \ldots, m$, *there exists a positive-definite function* $V_i(t, x_i)$ *that satisfies (2.100) and (2.101), and that* $g_i(t, x)$ *satisfies (2.102). If S defined by (2.103) is an M-matrix, then the trivial solution of (2.96) is uniformly asymptotically stable. Moreover, it is globally asymptotically stable if the assumptions hold globally and* $V_i(t, x_i)$ *is radially unbounded.*

The stability property of hybrid singularly perturbed systems will be addressed later in this book.

2.12 Miscellaneous Results

We conclude this chapter by presenting material that will be used throughout this book.

Jensen's Inequality. If $\varphi : \mathbb{R} \to \mathbb{R}$ is a convex function and $x : \Omega \to \mathbb{R}$ is a random variable on a probability space $(\Omega, \mathscr{F}, \mathbb{P})$ such that $\mathbb{E}[x] < \infty$, then

$$\varphi(\mathbb{E}[x]) \leq \mathbb{E}[\varphi(x)].$$

Tchebychev's Inequality. If $x : \Omega \to \mathbb{R}^n$ is a random variable such that $\mathbb{E}[\|x\|^p] < \infty$, for some $p > 0$, then

$$\mathbb{P}\big\{\omega \in \Omega \mid \|x\| \geq \varepsilon\big\} \leq \frac{\mathbb{E}[\|x\|^p]}{\varepsilon^p}, \qquad \text{for some } \varepsilon > 0.$$

Hölder's Inequality. Let x and y be $\mathbb{R}^n$-valued random processes. If $p, q \in (1, \infty)$ and $1/p + 1/q = 1$, then

$$\big|\mathbb{E}[x^T y]\big| \leq \mathbb{E}[\|x\|^p]^{1/p}\mathbb{E}[\|y\|^q]^{1/q}$$

holds provided that the pth moments on the right-hand side are finite.

Bihari's Inequality. For all $t \in [0, T]$ with $T > 0$, let $u(t) \geq 0$ be a Borel measurable function and $v(t) \geq 0$ be an integrable function. Suppose that $K : \mathbb{R}_+ \to \mathbb{R}_+$ is a continuous nondecreasing function such that $K(t) > 0$ for all $t > 0$. If, for some $c > 0$,

$$u(t) \leq c + \int_0^t v(s)K(u(s))ds, \qquad \forall t \in [0, T],$$

then

$$u(t) \leq G^{-1}\Big(G(c) + \int_0^t v(s)ds\Big)$$

holds for all $t \in [0, T]$ such that

$$G(c) + \int_0^t v(s)ds \in \mathrm{Dom}(G^{-1}),$$

where $G(r) = \int_{0+}^r \frac{ds}{K(s)}$, for $r > 0$, and G^{-1} is the inverse function of G.

Let x and y be two $\mathbb{R}^n$-valued random processes having probability measures $\mathbb{P}_x$ and $\mathbb{P}_y$, respectively. Then, the *Prokhorov distance* between the (probability) measures is denoted by $\mathscr{D}(x, y) = \mathscr{D}(\mathbb{P}_x, \mathbb{P}_y)$. Moreover, if $\mathscr{D}(x, y) = 0$, then x and y have the same probability measure. Also, if $\mathbb{P}\big\{\omega \in \Omega \mid \lim_{n\to\infty} \|x_n(\omega) - x(\omega)\| = 0\big\} = 1$, then $\{x_n\}$ is a $\mathscr{D}$-Cauchy sequence. The converse of this fact is true in the following sense.

Skorokhod's Theorem. Let $\{x_n\}$ be a $\mathscr{D}$-Cauchy sequence of random variables. Then, one can construct another sequence of random variables $\{y_n\}$ and a random variable y such that

$$\mathscr{D}(x_n, y_n) = 0 \quad \text{and} \quad \mathbb{P}\big\{\omega \in \Omega \mid \lim_{n\to\infty} \|y_n(\omega) - y(\omega)\| = 0\big\} = 1.$$

Definition 2.37 A collection of sequences of random variables $Q = \{x_r \mid r \in \Lambda\}$, for some index set Λ, is said to be totally $\mathscr{D}$-bounded if every infinite sequence $\{x_{n_r}\} \subset Q$ has a $\mathscr{D}$-Cauchy subsequence.

Prokhorov's Theorem. Q is totally $\mathscr{D}$-bounded if and only if, for every $\varepsilon > 0$, there exists a compact set K_ε of $\mathbb{R}^n$ such that

$$\mathbb{P}\{x \in K_\varepsilon\} > 1 - \varepsilon,$$

for every $x \in Q$.

2.13 Notes and Comments

This chapter serves as an introductory chapter for the rest of this book. The basic definitions of existence and uniqueness of solutions and stability notions of the trivial solution of ordinary differential equations are taken from [3], and the ISS definition and related theorems are taken from [17, 18]. The comparison functions used in the definitions of stability and ISS of nonlinear systems are taken from [3, 19]. Further reading about the ISS, one may consult the references [1–3, 17, 18, 20–23]. The comparison method for ordinary differential equations stated in Sect. 2.2 is taken from [3]. Section 2.3 is concerned with delay differential equations. The stability definitions and theorem are taken from [24–27], Lemma 2.1 and Theorem 2.5 are taken from [28], Lemmas 2.2 and 2.4 are taken from [29, 30], while Lemma 2.3 is taken from [31]. The impulsive systems of ordinary differential equations can be read in [32–36], Theorem 2.6 is taken from [14]. Also, one may read about ISS for impulsive systems, for instance, in [37]. The fundamental properties of impulsive system with time delay were initially developed in [4, 38, 39]. Section 2.7 deals with stochastic differential equations, where readers may refer to [40–50] and comparison method for stochastic system is taken from [51]. In fact, Theorems 2.11 and 2.12 have been slightly modified to fit our needs in this book. Section 2.9 addresses switched systems; the basic problems of the system, the definitions of dwell-time and average dwell-time switching signals can be read in [5–7], the Markovian switching can be read in books such as [12, 52], ISS with Markov switching can be read in [53], the state-dependent condition is taken from [8, 9] and initial-state-dependent dwell-time condition is taken from [10]. Further reading about switched systems can be found, for instance, in [6, 11, 54–83]. The singularly perturbed systems has been addressed in Sect. 2.11. One can read the theory of this system in, for instance, [3, 70, 71, 84–88]. As stated earlier, these systems are viewed as large-scale systems; so that, the theory of the latter systems can be read, for instance, in [3, 89–92]. The definition and properties of the M-matrix can be found in [3, 26]. Finally, in the miscellaneous section, the Jensen's inequality, Tchebychev's inequality and Hölder's inequality are taken from [12], Bihari's inequality is taken from [93] and Skorokhod's Theorem, Definition 2.37 and Prokhorov's Theorem are taken from [51].

References

1. Sontag ED (1989) Smooth stabilization implies coprime factorization. IEEE Trans Autom Control 34(4):435–443
2. Sontag ED (2002) The ISS philosophy as a unifying framework for stability-like behavior. Lectures notes control theory, vol 2. Springer, pp 443–468
3. Khalil HK (2002) Nonlinear systems. Prentice-Hall Inc., NJ
4. Ballinger GH (1999) Qualitative theory of impulsive delay differential equations. PhD thesis, University of Waterloo, ON, Canada
5. Liberzon D, Morse AS (1999) Basic problems in stability and design of switched systems. IEEE Control Syst Mag 19(5):59–70
6. Morse AS (1996) Supervisory control of families of linear set-point controllers-part I: exact matching. IEEE Trans Autom Control 41(10):1413–1431
7. Hespanha JP, Morse AS (1999) Stability of switched systems with average dwell-time. In: Proceeding of the 38th IEEE conference on decision and control. Phoenix, AR, pp 2655–2660
8. De Persis C, De Santis R, Morse AS (2002) Nonlinear switched systems with state dependent dwell-time. In: Proceedings of the 41st IEEE conference on decision and control, Las Vegas, Nevada, pp 4419–4424
9. De Persis C, De Santis R, Morse AS (2003) Switched nonlinear systems with state-dependent dwell-time. Syst Control Lett 50:291–302
10. Alwan MS, Liu XZ, Xie WC (2012) On input-to-state stability of nonlinear stochastic hybrid systems. Dyn Cont Discret Impuls Syst Ser A 19:513–533
11. Hu B, Xu X, Michel AN, Antsaklis PJ (1999) Stability analysis for a class of nonlinear switched systems. In: Proceedings of the 38th IEEE conference on decision and control. Phoenix, AR, pp 4374–4379
12. Mao X, Yuan C (2006) Stochastic differential equations with Markovian switching. Imperial College Press
13. Lakshmikantham V, Liu XZ (1998) Impulsive hybrid systems and stability theory. Dyn Syst Appl 7:1–10
14. Wang R, Liu X, Guan Z (2004) Robustness and stability analysis for a class of nonlinear switched systems with impulse effects. Dyn Syst Appl 14:233–248
15. Guan Z-H, Hill DJ, Shen X (2005) On hybrid impulsive and switching systems and application to nonlinear control. IEEE Trans Autom Control 50(7):1058–1062
16. Li Z, Soh Y, Wen C (eds) (2005) Switched and impulsive systems: analysis design and applications. Lecture notes in control and information sciences. Springer, Berlin
17. Sontag ED, Wang Y (1995) On characterization of input-to-state stability property. Syst Control Lett 24:351–359
18. Sontag ED, Wang Y (1996) New characterization of input-to-state stability property. IEEE Trans Autom Control 34(41):1283–1294
19. Hahn W (1967) Stability of motion. In: Baartz AP (Trans) Springer, New York
20. Angeli D, Sontag ED, Wang Y (2000) A characterization of integral input-to-state stability. IEEE Trans Autom Control 45(6):1082–1097
21. Sontag ED (1998) Comments on integral variants of ISS. Syst Control Lett 34:93–100
22. Teel AR, Moreau L, Nešic D (2001) A note on the robustness of ISS stability. In: Proceedings of the 40th IEEE on decision and control, Florida, pp 875–880
23. Teel AR, Moreau L, Neysic D (2003) A unified framework for input-to-state stability in systems with two time scales. IEEE Trans Autom Control 48(9):1526–1544

24. Driver RD (1977) Ordinary and delay differential equations. Springer, New York
25. El'sgol'ts LE, Norkin SB (1973) Introduction to the theory and application of differential equations with deviating arguments. Academic Press, New York
26. Gopalsamy K (1992) Stability and oscillations in delay differential equations of population dynamics. Kluwer Academic Puplishers, Netherlands
27. Hale JK, Lunel SMV (1993) Introduction to functional differential equations. Springer, New York
28. Halanay A (1966) Differential equations: stability, oscillations, time lags. Academic Press Inc., New York
29. Alwan MS (2006) Stability of hybrid singularly perturbed systems with time delay. Master's Thesis, University of Waterloo, ON, Canada
30. Alwan MS, Liu XZ (2008) On stability of linear and weakly nonlinear switched systems with time delay. Math Comput Model 48:1150–1157
31. Liu X, Shen X, Zhang Y (2003) Exponential stability of singularly perturbed systems with time delay. Appl Anal 82(2):117–130
32. Bainov DD, Simeonov PS (1989) Systems with impulsive effects. Ellis Horwood Ltd., England
33. Bainov DD, Simeonov PS (1993) Impulsive differential equations: periodic solutions and applications. Longman Scientific and Technical Group, England
34. Lakshmiknatham V, Bainov DD, Simeonov PS (1989) Theory of impulsive ifferential equations. World Scientific, Singapore
35. Samoilenko AM, Perestyuk NA (1995) Impulsive differential equations. World Scientific Publishing Co., Inc, River Edge, NJ
36. Yang T (2001) Impulsive control theory. Lecture notes in control and information sciences. Springer, Berlin
37. Hespanha JP, Liberzon D, Teel AR (2005) On input-to-state stability of impulsive systems. In: Proceedings of the 44th IEEE conference on decision and control, Spain, pp 3992–3997
38. Ballinger GH, Liu XZ (1990) Existence and uniqueness results for impulsive delay differential equations. Dyn Contin Discret Impuls Syst 6:579–591
39. Liu XZ, Ballinger GH (2000) Existence, uniqueness, and boundedness results for impulsive delay differential equations. Appl Anal 74(1–2):71–93
40. Gard TC (1988) Introduction to stochastic differential equations. Marcel Dekker
41. Gihman II, Skorohod AV (1972) Stochastic differential equations. In: Wickwire K (Trans Russian) Ergebnisse der Mathematik und ihrer Grenzgeiete, Band 72, Springer, New York
42. Hasm'iniskii RZ (1980) Stochastic stability of differential equations. In: Sijtjoff, Noordhoff (Trans Russian edn.)
43. Khasminiskii RZ (1980) Stochastic stability of differential equations. Sijthof & Noordhoff International Publishers, The Netherlands
44. Kloeden PE, Platen E (1999) Numerical solution of stochastic differential equations. Springer
45. Kushner HJ (1967) Stochastic stability and control. Academic Press Inc., New York
46. Mao X (1991) Stability of stochastic differential equations with respect to semimartingales. Longman Scientific and Technical
47. Mao X (1994) Exponential stability of stochastic differential equations. Marcel Dekker
48. Mao X (2008) Stochastic differential equations and applications. Horwood Publishing Ltd., England
49. Mohammed SEA (1984) Stochastic functional differential equations. Research notes in mathematics, vol 99. Pitman, Boston
50. Øksendal B (1995) Stochastic differential equations: an introduction with applications. Springer
51. Ladde GS, Lakshmikantham V (1980) Random differential inequalities. Academic Press, New York
52. Boukas EK (2006) Stochastic switching systems: analysis and design. Birkhauser
53. Huang L, Mao X (2009) On input-to-state stability of stochastic retarded systems with Markovian switching. IEEE Trans Autom Control 54(8):1898–1902
54. Alur R, Henzinger T, Sontag E (eds) Hybrid systems III-verification and control. Lecture notes in computer science. Springer, Berlin

55. Antsaklis PJ, Kohn W, Nerode N, Sastry S (eds) (1995) Hybrid systems. Lecture notes in computer science, vol 999. Springer, Berlin
56. Antsaklis PJ, Kohn W, Nerode N, Sastry S (eds) (1997) Hybrid systems. Lecture notes in computer science, vol 1273. Springer, Berlin
57. Antsaklis PJ, Koutsoukos X, Zaytoon J (1998a) On hybrid control of complex systems: a survey. Eur J Automat 32(9–10):1023–1045
58. Antsalkis PJ, Lemmon AD (1998b) J Discret Event Dyn Syst 8(2) (Special issue on hybrid control systems)
59. Antsalkis PJ, Nerode A (1998c) IEEE Trans Autom Control 43(Special issue on hybrid control systems)
60. Antsalkis PJ, Kohn W, Lemmon A, Sastry S (eds) (1999) Hybrid systems v. Lecture notes in computer sciences, vol 1567. Springer, Berlin
61. Antsalkis PJ (2000) Special issue on hybrid systems: theory and applications. Proc IEEE 88(7):879–887
62. Branicky MS, Borker VS (1994) A unified framework for hybrid control. In: Proceedings of the 33rd IEEE conference decision and control, Lake Buena Vista, FL, pp 4228–4234
63. Branicky MS (1998) Multiple Lyapunov functions and other analysis tools for switched and hybrid systems. IEEE Trans Autom Control 43(4):475–482
64. Branicky MS, Borker VS (1998) A unified framework for hybrid control: model and optimal control theory. IEEE Trans Autom Control 43(1):31–45
65. Dayawansa WP, Martin CF (1999) A Converse Lyapunov theorem for a class of dynamical systems which undergo switching. IEEE Trans Autom Control 44(4):751–759
66. DeCarlo RA, Branicky MS (2000) Perspective and results on the stability and stabilization of hybrid systems. Proc IEEE 88(7):1069–1082
67. Grossman RL, Nerode N, Ravn AP, Rischel H (eds) (1993) Hybrid systems. Lecture notes in computer science, vol 736. Springer, Berlin
68. Hespanha J, Tiwari A (2006) Hybrid systems: computation and control. In: 9th International Workshop HSCC 2006, Santa Barbara, CA. Lecture notes in computer science and general issues. Springer, USA
69. Henzinger T, Sastry S (eds) Hybrid systems computation and control. Lecture notes in computer sciences, vol 1386. Springer, Berlin
70. Hoppensteadt F (1968) Asymptotic stability in singular perturbation problems. J Differ Equ 4:350–358
71. Hoppensteadt F (1974) Asymptotic stability in singular perturbation problems II: problems having matched asymptotic solutions. J Differ Equ 15:510–521
72. Kim S (2005) Switching systems with delayed feedback control. PhD thesis, University of Waterloo, ON, Canada
73. Kim S, Campbell SA, Liu XZ (1999) Stability of a class of linear switching systems with time delay. IEEE Trans Circuits Syst I Regul Pap 53(2):384–393
74. Lemmon MD, He KX, Markovsky I (1999) Supervisory hybrid systems. Control Syst 19(44):42–55
75. Liberzon D (2003) Switching in systems and control. Brikhauser, Boston
76. Matveev AS, Savkin AV (2000) Qualitative theory of hybrid dynamical systems. Birkhauser, Cambridge, MA
77. Morse AS (ed) (1997) Control using logic-based switching. Lecture notes in control and information sciences, vol 222. Springer, New York
78. Morse A, Pantelides C, Sastry S, Schumacher J (1999) Automatica 35(Special issue on hybrid control systems)
79. Narendra KS, Balakrishnan J (1994) A common Lyapunov function for stable LTI systems with commuting A-matrices. IEEE Trans Autom Control 39(12):2469–2471
80. Peleties P, DeCarlo R (1993) A modeling strategy for hybrid systems based on event structors. Discret Event Dyn Syst 3:39–69
81. Peleties P, DeCarlo R (1988) Modeling of interacting continuous time and discrete events systems: an example. In: Proceedings of the 26th annual alleton conference, pp 1150–1159

82. Peleties P, DeCarlo R (1998) A modeling strategy for discrete events supervisory control of continuous time systems. In Proceedings of the 28th IEEE conference decision and control, pp 1308–1313
83. Sifakis J, Pnueli A (1995) J Theor Comput Sci 138(Special issue on hybrid control systems)
84. Kokotovic PV, Khalil HK (eds) (1986) Singular perturbations in systems and control. IEEE Press Inc., New York
85. Kokotovic PV, Khalil HK, O'Reilly J (1986) Singular perturbation methods in control: analysis and design. Academic Press Inc., London
86. Kokotovic P, Bensoussan A, Blankenship G (eds) (1987) Singular perturbations and asymptotic analysis in control systems. Lecture notes in control and information sciences. Springer, New York
87. Naidu DS (1988) Singular perturbation methodology in control systems. Peter Peregrinus Ltd.
88. Saberi A, Khalil H (1984) Quadratic-type Lyapunov functions for singularly perturbed system. IEEE Trans Autom Control 29(6):542–550
89. Michel AN, Miller RK (1977) Qualitative analysis of large scale dynamical systems. Academic Press Inc., New York
90. Šilijak DD (1991) Decentralized control of complex systems. Academic Press Inc
91. Šiljak DD (2007) Large-Scale dynamic systems: stability and structure. Dover Publications Inc
92. Zecevic A, Silijak DD (2010) Control of complex systems: structural constraints and uncertainty. Springer
93. Bihari I (1956) A generalization of a lemma of Bellman and its application to uniqueness problems of differential equations. Acta Math Hung 7:81–94

Chapter 3
Fundamental Properties of Stochastic Impulsive Systems with Time Delay

In this chapter, we address stochastic impulsive systems with time delay, where the impulse times are state-dependent. Using Itô calculus, we develop the essential foundation of the theory of the mentioned system, namely local and global existence, forward continuation and uniqueness of strong solutions. As a consequence of state-dependent impulses, nonidentical solutions can experience impulses at different times, but not at fixed times as in the case of state-independent impulse times. This also leads to an undesirable phenomenon, namely rhythmical beating upon impulsive hypersurfaces, which may arise when we attempt to extend the solution, unless some further conditions are imposed on these surfaces, as will be seen later.

Particularly, we start with establishing a local existence result assuming that the functionals f and g are bounded by a nonlinear random function having an integrable property. Also, we will state some conditions to ensure that when the solution hits the (impulsive) hypersurface, it will leave it immediately. Due to technical difficulties in extending the solution backwards, we concentrate on the classical forward solution that does not exhibit rhythmical beating phenomenon. A global solution is also obtained under the same bounded nonlinear estimate. Finally, we address the uniqueness problem supposing that a locally Lipschitz condition holds.

Consider the following stochastic impulsive system with time delay

$$dx(t) = f(t, x_t)dt + g(t, x_t)\,dW(t), \quad t \neq \tau_k(x(t^-)), \tag{3.1a}$$

$$\Delta x = \mathscr{I}(t, x_{t^-}), \quad t = \tau_k(x(t^-)), \tag{3.1b}$$

where $x \in \mathbb{R}^n$ is the system state random process, $f \in \mathbb{R}^n$, $g \in \mathbb{R}^{n\times m}$ and $\tau_k \in C^2(\mathbb{R}^n, \mathbb{R}_+)$ represents an impulsive hypersurface, for $k \in \mathbb{N}$, and satisfies $0 = \tau_0(x) < \tau_1(x) < \tau_2(x) < \cdots$ and $\lim_{k\to\infty} \tau_k(x) = \infty$ for $x \in \mathbb{R}^n$. We also assume that the solution of (3.1) is right-continuous (i.e., $x(t^+) = x(t)$). In the difference equation (3.1b), $\Delta x = x(t) - x(t^-)$ and the functional $\mathscr{I}(\cdot)$ is the impulse amount which is assumed to be $\mathscr{F}_{t_k}$-adapted.

M. S. Alwan and X. Liu, *Theory of Hybrid Systems: Deterministic and Stochastic*, Nonlinear Physical Science, https://doi.org/10.1007/978-981-10-8046-3_3

The initial condition is given by

$$x_{t_0} = \phi(s), \quad s \in [-r, 0] \tag{3.1c}$$

where $\phi \in \mathscr{L}^2_{\mathscr{F}_0}([-r, 0] \times \Omega; \mathbb{R}^n)$ (or $\phi \in \mathscr{L}^2_{\mathscr{F}_0}([-r, 0]; \mathbb{R}^n)$ after dropping the probability sample space Ω for simplicity of notation). That is, the initial state is assumed to be $\mathscr{F}_0$-adapted with finite pth moment.

Integrating the stochastic differential equation over (t_0, t) gives

$$x(t) = \phi(0) + \int_{t_0}^{t} f(s, x_s)ds + \int_{t_0}^{t} g(s, x_s)\, dW(s). \tag{3.2}$$

The first integral is a Riemann integral almost surely (a.s.), and the second one is an Itô integral satisfying

$$\mathbb{E}\Big[\int_{t_0}^{t} g(s, x_s)\, dW(s)\Big] = 0, \text{ and } \mathbb{E}\Big\|\int_{t_0}^{t} g(s, x_s)\, dW(s)\Big\|^2 = \int_{t_0}^{t} \mathbb{E}\|g(s, x_s)\|^2\, ds.$$

In the following, we define indistinguishable solutions and forward continuation of a solution.

Definition 3.1 The two random processes $x(t, \omega)$ and $y(t, \omega)$ are said to be indistinguishable if, for almost all $\omega \in \Omega$, $x(t, \omega) = y(t, \omega)$ for all $t \geq 0$, that is

$$\mathbb{P}\{\omega \mid x(t, w) = y(t, w) \quad \text{for all } t \geq 0\} = 1$$

or, for simplicity, we say $x = y$ (a.s.).

Definition 3.2 Let x and y be solutions of the impulsive stochastic system (3.1) that are defined on the intervals J_1 and J_2, respectively, where $J_1 \subset J_2$ and both intervals have the same closed left endpoints. If $x(t)$ and $y(t)$ are indistinguishable for all $t \in J_1$ (i.e., $x(t) = y(t)$ (a.s.) $\forall t \in J_1$), then y is said to be a *proper forward continuation of* x, or simply *continuation of* x. In this case, a solution x defined on J_1 is said to be *continuable*; otherwise, it said to be noncontinuable, and J_1 is called the *maximal interval of existence of* x.

We also need the following lemma to prove the existence result and whose proof is inspired by that of Lemma 2.1.1 in [1].

Lemma 3.1 *Let $\mathbb{N}$ be the set of natural numbers, $\mathscr{D} \subset \mathbb{R}^n$, $a, b \in \mathbb{R}_+$ with $a < b$, and c and ε are some positive constants. Then, the set*

$$Q' = \Big\{x^{(n)} \in C([a, b]; \mathscr{D}) \mid \mathbb{E}[\|x^{(n)}(t)\|^2] \leq c \text{ and }$$
$$\mathbb{E}[\|x^{(n)}(t_1) - x^{(n)}(t_2)\|^2] \leq \varepsilon, \quad \forall n \in \mathbb{N}, \forall t_1, t_2 \in [a, b]\Big\}$$

is totally $D-$bounded subset of $C([a, b]; \mathscr{D})$.

Proof By Tchebychev's inequality, one can find, for every $\varepsilon > 0$, $\gamma_1(\varepsilon)$ and $\gamma_2(\varepsilon)$ such that $\mathbb{P}\{\omega \in \Omega \mid \|x^{(n)}(t)\| > \gamma_1(\varepsilon)\} \leq \frac{\varepsilon}{2}$ and $\mathbb{P}\{\omega \in \Omega \mid \|x^{(n)}(t_1) - x^{(n)}(t_2)\| > \gamma_2(\varepsilon)\} \leq \frac{\varepsilon}{2}$. Hence, $\mathbb{P}\{\omega \in \Omega \mid \|x^{(n)}(t)\| > \gamma_1(\varepsilon)$ or $\|x^{(n)}(t_1) - x^{(n)}(t_2)\| > \gamma_2(\varepsilon)\} \leq \varepsilon$, which implies that $\mathbb{P}\{\omega \in \Omega \mid \|x^{(n)}(t)\| \leq \gamma_1(\varepsilon)$ or $\|x^{(n)}(t_1) - x^{(n)}(t_2)\| \leq \gamma_2(\varepsilon)\} > 1 - \varepsilon$ for every $x \in Q$. For some $\alpha > 0$, let

$$K_\varepsilon = \Big\{x^{(n)} \in \mathscr{C}([t_0, t_0 + \alpha], S) \Big| \, \|x^{(n)}(t)\| \leq \gamma_1(\varepsilon) \text{ and } \\ \|x^{(n)}(t_1) - x^{(n)}(t_2)\| \leq \gamma_2(\varepsilon), \; \forall t_1, t_2 \in [t_0, t_0 + \alpha]\Big\}.$$

Clearly, $\mathbb{P}\{x \in K_\varepsilon\} > 1 - \varepsilon$. By Arzela–Ascoli's theorem, the compactness of K_ε follows. Applying Prokhorov's theorem yields the totally $\mathscr{D}$–boundedness of the subset Q'.

Remark 3.1 Q' is a collection of sequences which are both uniformly bounded and equicontinuous in the m.s.

3.1 Existence of Solution

We start this section with establishing a local existence result of the initial-value problem in (3.1). We first show how the solution evolves between two impulsive hypersurfaces, and then, under the condition in (3.4), if this solution starts initially at a hypersurface, it will depart this surface in mean.

Theorem 3.1 *Let $J \subset \mathbb{R}_+$ and $\mathscr{D} \subset \mathbb{R}^n$ be an open set containing $\phi(0)$. Assume that $f \in \mathscr{L}_{ad}(\Omega; L[t_0, t_0 + \alpha])$ and $g \in \mathscr{L}_{ad}(\Omega; L^2[t_0, t_0 + \alpha])$, where $\alpha > 0$ and $[t_0, t_0 + \alpha] \subset J$, and are continuous in their second argument ψ. Moreover, there exists a (random) function $m(t)$ such that, for $(t, \psi) \in [t_0, t_0 + \beta] \times F$, for some positive $\beta \leq \alpha$ and compact set $F \subset \mathscr{D}$,*

$$\|f(t, \psi)\|^2 \vee \|g(t, \psi)\|^2 \leq m(t), \quad (a.s.) \tag{3.3}$$

where

$$\int_{t_0}^{t} m(s)\, ds < \infty, \qquad (a.s.).$$

Then, for almost all $\omega \in \Omega$ and each $(t, \phi) \in J \times \mathscr{L}^2_{\mathscr{F}_0}([-r, 0]; \mathbb{R}^n)$, there exists a (local) $\mathscr{F}_t$-adapted solution $x(t) = x(t; t_0, \phi)$ of (3.1) on $[t_0 - r, t_0 + \beta]$. Furthermore, assume that $\tau_k \in \mathscr{C}^2(\mathscr{D}; \mathbb{R}_+)$, for $k \in \mathbb{N}$, and, whenever $t^* = \tau_k(x^*)$ for some

$(t^*, x^*) \in J \times \mathscr{D}$ and some k, there exits a $\delta > 0$, where $[t^*, t^* + \delta] \subset J$, such that

$$\mathbb{E}[\mathscr{L}\tau_k(x(t))] \neq 1, \tag{3.4}$$

for all $t \in (t^*, t^* + \delta]$ and for all functions x that are $\mathscr{F}_t$-adapted $PC([t^* - r, t^* + \delta]; \mathscr{D})$ and continuous on $(t^*, t^* + \delta]$ and satisfy $x(t^*) = x^*$ and $\mathbb{E}[\|x(s) - x^*\|^2] < \lambda$ for $s \in [t^*, t^* + \delta]$ and $\lambda > 0$. Then, the solution x leaves the hypersurface $\tau_k(x)$ in mean; i.e., x exists on $[t_0 - r, t_0 + \beta]$ for some $\beta > 0$ for which x will not intersect any impulse hypersurface at any time $t \in (t_0, t_0 + \beta]$.

Proof Let $(t, \phi) \in J \times \mathscr{L}^2_{\mathscr{F}_0}([-r, 0] \times \Omega; \mathbb{R}^n)$ and choose $\alpha > 0$ such that $[t_0, t_0 + \alpha] \subset J$. Since for almost all $\omega \in \Omega$, $\phi(0) \in \mathscr{D}$ and $\mathscr{D}$ is an open set, one can choose $\lambda > 0$ such that

$$F := F(z, \lambda) = \{z \in \mathbb{R}^n \mid \|z - \phi(0)\| \leq \lambda\} \subset \mathscr{D}. \tag{3.5}$$

Clearly, F is a compact set. Set

$$M(t) = \int_{t_0}^{t} m(s)\, ds, \qquad t \in [t_0, t_0 + \alpha].$$

Then, $M(t)$ is absolutely continuous (a.s.) and nondecreasing (a.s.). Also, $M(t_0) = 0$ and $M(t)$ is bounded (a.s.). Therefore, there is a positive number, say $\widetilde{M}$, such that

$$M(t) = \int_{t_0}^{t} m(s)\, ds \leq \widetilde{M}, \qquad t \in [t_0, t_0 + \alpha].$$

Let $\beta = \min\{\alpha, \frac{\lambda}{2\widetilde{M}} - 1\} > 0$. For $0 < \beta_1 < \beta$, define

$$Q = \Big\{x \in PC([t_0 - r, t_0 + \beta_1], \mathscr{D}) \Big| \, x_{t_0} = \phi,\ x \text{ is continuous on } (t_0, t_0 + \beta_1]$$
$$\text{and } \mathscr{F}_t\text{-adapted and } \|x(t) - \phi(0)\|^2 \leq \lambda \ (a.s.)\ \forall t \in (t_0, t_0 + \beta_1]\Big\}.$$

If $x \in Q$, (i.e., x is continuous on $[t_0, t_0 + \beta_1]$ and $\mathscr{F}_t$-adapted), then the composite functions $f(t, x_t)$ and $g(t, x_t)$ are adapted and (a.s.) integrable (respectively square integrable) since $f(t, x_t) \in \mathscr{L}_{ad}(\Omega; L[t_0, t_0 + \beta_1])$ and $g(t, x_t) \in \mathscr{L}_{ad}(\Omega; L^2[t_0, t_0 + \beta_1])$.

For $n \in \mathbb{N}$, define the sequence of random processes

$$x^{(n)}(t) = \begin{cases} \phi(t - t_0), & t \in [t_0 - r, t_0], \\ \phi(0), & t \in (t_0, t_0 + \beta/n], \\ \phi(0) + \int_{t_0}^{t-\beta/n} f(s, x_s^{(n)})\, ds & \\ \quad + \int_{t_0}^{t-\beta/n} g(s, x_s^{(n)})\, dW(s), & t \in (t_0 + \beta/n, t_0 + \beta]. \end{cases} \tag{3.6}$$

By the above argument and $\phi \in \mathscr{L}^2_{\mathscr{F}_0}([-r, 0]; \mathbb{R}^n)$, the sequence $\{x^{(n)}\}$ is well defined and, for each n, $x^{(n)}(t)$ is $\mathscr{F}_t$-adapted. Moreover, for $t \in (t_0 + \beta/n, t_0 + 2\beta/n]$, we have

$$\|x^{(n)}(t) - \phi(0)\| \leq \Big\| \int_{t_0}^{t-\beta/n} f(s, x_s^{(n)})ds \Big\| + \Big\| \int_{t_0}^{t-\beta/n} g(s, x_s^{(n)})dW(s) \Big\|.$$

So that, in view of (3.3),

$$\begin{aligned}\mathbb{E}\Big[\|x^{(n)}(t) - \phi(0)\|^2\Big] &\leq 2\Big\{\mathbb{E}\Big\| \int_{t_0}^{t-\beta/n} f(s, x_s^{(n)})ds \Big\|^2 + \mathbb{E}\Big\| \int_{t_0}^{t-\beta/n} g(s, x_s^{(n)})dW(s) \Big\|^2\Big\} \\ &\leq 2\Big\{\frac{\beta}{n} \int_{t_0}^{t_0+\beta/n} \mathbb{E}\|f(s, x_s^{(n)})\|^2 ds + \int_{t_0}^{t_0+\beta/n} \mathbb{E}\|g(s, x_s^{(n)})\|^2 ds\Big\} \\ &\leq 2(\frac{\beta}{n} + 1)\widetilde{M} \leq \lambda,\end{aligned}$$

where we used $(a + b)^2 \leq 2(a^2 + b^2)$ and Caushy–Schwartz inequality. If a subsequence of $\{x^{(n)}\}$ is taken, then $\{x^{(n)}\} \in Q$ (a.s.), and by the mathematical induction, we can show that this is true for $t \in (t_0 + k\beta/n, t_0 + (k + 1)\beta/n]$, for $k = 1, 2, \ldots, n - 1$. Thus, for $n \geq 2$, $x^{(n)}$ belongs to Q. We also have, from (3.6),

$$\|x^{(n)}(t)\| \leq \|\phi(0)\| + \Big\| \int_{t_0}^{t-\beta/n} f(s, x_s^{(n)})ds \Big\| + \Big\| \int_{t_0}^{t-\beta/n} g(s, x_s^{(n)})dW(s) \Big\|.$$

So that

$$\begin{aligned}\mathbb{E}\Big[\|x^{(n)}(t)\|^2\Big] &\leq 3\Big\{\mathbb{E}\|\phi(0)\|^2 + \mathbb{E}\Big\| \int_{t_0}^{t-\beta/n} f(s, x_s^{(n)})ds \Big\|^2 + \mathbb{E}\Big\| \int_{t_0}^{t-\beta/n} g(s, x_s^{(n)})dW(s) \Big\|^2\Big\} \\ &\leq 3\Big\{c_1 + \frac{\beta}{n} \int_{t_0}^{t_0+\beta/n} \mathbb{E}\|f(s, x_s^{(n)})\|^2 ds + \int_{t_0}^{t_0+\beta/n} \mathbb{E}\|g(s, x_s^{(n)})\|^2 ds\Big\} \\ &\leq 3\Big\{c_1 + (\frac{\beta}{n} + 1)\widetilde{M}\Big\}.\end{aligned}$$

Namely, we have

$$\mathbb{E}\Big[\|x^{(n)}(t)\|^2\Big] \leq \lambda'. \tag{3.7}$$

where $\lambda' = 3\Big\{c_1 + (\frac{\beta}{n} + 1)\widetilde{M}\Big\}$. By Tchebychev's inequality, one can find, for $\varepsilon > 0$, $\gamma_1(\varepsilon)$ such that

$$\mathbb{P}\Big\{\|x^{(n)}(t)\| > \gamma_1(\varepsilon)\Big\} \leq \frac{\mathbb{E}\Big[\|x^{(n)}(t)\|^2\Big]}{\gamma_1(\varepsilon)^2} \leq \frac{\lambda'}{\gamma_1(\varepsilon)^2} = \frac{\varepsilon}{2}.$$

Now, for each n, let $y^{(n)}$ denote the restriction of $x^{(n)}$ to $[t_0, t_0 + \beta]$. Then, $y^{(n)}$ is continuous on $[t_0, t_0 + \beta]$, and moreover, for $t \in [t_0, t_0 + \beta]$, we have

$$\mathbb{P}\Big\{\|y^{(n)}(t)\| > \gamma_1(\varepsilon)\Big\} \le \frac{\varepsilon}{2}, \tag{3.8}$$

meaning that the sequence $\{y^{(n)}(t)\}$ is uniformly bounded (a.s.). We also have

$$y^{(n)}(t_1) - y^{(n)}(t_2) = \int_{t_2}^{t_1} f(s, y_s^{(n)})ds + \int_{t_2}^{t_1} g(s, y_s^{(n)})dW(s),$$

so that

$$\begin{aligned}\mathbb{E}\Big[\Big\|y^{(n)}(t_1) - y^{(n)}(t_2)\Big\|^2\Big] &\le 2\Big\{\mathbb{E}\Big\|\int_{t_2}^{t_1} f(s, y_s^{(n)})ds\Big\|^2 + \mathbb{E}\Big\|\int_{t_2}^{t_1} g(s, y_s^{(n)})\,dW(s)\Big\|^2\Big\}\\ &\le 2M^2|t_1 - t_2|(|t_1 - t_2| + 1) \le \varepsilon',\end{aligned}$$

namely

$$\mathbb{E}\Big[\Big\|y^{(n)}(t_1) - y^{(n)}(t_2)\Big\|^2\Big] \le \varepsilon'$$

which implies that, for a positive ε, there exists $\gamma_2(\varepsilon)$ such that

$$\mathbb{P}\Big\{\|y^{(n)}(t_1) - y^{(n)}(t_2)\| > \gamma_2(\varepsilon)\Big\} \le \frac{\varepsilon}{2}, \tag{3.9}$$

which shows that the sequence $\{y^{(n)}\}$ is equicontinuous (a.s.).

Combining (3.8) and (3.9) yields

$$\mathbb{P}\Big\{\|y^{(n)}(t)\| \le \gamma_1(\varepsilon) \text{ or} \|y^{(n)}(t_1) - y^{(n)}(t_2)\| \le \gamma_2(\varepsilon)\Big\} > 1 - \varepsilon,$$

Set

$$\begin{aligned}K_\varepsilon = \Big\{y^{(n)} \in C([t_0, t_0 + \beta], \mathscr{D}) \mid \|y^{(n)}(t)\| \le \gamma_1(\varepsilon)\\ \text{and } \|y^{(n)}(t_1) - y^{(n)}(t_2)\| \le \gamma_2(\varepsilon)\Big\}.\end{aligned}$$

The following part of the proof is aimed to prove the convergence of the sequence in (3.6).[1] Since K_ε is uniformly bounded and equicontinuous, by Arzela–Ascoli's theorem [1], it is a compact subset of $C([t_0, t_0 + \beta]; \mathscr{D})$. In addition, by Lemma 3.1, it satisfies $\mathbb{P}\{y^{(n)} \in K_\varepsilon\} > 1 - \varepsilon$. Thus, by Prokhorov's theorem, the collection of

[1] This part of the proof is inspired by that of Theorem 4.2.1 in [1] except the dynamics there are delay-free. We reproduced it here for self-contained proof reading.

continuous processes $\{y^{(n)}(t)\}$ is totally D–bounded. Thus, $\{(y^{(n)}(t), W^{(n)}(t), y_0^{(n)})\}$ is totally bounded, where $W^{(n)}(t) \equiv W(t)$ and $y_0^{(n)} \equiv \phi(0) =: y_0$. Therefore, one can find a D-Cauchy subsequence $\{(y^{(n_r)}(t), W^{(n_r)}(t), y_0^{(n_r)})\}$ of $\{(y^{(n)}(t), W^{(n)}(t), y_0^{(n)})\}$. By Skorohod's theorem [1], we can construct a sequence of random functions $\{(u^{(n_r)}(t), w^{(n_r)}(t), u_0^{(n_r)})\}$ and a random function $(u(t), w(t), u_0)$ such that the distance

$$D\Big((y^{(n_r)}(t), W^{(n_r)}(t), y_0^{(n_r)}), (u^{(n_r)}(t), w^{(n_r)}(t), u_0^{(n_r)})\Big) = 0, \tag{3.10}$$

for $n_1, n_2, n_3, \ldots$, and

$$\mathbb{P}\Big\{(u^{(n_r)}(t), w^{(n_r)}(t), u_0^{(n_r)}) \to (u(t), w(t), u_0)\Big\} = 1 \tag{3.11}$$

as $r \to \infty$.

Notation. Denote the *superscript* n_r by the *subscript* r; for example, the subsequence $\{u^{(n_r)}(t)\}$ becomes $\{u_r(t)\}$.

The subsequence $\{u_r(t)\}$ is a D–Cauchy sequence. By the definition of totally D–bounded set, one can construct or find (n-indexed) D–Cauchy subsequence $\{u_r^n(t)\}$ of $\{u_r(t)\}$ and construct a subsequence $\{u^n(t)\}$ of the (restricted) solution sequence $\{y^{(n)}\}$ as follows

$$u_r^n(t) = \begin{cases} u_{r_0}, & t \in (t_0, t_0 + \beta/n], \\ u_{r_0} + \int_{t_0}^{t-\beta/n} f(s, u_{r_s}^n)\, ds \\ \quad + \int_{t_0}^{t-\beta/n} g(s, u_{r_s}^n)\, dw_r(s), & t \in (t_0 + \beta/n, t_0 + \beta], \end{cases}$$

for every $r = 1, 2, \ldots$, and

$$u^n(t) = \begin{cases} u_0, & t \in (t_0, t_0 + \beta/n], \\ u_0 + \int_{t_0}^{t-\beta/n} f(s, u_s^n)\, ds \\ \quad + \int_{t_0}^{t-\beta/n} g(s, u_s^n)\, dw(s), & t \in (t_0 + \beta/n, t_0 + \beta]. \end{cases}$$

Set

$$I_r(t) = \int_{t_0}^{t} f(s, u_{r_s})\, ds + \int_{t_0}^{t} g(s, u_{r_s})\, dw_r(s), \tag{3.12a}$$

$$I_r^n(t) = \int_{t_0}^{t-\beta/n} f(s, u_{r_s}^n)\, ds + \int_{t_0}^{t-\beta/n} g(s, u_{r_s}^n)\, dw_r(s), \tag{3.12b}$$

$$I(t) = \int_{t_0}^{t} f(s, u_s)\, ds + \int_{t_0}^{t} g(s, u_s)\, dw(s), \tag{3.12c}$$

$$I^n(t) = \int_{t_0}^{t-\beta/n} f(s, u_s^n)\, ds + \int_{t_0}^{t-\beta/n} g(s, u_s^n)\, dw(s), \tag{3.12d}$$

$$I_r^r(t) = \int_{t_0}^{t-\beta/r} f(s, u_{r_s}^r)\, ds + \int_{t_0}^{t-\beta/r} g(s, u_{r_s}^r)\, dw_r(s). \tag{3.12e}$$

From (3.12a) and (3.12b), we have

$$\begin{aligned} I_r^n(t) - I_r(t) = &\overbrace{\int_{t_0}^{t-\beta/n} f(s, u_{r_s}^n)\, ds}^{l_0} - \int_{t_0}^{t} f(s, u_{r_s})\, ds \\ &+ \underbrace{\int_{t_0}^{t-\beta/n} g(s, u_{r_s}^n)\, dw_r(s)}_{l_1} - \int_{t_0}^{t} g(s, u_{r_s})\, dw_r(s) \end{aligned} \tag{3.13}$$

The stochastic integral l_0 and l_1 can be written as follows

$$\int_{t_0}^{t} f^n(s, u_{r_s}^n)\, ds, \qquad \int_{t_0}^{t} g^n(s, u_{r_s}^n)\, dw_r(s),$$

where $f^n(s, u_{r_s}^n)$ and $g^n(s, u_{r_s}^n)$ are sequences of step functions. As for f^n and g^n, the least we expect that they are piecewise continuous functions. Also, since the functionals f and g are continuous in the second argument and $u_r^n(t)$ is a $D-$Cauchy sequence which converges to $u_r(t)$, we have

$$\int_{t_0}^{t} \|f^n(s, u_{r_s}^n) - f(s, u_{r_s})\|^2\, ds \to 0$$

and

$$\int_{t_0}^{t} \|g^n(s, u_{r_s}^n) - g(s, u_{r_s})\|^2\, ds \to 0$$

in probability.[2] Therefore, the sequence of the deterministic integrals converges to

$$\int_{t_0}^{t} f(s, u_{r_s})\, ds$$

and by the definition of Itô integral, we have

$$\int_{t_0}^{t} g(s, u_{r_s})\, dw_r(s) = \int_{t_0}^{t} g^n(s, u_{r_s}^n)\, dw_r(s)$$

---

[2]In fact, if a subsequence is taken, the convergence holds with probability one.

in probability. Hence, $I_r^n(t)$ converges to $I_r(t)$ uniformly in probability as $n \to \infty$; namely, we have, for any $r \in \mathbb{N}$ and given $\varepsilon > 0$,

$$\mathbb{P}\{\|I_r^n(t) - I_r(t)\| > \varepsilon\} < \varepsilon \tag{3.14}$$

as $n \to \infty$. Similarly, from (3.12c) and (3.12d), we obtain

$$\mathbb{P}\{\|I^n(t) - I(t)\| > \varepsilon\} < \varepsilon. \tag{3.15}$$

From (3.12b) and (3.12d), we get

$$\mathbb{P}\{I_r^n(t) \to I^n(t)\} = 1 \tag{3.16}$$

as $r \to \infty$, because we have a sequence of stochastic integrals $\{I_r^n(t)\}_{r=1}^{\infty}$ which, by (3.11), converges to the stochastic integral $I^n(t)$ as $r \to \infty$. Also, (3.16) implies that, for any $\varepsilon > 0$, there exists a positive number r such that $r \geq r_0 = r_0(\varepsilon)$,

$$\|f(s, u_{r_s}^n) - f(s, u_s^n)\| < \sqrt{\frac{\varepsilon^3}{4\beta^2}}$$

and

$$\|g(s, u_{r_s}^n) - g(s, u_s^n)\| < \sqrt{\frac{\varepsilon^3}{4\beta}}.$$

Hence

$$\begin{aligned}
\mathbb{E}[\|I_r^n(t) - I_r(t)\|^2] &\leq 2\mathbb{E}\Big[\beta \int_{t_0}^{t_0+\beta} \|f(s, u_{r_s}^n) - f(s, u_s^n)\|^2\, ds\Big] \\
&\quad + 2\mathbb{E}\Big[\int_{t_0}^{t_0+\beta} \|g(s, u_{r_s}^n) - g(s, u_s^n)\|^2\, ds\Big] \\
&\leq 4\mathbb{E}\Big[\int_{t_0}^{t_0+\beta} \frac{\varepsilon^3}{4\beta}\, ds\Big] = \varepsilon^3
\end{aligned}$$

and by Tchebychev's inequality, we get

$$\mathbb{P}\{\|I_r^n(t) - I^n(t)\| > \varepsilon\} < \varepsilon, \qquad r \geq r_0(\varepsilon). \tag{3.17}$$

We want now to show that

$$u(t) = \phi(0) + \overbrace{\int_{t_0}^{t} f(s, u_s)\, ds + \int_{t_0}^{t} g(s, u_s)\, dW(s)}^{=:I(t)}$$

holds. Note that

$$
\begin{aligned}
&\mathbb{P}\Big\{\|u(t)-\phi(0)-I(t)\|>6\varepsilon\Big\}\\
&=\mathbb{P}\Big\{\|u(t)-\overbrace{u_r^r(t)+u_{r0}+I_r^r(t)}^{=0}-\phi(0)-I(t)+I^n(t)-I^n(t)+I_r^n(t)-I_r^n(t)\\
&\qquad+I_r(t)-I_r(t)\|>6\varepsilon\Big\}\\
&=\mathbb{P}\Big\{\|(u(t)-u_r^r(t))+(u_{r0}-\phi(0))-(I(t)-I^n(t))+(I_r^n(t)-I^n(t))\\
&\qquad+(I_r(t)-I_r^n(t))+(I_r^r(t)-I_r(t))\|>6\varepsilon\Big\}\\
&\le\mathbb{P}\Big\{\|y(t)-y_r^r(t)\|>\varepsilon\Big\}+\mathbb{P}\Big\{\|u_{r0}-\phi(0)\|>\varepsilon\Big\}+\mathbb{P}\Big\{\|I(t)-I^n(t)\|>\varepsilon\Big\}\\
&\qquad+\mathbb{P}\Big\{\|I_r^n(t)-I^n(t)\|>\varepsilon\Big\}+\mathbb{P}\Big\{\|I_r(t)-I_r^n(t)\|>\varepsilon\Big\}\\
&\qquad+\mathbb{P}\Big\{\|I_r^r(t)-I_r(t)\|>\varepsilon\Big\}<6\varepsilon,
\end{aligned}
$$

namely

$$\mathbb{P}\Big\{\|u(t)-\phi(0)-I(t)\|>6\varepsilon\Big\}<6\varepsilon.$$

Since $\varepsilon>0$ is arbitrary, this implies that

$$u(t)=\phi(0)+\int_{t_0}^{t}f(s,u_s)\,ds+\int_{t_0}^{t}g(s,u_s)\,dW(s),\qquad (a.s.).$$

Hence, $y\equiv u$. Finally, define

$$x(t)=\begin{cases}\phi(t-t_0), & t\in[t_0-r,t_0],\\ y(t), & t\in(t_0,t_0+\beta].\end{cases}\tag{3.18}$$

Thus, x is the required solution of (3.1). In the rest of the proof, we show that, under the condition in (3.4), the solution x cannot continue along the hypersurface $t=\tau_k(x)$ after it initially starts on it. Define the random function

$$g(t)=t-\tau_k(x(t)),\tag{3.19}$$

for all $t\in[t_0,t_0+\beta]$. Then, $g(t_0)=t_0-\tau_k(\phi(0))=0$ and $g(t)$ has a derivative in mean. Differentiating with respect to t, applying Itô formula and taking the mathematical expectation lead to

$$\frac{d}{dt}\mathbb{E}[g(t)]=1-\mathbb{E}[\mathscr{L}\tau_k(x(t))].\tag{3.20}$$

Clearly, the right-hand side of (3.20) is continuous at least in a small neighborhood of t_0, so that if we let $t^* = t_0$ and $x^* = \phi(0)$ and apply (3.4), $g(t)$ is either strictly increasing or strictly decreasing in mean on $(t^*, t^* + \delta)$. Hence, the solution will depart any hypersurface $t = \tau_k(x(t))$ in mean for some amount of time after t_0 (or t^*) before it hits another hypersurface. This completes the proof.

3.2 Forward Continuation

Having seen how the solution x evolves between two hypersurfaces, regardless of where it initially starts, we address now the problem of forward continuation of solution of (3.1) which, at the same time, does not exhibit the beating phenomenon upon an impulse hypersurface. These extensions require further conditions on the impulsive moments τ_k and function $\mathscr{I}$ as stated in the following theorem.

Theorem 3.2 *Suppose that the functionals f and g satisfy the conditions in Theorem 3.1, $\tau_k \in \mathscr{C}^2(\mathscr{D}; \mathbb{R}_+)$, for some $k \in \mathbb{N}$, and $\lim_{k\to\infty} \tau_k(x) = \infty$ uniformly in x. Assume that*

$$\mathbb{E}[\mathscr{L}\tau_k(\psi(0))] < 1, \tag{3.21}$$

for all $(t, \psi) \in J \times \mathscr{PC}([-r, 0]; \mathscr{D})$ and $k \in \mathbb{N}$, and the relations

$$\begin{aligned} &\psi(0) + \mathscr{I}(\tau_k(\psi(0)), \psi) \in \mathscr{D}; \ \text{and} \\ &\tau_k(\psi(0) + \mathscr{I}(\tau_k(\psi(0)), \psi)) \le \tau_k(\psi(0)) \end{aligned} \tag{3.22}$$

hold almost surely for all $\psi \in \mathscr{PC}([-r, 0]; \mathscr{D})$ for which $\psi(0^-) = \psi(0)$ (a.s.) and for all $k \in \mathbb{N}$. Then, for every continuable solution x of (3.1), there exists a continuation y of x that is noncontinuable. Moreover, any solution x of (3.1) can intersect each impulse hypersurface at most once.

Proof Let $x(t)$ be any solution of (3.1) that is defined on $[t_0 - r, t_0 + \beta_1)$ or $[t_0 - r, t_0 + \beta_1]$, where $0 < \beta_1 < \infty$. Denote by X the set of all solutions x with their continuations. For any $y, z \in X$, we define the partial ordering $\prec$ by $y \prec z$ if, for almost all $\omega \in \Omega$, either $y = z$ or z is a continuation of y. Let S be a totally ordered subset of X. Now for $y \in S$, we associate $\beta(y)$ such that $\beta_1 \le \beta(y) \le \infty$ and by which the solution y is defined on $[t_0 - r, t_0 + \beta(y))$ or $[t_0 - r, t_0 + \beta(y)]$.

Define

$$\beta_2 = \sup\{\beta(y) \mid y \in S\}.$$

Clearly, $\beta_1 \le \beta_2 \le \infty$ and y is defined on a subset of $[t_0 - r, t_0 + \beta_2]$ if $\beta_2 < \infty$ or $[t_0 - r, t_0 + \beta_2)$ if $\beta_2 = \infty$. At this stage, one considers two cases. The trivial case is when $\beta_2 < \infty$, and there is a solution y defined on $[t_0 - r, t_0 + \beta_2]$. Consequently, this solution y of (3.1) is an upper bound on S, and at the same time, it is the required solution continuation. In the other case, we will show that there is a solution z defined on $[t_0 - r, t_0 + \beta_2)$ such that for all $y \in S$, $y \prec z$, that is z will be an upper bound on S. Hence, by Zorn's lemma, the set X has a maximal element. To show this fact, for $t \in [t_0 - r, t_0 + \beta_2)$, we define the following function

$$z(t) = y(t), \quad \text{(a.s.)} \tag{3.23}$$

where y is any solution in S for which $t < t_0 + \beta(y)$. The new function z is well defined, it is right-continuous (i.e., $z(t^+) = z(t)$ (a.s.)) for all $t \in [t_0 - r, t_0 + \beta_2)$, the left limit $z(t^-)$ exits for all $t \in (t_0 - r, t_0 + \beta_2)$ and $z(t^-) = z(t)$ (a.s.) for all but at most finite number of points in $(t_0 - r, t_0)$. Moreover, if z has a finite number of simple jump discontinuities in any finite interval of $(t_0, t_0 + \beta_2)$, then z is a solution of (3.1) (i.e., $z \in \mathscr{PC}([t_0 - r, t_0 + \beta_2); \mathscr{D})$ and $\mathscr{F}_t$-adapted). To show this is the only possible case, for $\beta_2 < \infty$, define

$$\mathbb{T} = \{t \in (t_0, t_0 + \beta_2) \,|\, t = \tau_k(z(t^-)) \text{ for some } k\}.$$

Then, except at these points, $z(t^-) = z(t)$ (a.s.). At this point, we also consider two case; the first one is when $\mathbb{T}$ is finite. By the assumptions imposed on the functionals f and g from the last theorem, $f(t, z_t)$ and $g(t, z_t)$ can only have a finite number of simple jump discontinuities on the interval $(t_0, t_0 + \beta_2)$, and except at these point or at the points of $\mathbb{T}$, the solution z is continuous and has the solution form given in (2.53). This is because the functionals $f(t, z_t)$ and $f(t, y_t)$ have the same properties. We conclude that, if $y \in \mathscr{PC}([t_0 - r, t_0 + \beta_2); \mathscr{D})$ and $\mathscr{F}_t$-adapted, so is z. The more challenging case is when $\beta_2 < \infty$ and $\mathbb{T}$ has an infinite number of discontinuities in $(t_0, t_0 + \beta_2)$. In this case, $\mathbb{T}$ has an increasing sequence of impulse times $\mathbb{T} = \{t_k\}_{k=1}^{\infty}$, where $t_0 < t_1 < t_2 < \cdots < t_k < \cdots < t_0 + \beta_2$ and $\lim_{k\to\infty} t_k = t_0 + \beta_2$. For $k \in \mathbb{N}$, denote by j_k the index of the unique impulse hypersurface τ_{j_k} that the solution z reaches at t_k, i.e., $t_k = \tau_{j_k}(z(t_k^-))$. For some finite integer number $N > 0$, if $j_k < N$, then z can reach only a finite number of impulse hypersurfaces. Since, as assumed, there is an increasing number of impulse times, the solution z must reach at least one impulse hypersurface more than once. In other words, $j_k = j_{k+m}$, and hence, $t_k = \tau_{j_k}(z(t_k^-))$ and $t_{k+m} = \tau_{j_k}(z(t_{k+m}^-))$ for some positive integers k and m (i.e., the hypersurface τ_{j_k} is being hit at times t_k and t_{k+m}). This also implies that if $y \in S$, then $t_k = \tau_{j_k}(y(t_k^-))$ and $t_{k+m} = \tau_{j_k}(y(t_{k+m}^-))$ where $t_{k+m} < t_0 + \beta(y)$. We will show that, according to our assumptions, this cannot happen for the solution y to reach the same hypersurface more than once. For this purpose, for $i = 0, 1, 2, \ldots, m$, we define

$$h_{k+i}(t) = t - \tau_{j_{k+i}}(y(t)), \qquad \text{(a.s.)} \tag{3.24}$$

for $t \in [t_0 - r, t_{k+m}]$. Note that $h_{k+i}(t_{k+i}^-) = 0$ for all i. Suppose for the sake of contradiction that, for some $0 \le i \le m - 1$, we have $j_{k+i} > j_{k+i+1}$ and hence $\tau_{j_{k+i}}(\nu) > \tau_{j_{k+i+1}}(\nu)$ for all $\nu \in \mathscr{D}$. This implies

$$h_{k+i+1}(t_{k+i}) \ge 0, \qquad \text{(a.s.)} \tag{3.25}$$

On the other hand, differentiating $h_{k+i+1}(t)$ with respect to t, for all $t \in (t_{k+i}, t_{k+i+1})$, applying Itô formula and taking the mathematical expectation give

$$\frac{d}{dt}\mathbb{E}[h_{k+i+1}(t)] = 1 - \mathbb{E}[\mathscr{L}\tau_{j_{k+i+1}}(y(t))], \tag{3.26}$$

for all $t \in (t_{k+i}, t_{k+i+1})$. By (3.21) and the fact that $h_{k+i}(t_{k+i}^-) = 0$, we conclude that $h_{k+i+1}(t_{k+i}) < 0$ in mean, which contradicts with what we got in (3.25). Thus, $j_{k+i} < j_{k+i+1}$ and hence $j_k < j_{k+1} < \cdots < j_{k+m}$, which also contradicts with our supposition $j_k = j_{k+m}$. Therefore, the solution y and hence z must intersect a given impulse hypersurface at most once in mean. This completes the proof.

Before developing our global existence result, we would like to address the case where the solution is noncontinuable in the sense that the solution cannot be entirely contained in any compact set.

Theorem 3.3 *Let x be a solution of (3.1) that is defined for all $t \in [t_0 - r, t_0 + \beta)$, where $0 < \beta < \infty$ and $[t_0, t_0 + \beta] \subset J$. If x is noncontinuable, then there is a sequence $\{s_k\}_{k=1}^{\infty}$, with $t_0 < s_1 < s_2 < \cdots < s_k < \cdots < t_0 + \beta$ and $\lim_{k\to\infty} s_k = t_0 + \beta$ (a.s.) such that $x(s_k) \notin F$, for any compact set $F \subset \mathscr{D}$.*

Proof Assume, for contradiction, that there is a compact set $F_1 \subset \mathscr{D}$ and $\beta_1 > 0$ for which $x(t) \in F_1$ for all $t \in [t_0 + \beta_1, t_0 + \beta)$. Let F_2 be the closure of the range of the solution x when t is restricted to $[t_0 - r, t_0 + \beta_1]$. Then, the set $F = F_1 \cup F_2 \subset \mathscr{D}$ is also compact and $x(t) \in F$ for all $t \in [t_0 - r, t_0 + \beta)$. Now, for any $t, \bar{t} \in [t_0 + \beta_1, t_0 + \beta)$, we have from (2.53)

$$\|x(t) - x(\bar{t})\| \leq \left\| \int_{\bar{t}}^{t} f(s, x_s)ds \right\| + \left\| \int_{\bar{t}}^{t} g(s, x_s)dW(s) \right\|. \tag{3.27}$$

Hence

$$\begin{aligned}
\mathbb{E}\Big[\|x(t) - x(\bar{t})\|^2\Big] &\leq 2\Big\{\mathbb{E}\Big\| \int_{\bar{t}}^{t} f(s, x_s)ds \Big\|^2 + \mathbb{E}\Big\| \int_{\bar{t}}^{t} g(s, x_s)dW(s) \Big\|^2\Big\} \\
&\leq 2\Big\{[t - \bar{t}\,] \int_{\bar{t}}^{t} \mathbb{E}\|f(s, x_s)\|^2 ds + \int_{\bar{t}}^{t} \mathbb{E}\|g(s, x_s)\|^2 ds\Big\} \\
&\leq 2\widetilde{M}^2 |t - \bar{t}\,| \{|t - \bar{t}\,| + 1\} < \varepsilon,
\end{aligned} \tag{3.28}$$

for some arbitrary positive $\varepsilon > 0$ and $\widetilde{M} > 0$, which is guaranteed by Theorem 3.1. Then, by Tchebychev's inequality, we obtain

$$\mathbb{P}\Big\{\|x(t) - x(\bar{t})\| > \eta\Big\} \leq \frac{\varepsilon}{\eta^2},$$

for some $\eta > 0$. Also, by Cauchy criterion, the limit $\lim_{t \to (t_0+\beta)} x(t)$ exits with probability one and its limit point, say ζ, is in F. That is, the solution x can be continued by defining $x(t_0 + \beta) = \zeta$. But this contradicts with our supposition that x is noncontinuable. Thus, the conclusion of the theorem follows.

3.3 Global Existence

Having shown the evolution of a local solution of (3.1), we address the problem of the global existence of the solution. This demands imposing further assumptions on the functionals f and g.

Theorem 3.4 *Let $J = \mathbb{R}_+$, $\mathscr{D} = \mathbb{R}^n$ be an open set containing $\phi(0)$ and the functionals $f \in \mathscr{L}_{ad}(\Omega; L[t_0, t_0 + \alpha])$ and $g \in \mathscr{L}_{ad}(\Omega; L^2[t_0, t_0 + \alpha])$, where $\alpha > 0$ and $[t_0, t_0 + \alpha] \subset J$ are continuous in their second argument, say ψ. Assume further that there are two measurable functions h_1, h_2 (or $h_1, h_2 \in \mathscr{PC}(\mathbb{R}_+; \mathbb{R}_+)$) and a continuous increasing concave function $\kappa : \mathbb{R}_+ \to \mathbb{R}_+$ such that*

$$\|f(t, \psi)\|^2 \vee \|g(t, \psi)\|^2 \leq h_1^2(t) + h_2^2(t)\kappa(\|\psi\|_r^2)$$

for all $(t, \psi) \in \mathbb{R}_+ \times \mathscr{L}^2_{\mathscr{F}_t}([-r, 0]; \mathbb{R}^n)$ (i.e., ψ is an $\mathscr{F}_t$-adapted and $\mathbb{E}[\|\psi\|_r^2] < \infty$). Then, for each $(t, \phi) \in \mathbb{R}_+ \times \mathscr{L}^2_{\mathscr{F}_0}([-r, 0]; \mathbb{R}^n)$, there exists a local $\mathscr{F}_t$-adapted solution $x = x(t;\ t_0,\ \phi(0))$ for (3.1) that can be continued to $[t_0 - r, \infty)$.

Proof For all $(t, \phi) \in \mathbb{R}_+ \times \mathscr{L}^2_{\mathscr{F}_0}([-r, 0]; \mathbb{R}^n)$, let $x(t) = x(t; t_0, \phi(0))$ be a local solution of (3.1) that is guaranteed by Theorem 3.2. Suppose that, for contradiction, for a finite β the solution x is noncontinuable in the sense of Theorem 3.3. We will show that based on the theorem assumptions this supposition would be impossible.

Let $a = \mathbb{E}[\|\phi(0)\|^2] + \mathbb{E}\left[\left(\sum_{\{k:t_k \in (t_0,t]\}} \left\|I(t_k, x_{t_k^-})\right\|\right)^2\right]$, $b = (\beta + 1)\beta\hbar^2$, where $\hbar = \sup\{h_1(t) \mid \forall t \in [t_0, t_0 + \beta]\}$ and $c = \mathbb{E}[\|\phi\|_r^2]$.

Then, for all $t \in (t_0, t_0 + \beta)$,

$$\begin{aligned}
\mathbb{E}[\|x(t)\|^2] &\leq 4\Big\{\mathbb{E}[\|\phi(0)\|^2] + \mathbb{E}\Big[\Big(\sum_{\{k:t_k \in (t_0,t]\}} \left\|I(t_k, x_{t_k^-})\right\|\Big)^2\Big] + \beta \int_{t_0}^t \mathbb{E}\|f(s, x_s)\|^2 ds \\
&\quad + \int_{t_0}^t \mathbb{E}\|g(s, x_s)\|^2 ds\Big\} \\
&\leq 4\Big\{a + b + (\beta + 1) \int_{t_0}^t h_2^2(s)\kappa(\mathbb{E}[\|x_s\|_r^2])ds\Big\}
\end{aligned} \tag{3.29}$$

which implies that

$$\begin{aligned}\mathbb{E}[\|x_t\|_r^2] &\le c + 4(a+b) + 4(\beta+1)\int_{t_0}^{t} h_2^2(s)\kappa(\mathbb{E}[\|x_s\|_r^2])\,ds\\ &= B + 4(\beta+1)\int_{t_0}^{t} h_2^2(s)\kappa(\mathbb{E}[\|x_s\|_r^2])\,ds,\end{aligned}$$

where $B = c + 4(a+b)$. Using Bihari's Lemma yields

$$\mathbb{E}[\|x_t\|_r^2] \le G^{-1}\Big(G(B) + 4(\beta+1)\int_{t_0}^{t} h_2^2(s)ds\Big),$$

where

$$G(u) = \int_{0+}^{u} \frac{ds}{\kappa(s)}, \qquad u > 0$$

and $G(B) + 4(\beta+1)\int_{t_0}^{t} h_2^2(s)ds \in \mathrm{Dom}(G^{-1})$. If $B \to 0$, then $G(B) \to -\infty$ and, hence, $G^{-1} \to 0$. That is to say, if $B \to 0$, then $\mathbb{E}[\|x_t\|_r^2] \le 0 < \infty$. Thus, $\mathbb{E}[\|x(t)\|^2] < \infty$. This contradicts with that x is noncontinuable. Therefore, the solution must be bounded when $t \to (t_0+\beta)^-$ and the global existence result follows.

3.4 Uniqueness of Solution

Having established the local/global existence result for impulsive system (3.1), we are in a position to prove the uniqueness of the solution.

Theorem 3.5 *Suppose that the assumptions of Theorem 3.4 hold and that the functionals $f(t,\psi)$ and $g(t,\psi)$ are locally Lipschitz in ψ for all $t \in J$. Then, system (3.1) has a unique solution defined on $[t_0 - r, t_0 + \beta)$, where $0 < \beta \le \infty$ and $[t_0, t_0+\beta) \subset J = \mathbb{R}_+$.*

Proof For all $t \in [t_0 - r, t_0 + \beta)$ with $0 < \beta \le \infty$ and $[t_0, t_0+\beta) \subset J$, let $x = x(t; t_0, \phi(0))$ and $y = y(t; t_0, \phi(0))$ be two solutions of (3.1). So that, $x(t) = y(t) = \phi(t - t_0)$ for all $t \in [t - t_0, t_0]$. For contradiction, assume that $x \not\equiv y$ (a.s.) (i.e., $x(t) \ne y(t)$ (a.s.) for all $t \in J$). Then, there would be some $t \in (t_0, t_0+\beta)$ such that $x(t) \ne y(t)$ (a.s.). Define the stopping time $t_1 = \inf\{t \in (t_0, t_0+\beta) \mid x(t) \ne y(t)\}$. If t_1 is not an impulsive time (i.e., $t_1 \ne \tau_k(x(t_1^-))$ or equivalently $t_1 \ne \tau_k(y(t_1^-))$ for all k), then $x(t_1) = x(t_1^-) = y(t_1^-) = y(t_1)$ (a.s.); otherwise, $x(t_1) = x(t_1^-) + \mathscr{I}(t_1, x_{t_1^-}) = y(t_1^-) + \mathscr{I}(t_1, y_{t_1^-}) = y(t_1)$. Hence, in both cases, we have $x(t_1) = y(t_1)$ (a.s.). Let $\varepsilon > 0$ be sufficiently small such that $t_1 + \varepsilon < t_0 + \beta$ and the solutions x and y do not reach any hypersurface over $(t_1, t_1+\varepsilon]$. Let $\delta > 0$ be a sufficiently small number such that $\delta < \varepsilon$ and $\delta(\delta+1)L^2 \le \frac{1}{4}$, where $L > 0$, such that $\|f(t,\psi_1)-f(t,\psi_2)\| \vee \|g(t,\psi_1)-g(t,\psi_2)\| \le L\|\psi_1-\psi_2\|_r$, for all $t\in[t_0, t_1+\epsilon]$ and all $\psi_1,\ \psi_2$ in some compact set $F\subset\mathscr{D}$ with $\mathscr{D}$ being an open subset of $\mathbb{R}^n$, where the last inequality is guaranteed as f and g are locally Lipschitz in ψ. Then, for all $t\in[t_1, t_1+\delta]$ (where x and y do not intersect with any impulsive hypersurface), we have from (2.53)

$$\begin{aligned}
\mathbb{E}[\|x(t)-y(t)\|^2] &= 2\Big\{\mathbb{E}\Big\|\int_{t_1}^{t}(f(s,x_s)-f(s,y_s))\,ds\Big\|^2 \\
&\quad + \mathbb{E}\Big\|\int_{t_1}^{t}(g(s,x_s)-g(s,y_s))\,dW(s)\Big\|^2\Big\} \\
&\le 2\Big\{\delta\int_{t_1}^{t}\mathbb{E}\|f(s,x_s)-f(s,y_s))\|^2\,ds \\
&\quad + \int_{t_1}^{t}\mathbb{E}\|g(s,x_s)-g(s,y_s))\|^2\,ds\Big\} \\
&\le 2(\delta+1)\int_{t_1}^{t}\mathbb{E}[L^2\|x_s-y_s\|_r^2]\,ds \\
&\le 2(\delta+1)L^2\int_{t_1}^{t}\sup_{u\in[t_1,s]}\mathbb{E}\|x(u)-y(u)\|^2 ds \\
&\le 2(\delta+1)L^2\int_{t_1}^{t_1+\delta}\sup_{u\in[t_1,t_1+\delta]}\mathbb{E}\|x(u)-y(u)\|^2 ds \\
&\le 2(\delta+1)L^2\delta\sup_{u\in[t_1,t_1+\delta]}\mathbb{E}\|x(u)-y(u)\|^2 \\
&\le \frac{1}{2}\sup_{u\in[t_1,t_1+\delta]}\mathbb{E}\|x(u)-y(u)\|^2
\end{aligned}$$

for all $t\in[t_1, t_1+\delta]$. The last inequality implies that $\sup_{[t_1,t_1+\delta]}\mathbb{E}[\|x(t)-y(t)\|^2]=0$. Since x and y are continuous functions for all $t\in[t_1,t_1+\delta]$, then

$$\mathbb{P}\Big(\sup_{[t_1,t_1+\delta]}\|x(t)-y(t)\|>0\Big)=0, \tag{3.30}$$

which implies that $x(t)=y(t)$ (a.s.) for all $t\in[t_1,t_1+\delta]$ [2]. But this contradicts with our supposition that $x\not\equiv y$ (a.s.). Thus, it must be true that (3.1) has a unique solution.

3.5 Notes and Comments

In this chapter, we have considered a general nonlinear stochastic system with time delay and impulsive effects occurring at state-dependent variable times. We have first addressed a local existence result for the system over a space of piecewise continuous,

$\mathscr{F}_t$-adapted functions. The material of this chapter is taken from [3]. Also, if the stochastic diffusion process is completely set to zero (i.e., $g \equiv 0$), then system (3.1) reduces to impulsive system (2.31). So that, one may refer to [4, 5] to further study the fundamental properties system (2.31). We should mention that, in proving the equi-continuity property of the solution sequence, one may get the same result by following another, but lengthy, approach and then employing Kolmogorov's theorem for continuity. As mentioned earlier, the proof of the convergence of sequence of SIEs is inspired by that of Theorem 4.2.1 in [1]; instead, one can obtain the same convergence property if the functionals satisfy the Lipschitz condition. We have also shown that, by imposing further restriction on the impulsive hypersurface, solutions leave this surface in mean. Due to some technical difficulties in backward extending a given solution of an impulsive system with or without time delay, we have focused on forward continuation, which meets our interest in this book when studying the qualitative properties of the stochastic delay system with fixed impulses. Later, under further conditions on the impulse function and impulses, solutions evolve without exhibiting rhythmical beating upon a hypersurface. Supposing that the drift and diffusion coefficients (i.e., f and g) are bounded by some nonlinear estimate in their delayed-state argument, a global result has been achieved. In fact, one can reach the same finding if the coefficients are assumed to grow linearly; however, the result will be analyzed differently. Finally, a unique solution is guaranteed if f and g be locally Lipschitz.

References

1. Ladde GS, Lakshmikantham V (1980) Random differential inequalities. Academic Press, New York
2. Gard TC (1988) Introduction to stochastic differential equations. Marcel Dekker
3. Alwan MS, Liu XZ, Xie W-C (2010) Existence, continuation, and uniqueness problems of stochastic impulsive systems with time delay. J Frankl Inst 347:1317–1333
4. Ballinger GH, Liu XZ (1990) Existence and uniqueness results for impulsive delay differential equations. Dyn Contin Discret Impuls Syst 6:579–591
5. Liu XZ, Ballinger GH (2000) Existence, uniqueness, and boundedness results for impulsive delay differential equations. Appl Anal 74(1–2):71–93

Chapter 4
Stability of Stochastic Impulsive Systems with Time Delay

In this chapter, we consider nonlinear stochastic impulsive systems with time delay. Particularly, the time delay here is finite; the stochastic noise is represented by a Wiener process, and the impulses are state-independent and are of types diminishing and unbounded in total. Namely, we consider the following system

$$dx(t) = f(t, x_t)dt + g(t, x_t)\, dW(t), \qquad t \neq \tau_k, \tag{4.1a}$$

$$\Delta x = \mathscr{I}(t, x_{t^-}), \qquad t = \tau_k, \tag{4.1b}$$

$$x_{t_0} = \phi(s), \quad s \in [-r, 0] \tag{4.1c}$$

The main interest here is to address the problem of mean square (m.s.) global asymptotic stability and the problem of stabilization by impulsive controller. Precisely, we develop Lyapunov-like sufficient conditions to ensure the stability property using the classical Lyapunov-based approach and the comparison method.

Similar to those stated in Chap. 2, while investigating the stability properties of impulsive systems, we make for convenient reading two assumptions, taking into account the random noise affecting the system in Assumption A1.

Assumption A1 There exist $0 \leq \varrho_1 \leq \varrho$ such that, for all $\tau_k \in \mathbb{R}_+$ and x defined on $\mathscr{PC}([-r, 0]; \mathscr{D})$, for some open set $\mathscr{D} \in \mathbb{R}^n$, if

$$\mathbb{E}[\|x(\tau_k^-)\|^2] < \varrho_1, \quad \text{then} \quad \mathbb{E}[\|x(\tau_k)\|^2] < \varrho.$$

Assumption A2 For any $k \in \mathbb{N}$, we have

$$\tau_{\sup} = \sup\{\tau_k - \tau_{k-1}\} < \infty \qquad \text{and} \qquad \tau_{\inf} = \inf\{\tau_k - \tau_{k-1}\} > 0.$$

M. S. Alwan and X. Liu, *Theory of Hybrid Systems: Deterministic and Stochastic*, Nonlinear Physical Science, https://doi.org/10.1007/978-981-10-8046-3_4

Definition 4.1 The trivial solution of system (4.1) is said to be

(i) *stable in the m.s.*, if for every $\varepsilon > 0$ and $t_0 \in \mathbb{R}_+$, there exists a $\delta = \delta(t_0, \varepsilon) > 0$ such that

$$\mathbb{E}[\|\phi\|_r^2] \leq \delta \quad \text{implies} \quad \mathbb{E}[\|x(t)\|^2] < \varepsilon, \quad \forall t \geq t_0,$$

where $x(t) = x(t; t_0, \phi)$ is any solution of (4.1), with $x \in \mathscr{PC}([t_0 - r, t_0 + \alpha]; \mathscr{D})$ for some $\alpha > 0$ and $\phi \in \mathscr{L}^2_{\mathscr{F}_0}([-r, 0]; \mathscr{D})$;

(ii) *uniformly stable in the m.s.* if δ in (i) is independent of t_0;

(iii) *asymptotically stable in the m.s.* if it is stable and for any $t_0 \in \mathbb{R}_+$, there exists $\eta = \eta(t_0) > 0$ such that

$$\mathbb{E}[\|\phi\|_r^2] \leq \eta \quad \text{implies} \quad \lim_{t\to\infty} \mathbb{E}[\|x(t)\|^2] = 0;$$

(iv) *uniformly asymptotically stable in the m.s.* if it is uniformly stable in the m.s. and there exists some $\eta > 0$ such that, for every $\gamma > 0$, there exists a constant $T = T(\eta, \gamma) > 0$ for which

$$\mathbb{E}[\|\phi\|_r^2] \leq \eta \quad \text{implies} \quad \mathbb{E}[\|x(t)\|^2] < \gamma, \quad \forall t \geq t_0 + T;$$

(v) *uniformly attractive in the m.s.* if, for any $\eta > 0$, there exists a $\delta_0 = \delta_0(\eta)$ and $T = T(\eta) > 0$ for which $\mathbb{E}[\|x(t)\|^2] < \eta$, for all $t \geq T$, whenever $\mathbb{E}[\|\phi\|^2] < \delta_0$. It is said to be *uniformly asymptotically stable* if it is uniformly attractive and (ii) holds simultaneously.;

(vi) *exponentially stable in the m.s.* if there exist positive constants K and λ such that

$$\mathbb{E}[\|x(t)\|^2] \leq K\mathbb{E}[\|\phi\|_r^2]e^{-\lambda(t-t_0)}, \quad \forall t \geq t_0;$$

(vii) *unstable in m.s.* if (i) fails to hold.

4.1 Stability Analysis by Classical Lyapunov Technique

In this section, we address the m.s. stability properties of (4.1) using Lyapunov-based theorems together with Razumikhin technique. Particularly, in Theorem 4.1, the underlying continuous system is assumed to have unstable trivial solution that is stabilized by the action of impulsive effects, which are not necessarily bounded. Later in Corollary 4.1, the underlying continuous systems is assumed to be stable that is perturbed by impulses. To maintain the stability property, the impulses are treated as perturbation to the continuous system.

Theorem 4.1 *Assume that Assumptions A1 and A2 are satisfied and there exist functions $a, c \in \mathscr{K}_c$, $b \in \mathscr{K}_v$ and $p \in \mathscr{PC}(\mathbb{R}_+; \mathbb{R}_+)$. Let $V \in \mathscr{C}^{1,2}([-r, \infty) \times \mathbb{R}^n; \mathbb{R}_+)$ such that the following conditions hold:*

(i) for all $(t, \psi(0)) \in [-r, \infty) \times S(\varrho)$,

$$b(\|\psi(0)\|^2) \le V(t, \psi(0)) \le a(\|\psi(0)\|^2), \qquad (a.s.);$$

(ii) for all $t \neq \tau_k \in \mathbb{R}_+$ and $\psi \in \mathscr{PC}([-r, \infty); S(\varrho))$,

$$\mathscr{L}V(t, \psi) \le p(t)c(V(t, \psi(0))), \qquad (a.s.)$$

provided that $\bar{g}(V(t+s, \psi(s))) \le V(t, \psi(0))$ for some $s \in [-r, 0]$, where $\bar{g} \in \mathscr{K}_3$;

(iii) at any impulsive moment $\tau_k \in \mathbb{T}$ and $\psi \in \mathscr{PC}([-r, \infty); S(\varrho))$,

$$V(\tau_k, \psi(0) + \mathscr{I}(\tau_k, \psi(\tau_k^-))) \le \bar{g}(V(\tau_k^-, \psi(0))), \qquad (a.s.)$$

with $\psi(0^-) = \psi(0)$, where $(\tau_k, \psi(\tau_k^-)) \in \mathbb{R}_+ \times \mathscr{PC}([-r, 0]; S(\varrho_1))$;

(iv) $M_1 = \sup_{t \ge 0} \int_t^{t+\tau} p(s)ds < \infty$, with $\tau = \sup_{k \in \mathbb{N}}\{\tau_k - \tau_{k-1}\} < \infty$ and $M_2 = \inf_{q>0} \int_{\bar{g}(q)}^{q} ds/c(s) > M_1$.

Then, the trivial solution $x \equiv 0$ of (4.1) is uniformly asymptotically stable in the m.s.

Proof From condition (i), we have for $s \in [0, \varrho]$, $b(s) \le a(s)$, so that we can find two functions $\hat{b} \in \mathscr{K}_v$ and $\hat{a} \in \mathscr{K}_c$ such that $\hat{b}(s) \le b(s) \le a(s) \le \hat{a}(s)$ for all $s \in [0, \varrho]$. This implies

$$\hat{b}(\|\psi(0)\|^2) \le V(t, \psi(0)) \le \hat{a}(\|\psi(0)\|^2), \qquad \text{(a.s.)},$$

for all $t \in \mathbb{R}_+$ and $\psi \in \mathscr{PC}\big([-r, 0]; S(\varrho)\big)$.

We first prove uniform stability in the m.s. Let $0 < \varepsilon < \rho_1$ and $x(t) = x(t; t_0, \phi)$ be a solution of (4.1) with its maximal interval of existence $[t_0, t_0 + \beta)$. Choose $\delta = \delta(\varepsilon)$ such that $\delta < \hat{a}^{-1}(\bar{g}(\hat{b}(\varepsilon)))$. Since, by the definition of M_2, $0 < \bar{g}(q) < q$, we have $0 < \delta < \varepsilon$.

Claim 1 *Let ϕ be the initial function such that $\mathbb{E}[\|\phi\|_r^2] \le \delta$. Then, $x \equiv 0$ is uniformly stable in the m.s.*

Proof of Claim 1. If our claim were not true, there would be some $t \in [t_0, t_0 + \beta)$ for which $\mathbb{E}[\|x(t)\|^2] > \varepsilon$. Then, define $\hat{t} = \inf\{t \in [t_0, t_0 + \beta) \big| \mathbb{E}[\|x(t)\|^2] > \varepsilon\}$. Clearly that $\mathbb{E}[\|x(t)\|^2] \le \mathbb{E}[\|\phi\|_r^2] \le \delta < \varepsilon$ for all $t \in [t_0 - r, t_0]$, and particularly, $\mathbb{E}[\|x(t_0)\|^2] < \varepsilon$. Therefore, $\hat{t} \in (t_0, t_0 + \beta)$, $\mathbb{E}[x(t)] \le \varepsilon \le \varrho_1$ for all $t \in [t_0 - r, \hat{t})$ which is guaranteed by Assumption A1 and either $\mathbb{E}[\|x(t)\|^2] = \varepsilon$ or $\mathbb{E}[\|x(t)\|^2] > \varepsilon$ at $\hat{t} = \tau_k$ for some k. Therefore, $V(t, x(t))$ is defined for all $t \in [t_0, \hat{t}]$. Thus, define

$m(t) = \mathbb{E}[V(t, x(t))]$. With the aid of Itô formula and the property of the function c, we get, after taking the mathematical expectation,

$$\begin{aligned} m(t) &\leq m(s) + \mathbb{E} \int_s^t \mathscr{L}V(u, x_u)du, \qquad \forall\, t_0 \leq s \leq t \leq \hat{t} \\ &\leq m(s) + p(t)c(m(t)), \end{aligned}$$

which implies that the Dini derivative of m is given by

$$D^+m(t) = \lim_{h \to 0^+} \sup \frac{1}{h}[m(t+h) - m(t)] \leq p(t)c(m(t)), \tag{4.2}$$

for all $t \neq \tau_k$ in $(t_0, \hat{t}]$, provided that $m(t) \geq \bar{g}(\|m(t)\|_r)$ and, at the impulsive moments, we have

$$m(\tau_k) \leq \bar{g}(m(\tau_k^-)), \tag{4.3}$$

for all $\tau_k \in (t_0, \hat{t}]$. Let $t^* = \inf\{t \in [t_0, \hat{t}] \big| m(t) \geq \hat{b}(\varepsilon)\}$. Since $m(t_0) \leq \hat{a}(\|\phi\|_r^2) \leq \hat{b}(\delta) \leq \bar{g}(\hat{b}(\varepsilon)) \leq \hat{b}(\varepsilon)$ and $m(\hat{t}) \geq \hat{b}(\varepsilon)$, which is guaranteed by $\hat{b}(\mathbb{E}[\|x(t)\|^2]) \leq m(t) \leq \hat{a}(\mathbb{E}[\|x(t)\|^2])$, we conclude that $t^* \in (t_0, \hat{t}]$. Furthermore, $m(t) < \hat{b}(\varepsilon)$ for all $t \in [t_0 - r, t^*)$. This is because, as seen above, $m(t) \leq \hat{b}(\varepsilon)$ for all $t \in [t_0 - r, t_0]$, $m(\hat{t}) \geq \hat{b}(\varepsilon)$ and the definition of t^*. Before finishing the proof of Claim 1, we need to prove the following result.

Claim 2 *For $\tau_k \in (t_0, \hat{t}]$ (for any k), $m(t^*) = \hat{b}(\varepsilon)$ and $t^* \neq \tau_k$.*

Proof of Claim 2. Note that, from the definition of t^*, $m(t^*) \geq \hat{b}(\varepsilon) > 0$. Now, if $t^* = \tau_k$ for some k, then

$$0 < \hat{b}(\varepsilon) \leq m(t^*) \leq \bar{g}(m(t^{*-})) \leq m(t^{*-}) < \hat{b}(\varepsilon)$$

which is impossible. Thus, $t^* \neq \tau_k$ for any k. This also implies that $m(t^*) = \hat{b}(\varepsilon)$ because $m(t)$ is continuous at t^*. This completes the proof of Claim 2.

To pursue the proof of Claim 1, consider that $\tau_{l-1} \leq t_0 < t^* < \tau_l$. Let $\bar{t} = \sup\{t \in [t_0, t^*] \big| m(t) \leq \bar{g}(\hat{b}(\varepsilon))\}$. We have seen $m(t_0) \leq \bar{g}(\hat{b}(\varepsilon))$, $m(t^*) = \hat{b}(\varepsilon) > \bar{g}(\hat{b}(\varepsilon))$ and $m(t)$ is continuous on $[t_0, t^*]$. Then, $\bar{t} \in (t_0, t^*)$, $m(\bar{t}) = \bar{g}(\hat{b}(\varepsilon))$ and $m(t) \geq \bar{g}(\hat{b}(\varepsilon))$ for all $t \in [\bar{t}, t^*]$. Hence, for $t \in [\bar{t}, t^*]$ and $s \in [-r, 0]$, we have $\bar{g}(m(t+s)) \leq \bar{g}(\hat{b}(\varepsilon)) \leq m(t)$. Thus, the inequality

$$D^+m(t) \leq p(t)c(m(t))$$

holds for all $t \in (\bar{t}, t^*]$. Integrating this differential inequality yields

$$\int_{m(\bar{t})}^{m(t^*)} \frac{ds}{c(s)} \leq \int_{\bar{t}}^{t^*} p(s)ds \leq \int_{\bar{t}}^{\bar{t}+\tau} p(s)ds \leq \sup \int_{\bar{t}}^{\bar{t}+\tau} p(s)ds = M_1. \quad (4.4)$$

On the other hand, we have

$$\int_{m(\bar{t})}^{m(t^*)} \frac{ds}{c(s)} = \int_{\bar{g}(\hat{b}(\varepsilon))}^{\hat{b}(\varepsilon)} \frac{ds}{c(s)} \geq M_2, \quad (4.5)$$

which contradicts with the assumption $M_2 > M_1$.

Next, we consider the case where $\tau_k < t^* < \tau_{k+1}$ for some $k \geq l$. Then, we have

$$m(\tau_k) \leq \bar{g}(m(\tau_k^-)) \leq \bar{g}(\hat{b}(\varepsilon)),$$

where the second inequality is guaranteed because $m(t_k^-) \leq \hat{b}(\varepsilon)$ and $\bar{g}$ is a nondecreasing function. Define $\bar{t} = \sup\{t \in [\tau_k, t^*] \big| m(t) \leq \bar{g}(\hat{b}(\varepsilon))\}$. Then, as achieved in the previous analysis, we get $\bar{t} \in [\tau_k, t^*)$, $m(\bar{t}) = \bar{g}(\hat{b}(\varepsilon))$ and $m(t) \geq \bar{g}(\hat{b}(\varepsilon))$ for all $t \in [\bar{t}, t^*)$. Following the same argument applied to the differential inequality over the interval $[\bar{t}, t^*]$, we reach the same contradiction. Thus, we conclude that the trivial solution $x \equiv 0$ is uniformly stable in the m.s. This completes the Proof of Claim 1.

Now, we are aiming to prove that $x \equiv 0$ is uniformly asymptotically stable in the m.s. Since it is uniformly stable in the m.s., there exists $\eta > 0$ such that $\mathbb{E}[\|\phi\|_r^2] \leq \eta$ implies $\mathbb{E}[\|x(t)\|^2] \leq \varrho_1$ for all $t \geq t_0 - r$. We also have that, with the aid of (i), $\mathbb{E}[V(t, x(t))] \leq \hat{a}(\mathbb{E}[\|x(t)\|^2]) \leq \hat{a}(\varrho_1)$ for all $t \geq t_0 - r$. Let $0 < \gamma < \varrho_1$ and define

$$0 < M = M(\gamma) = \sup\Big\{\frac{1}{c(s)}\Big| s \in [\bar{g}(\hat{b}(\gamma)), \hat{a}(\varrho_1)]\Big\}.$$

For $\hat{b}(\gamma) \leq q \leq \hat{a}(\varrho_1)$, we have $\bar{g}(\hat{b}(\gamma)) \leq \bar{g}(q) \leq q \leq \hat{a}(\varrho_1)$ and so that

$$M_2 \leq \int_{\bar{g}(q)}^{q} \frac{ds}{c(s)} \leq M[q - \bar{g}(q)],$$

from which $q - \bar{g}(q) \geq M_2/M$ or $\bar{g}(q) \leq q - \frac{M_2}{M} < q - d$, where $d = d(\gamma)$ is chosen so that $d < (M_2 - M)/M < M_2/M$.

Let $N = N(\gamma)$ be the smallest positive integer for which $\hat{a}(\varrho_1) \leq \hat{b}(\gamma) + Nd$ and define $T = T(\gamma) = \tau + (r + \tau)(N - 1)$. For the given γ, choose η such that $\mathbb{E}[\|\phi\|_r^2] \leq \eta$ implies $\mathbb{E}[\|x(t)\|^2] \leq \gamma$ for any solution $x(t) = x(t, t_0, \phi)$ of (12.4), $t_0 \in [\tau_{l-1}, \tau_l]$ and $t \geq t_0 + T$. From the previous analysis, we have shown that $m(t) \leq \hat{a}(\varrho_1)$ for all $t \geq t_0 - r$. Given, $0 < A < \hat{a}(\varrho_1)$ and $j \geq l$, we will show that

(1) if $m(t) \leq A$ for $t \in [\tau_j - r, \tau_j)$, then $m(t) \leq A$ for all $t \geq \tau_j$; and
(2) if, in addition, $\hat{b}(\gamma) \leq A$, then $m(t) \leq A - d$ for all $t \geq \tau_j$.

To prove (1), suppose for contradiction that there exists some $t \geq \tau_j$ such that $m(t) > A$. Define $t^* = \inf\{t \geq \tau_j | m(t) > A\}$. Then, $t^* \in [\tau_k, \tau_{k+1})$ for some $k \geq j$. Because $m(\tau_k) \leq \bar{g}(m(\tau_k^-)) \leq \bar{g}(A) < A$ (i.e., $m(\tau_k) < A$), then $t^* \in (\tau_k, \tau_{k+1})$. Moreover, $m(t^*) = A$ and, by the definition of t^*, $m(t) \leq A$ for all $t \in [\tau_j - r, t^*]$. Define $\bar{t} = \sup\{t \in [\tau_k, t^*] \big| m(t) \leq \bar{g}(A)\}$. Note that $\bar{t} \neq t^*$ (i.e., $\bar{t} \in [\tau_k, t^*)$) because $m(\tau_k) \leq \bar{g}(A) < A = m(t^*)$. We can also see that $m(\bar{t}) = \bar{g}(A)$, and $m(t) \geq \bar{g}(A)$ for all $t \in [\bar{t}, t^*]$. Thus, for $t \in [\bar{t}, t^*]$ and $s \in [-r, 0]$, we have $\bar{g}(m(t+s)) \leq \bar{g}(A) \leq m(t)$, where we have used the fact $m(t) \leq A$ for all $t \in [\tau_j - r, t^*]$. Hence, the differential inequality

$$D^+ m(t) \leq p(t)c(m(t))$$

holds for all $t \in (\bar{t}, t^*]$, and by integration over the last interval, we obtain

$$\int_{m(\bar{t})}^{m(t^*)} \frac{ds}{c(s)} \leq \int_{\bar{t}}^{t^*} p(s)ds \leq \int_{\bar{t}}^{\bar{t}+\tau} p(s)ds \leq M_1.$$

On the other hand, we have

$$\int_{m(\bar{t})}^{m(t^*)} \frac{ds}{c(s)} = \int_{\bar{g}(A)}^{A} \frac{ds}{c(s)} \geq M_2,$$

which contradicts with the assumption $M_2 > M_1$. This proves $m(t) \leq A$ for all $t \geq \tau_j$.

The proof of (2) can be carried over similarly. Assume there is some $t \geq \tau_j$ such that $m(t) > A - d$. Define $t^* = \inf\{t \geq \tau_j | m(t) > A - d\}$ and let $k \geq j$ be chosen so that $t^* \in [\tau_k, \tau_{k+1})$. Note that $t^* \neq \tau_k$ (or $t \in (\tau_k, \tau_{k+1})$). This is because $\bar{g}(A) < A - d$, the fact that $\hat{b}(\gamma) \leq A \leq \hat{a}(\varrho_1)$ and $m(\tau_k) \leq \bar{g}(m(\tau_k^-)) \leq \bar{g}(A) < A - d$. Moreover, as achieved before, $m(t^*) = A - d$ and $m(t) \leq A - d$ for all $t \in [\tau_k, t^*]$. Define $\bar{t} = \sup\{t \geq \tau_k | m(t) \leq \bar{g}(A - d) \leq \bar{g}(A)\}$. Since $m(t^*) = A - d > \bar{g}(A) \geq m(\tau_k)$, then $\bar{t} \in [\tau_k, t^*)$, $m(\bar{t}) = \bar{g}(A)$ and $m(t) \geq \bar{g}(A)$ for all $t \in [\bar{t}, t^*]$. By the same manner, we obtain

$$\int_{m(\bar{t})}^{m(t^*)} \frac{ds}{c(s)} \leq M_1$$

and

$$\int_{m(\bar{t})}^{m(t^*)} \frac{ds}{c(s)} = \int_{\bar{g}(A)}^{A-d} \frac{ds}{c(s)} = \int_{\bar{g}(A)}^{A} \frac{ds}{c(s)} - \int_{A-d}^{A} \frac{ds}{c(s)} \geq M_2 - \int_{A-d}^{A} \frac{ds}{c(s)}.$$

Let $M = \sup\{\frac{1}{c(s)} | s \in [A - d, A]\}$. Then

$$\int_{m(\bar{t})}^{m(t^*)} \frac{ds}{c(s)} \geq M_2 - Md > M_2 + M(M_1 - M_2)/M = M_1.$$

Thus, we have arrived a contradiction which completes the proof of $m(t) \leq A - d$ for all $t \geq \tau_j$.

Finally, define the indices $k^{(i)}$ for $i = 1, 2, \ldots, N$ as follows. For $i = 1, k^{(1)} = \tau_l$ and for the rest of i, we have $\tau_{k^{(i)}-1} < \tau_{k^{(i-1)}} + r \leq \tau_{k^{(i)}}$, which implicitly implies that $\tau_{k^{(i)}} - \tau_{k^{(i-1)}} \geq r$ and $\tau_{k^{(i)}-1} - \tau_{k^{(i-1)}} < r$. Then, for $i = 1$, we have $\tau_{k^{(1)}} = \tau_l \leq t_0 + \tau$ and, for $i = 2, 3, \ldots, N$, we have $\tau_{k^{(i)}} \leq \tau_{k^{(i)}-1} + \tau \leq \tau_{k^{(i-1)}} + r + \tau$. In general, one may get, after combining the two inequalities, $\tau_{k^{(N)}} \leq t_0 + \tau + (r + \tau)(N - 1) = t_0 + T$. Now, we claim that, for $i = 2, 3, \ldots, N$ and $t \geq \tau_{k^{(i)}}$, $m(t) \leq A - id$. To justify our claim, note that, for $t \in [t_0 - r, \tau_{k^{(1)}})$, we have $m(t) \leq \hat{a}(\varrho_1)$. For $t \geq \tau_{k^{(1)}}$, we have shown $m(t) \leq \hat{a}(\varrho_1) - d$, where we have set $A = \hat{a}(\rho_1)$. Assume the inequality is true for $t \geq \tau_{k^{(j)}}$ for some $1 \leq j \leq N - 1$; i.e., $m(t) \leq \hat{a}(\varrho_1) - jd$. Let $A = \hat{a}(\varrho_1) - jd$. From the definition of $\tau_{k^{(i)}}$ for $i \geq 2$, we get $\tau_{k^{(j)}} \leq \tau_{k^{(j+1)}} - r$. Then, $m(t) \leq A$ for $t \in [\tau_{k^{(j+1)}} - r, \tau_{k^{(j+1)}})$ and $m(t) \leq A - d$ for all $t \geq \tau_{k^{(j+1)}} = \hat{a}(\varrho_1) - Nd$. Thus, we have proved our claim by induction. In particular, we have $m(t) \leq (\hat{b}(\gamma) + Nd) - Nd = \hat{b}(\gamma)$ for all $t \geq t_0 + T \geq \tau_{k^{(N)}}$. To conclude the proof, we use assumption (i) to get $\hat{b}(\mathbb{E}[\|x(t)\|^2]) \leq m(t) \leq \hat{b}(\gamma)$, which implies that $\mathbb{E}[\|x(t)\|^2] \leq \gamma$, for all $t \geq t_0 + T$. This shows that $x \equiv 0$ is uniformly asymptotically stable in the m.s. This completes the proof of Theorem 4.1.

Remark 4.1 The importance of Theorem 4.1 is its applicability to unstable continuous systems that can be stabilized by impulsive effects. In condition (iv), the requirement $M_2 > M_1$ is made to ensure that any possible growth in V between impulses is reduced by V at the impulses. Furthermore, the definition of M_1 can be weakened by redefining M_1 as follows

$$M_1 = \sup_{k \in \mathbb{N}} \int_{\tau_{k-1}}^{\tau_k} p(s)ds.$$

Moreover, if we were interested in establishing only m.s. uniform stability, we could drop the requirement $\tau < \infty$. Another interesting finding in this theorem is that the condition in (iii) is independent of the time delay. Furthermore, the proof of Theorem 4.1 can be use to establish the pth moment stability of the trivial solution. This requires modifying the inequalities in Theorem 4.1(i) as follows: $b(\|\psi(0)\|^p) \leq V(t, \psi(0)) \leq a(\|\psi(0)\|^p)$, (a.s.). Furthermore, as a special result of Theorem 4.1 is exponential stability in the m.s. or pth moment. In this case, the nonlinear class $\mathscr{K}_c$ and $\mathscr{K}_v$ functions reduce to linear functions as follows: $b(\|\psi(0)\|^p) = b\|\psi(0)\|^p$, $a(\|\psi(0)\|^p) = a\|\psi(0)\|^p$, and $c(V(t, \psi(0))) = cV(t, \psi(0))$ for some positive constants a, b and c, where for the m.s result, $p = 2$.

In the following corollary, we only state the sufficient conditions that guarantee the stability property for system (4.1), where the proof can be obtained from Theorems 4.1 and 3.2 in [1].

Corollary 4.1 *Assume that Assumptions A1 and A2 are satisfied and there exist functions $a \in \mathscr{K}_c$, $b, c \in \mathscr{K}_v$ and $p \in \mathscr{PC}(\mathbb{R}_+; \mathbb{R}_+)$. Let $V \in \mathscr{C}^{1,2}([-r, \infty) \times \mathbb{R}^n; \mathbb{R}_+)$ such that the following conditions hold:*

(i) *assumption (i) of Theorem 4.1;*

(ii) *for all* $t \neq \tau_k \in \mathbb{R}_+$ *and* $\psi \in \mathscr{PC}([-r, \infty); S(\varrho))$,

$$\mathscr{L}V(t, \psi) \leq -p(t)c(V(t, \psi(0))), \qquad (a.s.)$$

provided that $V(t + s, \psi(s)) \leq \bar{g}(V(t, \psi(0)))$ *for some* $s \in [-r, 0]$, *where* $\bar{g} \in \mathscr{K}_3$;

(iii) *at any impulsive moment* $\tau_k \in \mathbb{T}$ *and* $\psi \in \mathscr{PC}([-r, \infty); S(\varrho))$,

$$V(\tau_k, \psi(0) + \mathscr{I}(\tau_k, \psi(\tau_k^-))) \leq \hat{g}(V(\tau_k^-, \psi(0))), \qquad (a.s.)$$

with $\psi(0^-) = \psi(0)$, *where* $(\tau_k, \psi(\tau_k^-)) \in \mathbb{R}_+ \times \mathscr{PC}([-r, 0]; S(\varrho_1))$ *and* $\hat{g} \in \mathscr{K}_3$;

(iv) $M_2 = \sup_{q \geq 0} \int_q^{\bar{g}(q)} \frac{ds}{c(s)} < \infty$ *and* $M_1 = \inf_{t>0} \int_t^{t+\mu} p(s)ds > M_2$, *with* $\mu = \inf\{\tau_k - \tau_{k-1}\} > 0$.

Then, the trivial solution $x \equiv 0$ *of (4.1) is uniformly asymptotically stable in the m.s.*

Example 4.1 Consider the following impulsive system

$$\begin{aligned} dx &= \big(-ax + bx(t-1)e^{-x^2}\big)dt + 1/2x(t-1)dW, && t \neq \tau_k, \\ \Delta x(t) &= -0.8x(t^-), && t = \tau_k, \end{aligned}$$

where $b > (2a - 1/2\lambda)/(1 + \lambda)$ for some positive constants a and $0 < \lambda < 1$. Defining $V(x) = x^2$ implies that $\mathscr{L}V(x) \leq pc(x)$ where $p = (-2a + (1 + \lambda)b + 1/2\lambda)/2 > 0$, $c(s) = s$ and $\bar{g}(s) = \lambda s$. From (iv), we get $M_1 = p\tau$, $M_2 = -\ln(\lambda)$ and the condition $M_2 > M_1$ implies that $\tau < -\frac{1}{p}\ln(\lambda)$. Choosing $a = 1/2$ and $\lambda = 0.8$ leads to $b > 0.33$, so that if $b = 1$, the $\tau < 0.3719$ which represents the upper bound on the time between impulses. Figure 4.1 shows the stabilization of the trivial solution, where the time between impulses is taken $\tau = 0.2$.

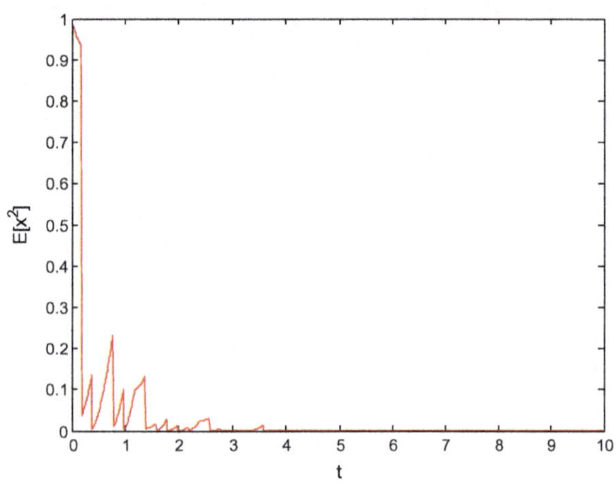

Fig. 4.1 Mean square asymptotic stability of $x \equiv 0$

4.2 Stability Analysis by Comparison Method

We continue to address the stability problem for system (4.1) by using the Lyapuno–Razumikhim technique. In this section, we use the comparison method to establish some m.s. stability properties for this system. As stated earlier, the comparison method enables one to compare multivariable systems with an auxiliary scalar system, and hence, the features of the latter system imply the corresponding features of the compared systems.

Theorem 4.2 *Assume that Assumptions A1 and A2 are satisfied and there exists a function* $a \in \mathscr{K}_2$. *Let* $V \in \mathscr{C}^{1,2}\big([-r, \infty) \times \mathbb{R}^n; \mathbb{R}_+\big)$ *such that the following assumptions hold:*

(i) $V(t, \psi(0)) \le a(\|\psi(0)\|^2) \le a(\|\psi\|_r^2)$, *(a.s.),* $\forall (t, \psi(0)) \in [-r, \infty) \times S(\varrho)$;

(ii) $\mathscr{L}V(t, \psi(t)) \le h(t, V(t, \psi(0)))$, *(a.s.),* $\forall t \ne \tau_k$ *and* $\psi \in \mathscr{PC}\big([-r, 0]; S(\varrho)\big)$ *provided that* $V(t+s, \psi(s)) \le q(V(t, \psi(0)))$ *for all* $s \in [-r, 0]$, *with* $q \in \mathscr{K}_3$, *where* $h : \mathbb{R}_+ \times \mathbb{R}_+ \to \mathbb{R}$ *is continuous on* $[\tau_{k-1}, \tau_k)$, $h(t, z)$ *is concave in* z *for any* $t \in \mathbb{R}_+$, *and, for each* $x \in \mathbb{R}^n$ *and* $k \ge 1$,

$$\lim_{(t,y)\to(\tau_k^-,x)} h(t, y) = h(\tau_k^-, x)$$

exists;

(iii) $\forall \tau_k \in \mathbb{T}$ *and* $\psi \in \mathscr{PC}\big([t_0 - r, \infty); S(\varrho)\big)$,

$$V(\tau_k, \psi(0) + \mathscr{I}(\tau_k, \psi(\tau_k^-))) \le \alpha_k(V(\tau_k^-, \psi(0^-))), \qquad \text{(a.s.)}$$

where $\psi(0^-) = \psi(0)$, $(\tau_k, \psi(\tau_k^-)) \in \mathbb{R}_+ \times \mathscr{PC}\big([-r, 0]; S(\varrho_1)\big)$ *(with* $\varrho_1 < \varrho$*), and* $\alpha_k : \mathbb{R}_+ \to \mathbb{R}_+$ *is a non-decreasing, concave function;*

(iv) the auxiliary scalar impulsive system

$$\begin{cases} D^+ v(t) = h(t, v(t)), & t \ne \tau_k, \\ v(t) = \alpha_k(v(t^-)), & t = \tau_k, \\ v(t_0) = v_0 \ge 0 \end{cases} \tag{4.6}$$

has a maximal solution $r(t) = r(t, t_0, v_0)$.

Then, $\mathbb{E}[V(t_0, x_0)] < v_0$ *implies* $\mathbb{E}[V(t, x(t))] < r(t)$ *for all* $t \ge t_0$.

Proof Let $x(t) = x(t; t_0, \phi)$ be any solution of system (4.1). From (i), we have $\mathbb{E}[V(t, x(t))] < \infty$. Also, by Itô formula and condition (ii), we have, for all $t \in [\tau_{k-1}, \tau_k)$,

$$\mathbb{E}[V(t, x(t))] \le \mathbb{E}[V(\tau_{k-1}, x(\tau_{k-1})] + \int_{\tau_{k-1}}^{t} h(s, \mathbb{E}[V(s, x(s))])ds,$$

from which we get

$$D^+m(t) \le h(t, m(t)), \qquad t \ne \tau_k,$$

where $m(t) = \mathbb{E}[V(t, x(t))]$ for all $t \in [\tau_{k-1}, \tau_k)$. At the impulsive moments, we have from condition (iii), $m(\tau_k) \le \alpha_k(m(\tau_k^-))$. In summary, we have

$$\begin{cases} D^+m(t) \le h(t, m(t)), & t \ne \tau_k, \\ m(t) \le \alpha_k(m(t^-)), & t = \tau_k, \\ m(t_0) = \mathbb{E}[V(t_0, x_0)]. \end{cases}$$

Therefore, comparing with (4.6) leads to (see Theorem 1.6.1 in [2])

$$m(t) = \mathbb{E}[V(t, x(t))] < r(t) = v(t), \qquad \forall t \ge t_0.$$

This completes the proof.

In this following, we make use of this comparison result to show how the stability properties of the auxiliary scalar impulsive system (4.6) imply those of (4.1).

Theorem 4.3 *Assume that Assumptions A1 and A2 hold, and there exist functions $a \in \mathscr{K}_c$ and $b \in \mathscr{K}_v$. Assume further that $V \in \mathscr{C}^{1,2}([-r, \infty) \times \mathbb{R}^n; \mathbb{R}_+)$ such that the following hold:*

(i) for all $(t, \psi(0)) \in [-r, \infty) \times S(\varrho)$,

$$b(\|\psi(0)\|^2) \le V(t, \psi(0)) \le a(\|\psi(0)\|^2), \qquad (a.s.);$$

(ii) for all $t \ne \tau_k$ and $\psi \in \mathscr{PC}([-r, 0]; S(\varrho))$,

$$\mathscr{L}V(t, \psi(t)) \le h(t, V(t, \psi(0))), \qquad (a.s.)$$

provided that $V(t+s, \psi(s)) \le q(V(t, \psi(0)))$ with $s \in [-r, 0]$, where $q \in \mathscr{K}_3$, $h : \mathbb{R}_+ \times \mathbb{R}_+ \to \mathbb{R}$ is continuous in its variables, $h(t, 0) = 0$ and $h(t, z)$ is concave in z for any $t \in \mathbb{R}_+$, and, for each $x \in \mathbb{R}^n$ and $k \ge 1$,

$$\lim_{(t,y)\to(\tau_k^-,x)} h(t, y) = h(\tau_k^-, x)$$

exists;

(iii) $\forall \tau_k \in \mathbb{T}$ and $\psi \in \mathscr{PC}([-r, 0]; S(\varrho))$,

$$V(\tau_k, \psi(0) + \mathscr{I}_k(\tau_k, \psi(\tau^-))) \le \alpha_k(V(\tau_k^-, \psi(0^-))), \qquad (a.s.)$$

where $\psi(0^-) = \psi(0)$, $(\tau_k, \psi(\tau_k^-)) \in \mathbb{R}_+ \times \mathscr{PC}([-r, 0]; S(\varrho_1))$, and $\alpha_k \in \mathscr{K}_3$.

Then, the stability properties of the trivial solution of auxiliary scalar impulsive system (4.6) imply the corresponding stability properties of the trivial solution of system (4.1).

Proof Let $0 < \varepsilon < \varrho_1 < \varrho$ and $t_0 \in \mathbb{R}_+$. Assume that the auxiliary scalar comparison system is stable. Then, for given $b(\varepsilon) > 0$ and $t_0 \in \mathbb{R}_+$, there exists a $\delta = \delta(t_0, \varepsilon) > 0$ such that

$$v_0 < \delta \quad \text{implies} \quad v(t; t_0, v_0) < b(\varepsilon), \quad \forall t \geq t_0,$$

where $v(t; t_0, v_0)$ is any solution of the comparison system.

Choose $v_0 = a(\|\phi\|_r^2)$ and $\delta_1 = \delta_1(\varepsilon) > 0$ for which $a(\delta_1) < b(\varepsilon)$. Define $\hat{\delta} = \min\{\delta, \delta_1\}$. We claim that, if $\mathbb{E}[\|\phi\|_r^2] \leq \hat{\delta}$, then

$$\mathbb{E}[\|x(t)\|^2] < \varepsilon, \qquad \forall t \geq t_0.$$

If our claim were not true, there would be a $\bar{t} \in [\tau_k, \tau_{k+1})$ for some k such that

$$\varepsilon \leq \mathbb{E}[\|x(\bar{t})\|^2]$$

and

$$\mathbb{E}[\|x(t)\|^2] < \varepsilon, \qquad \forall t \in [\tau_k, \bar{t}).$$

By Assumption A1 (i.e., if $\mathbb{E}[\|x(\tau_k^-)\|^2] < \varepsilon < \varrho_1$, then $\mathbb{E}[\|x(\tau_k)\|^2] = \mathbb{E}[\| x(\tau_k^-) + \mathscr{I}(\tau_k, x_{\tau_k^-})\|^2] < \varrho$), there exists a $\underline{t}$ such that $\tau_k < \underline{t} \leq \bar{t}$ satisfying

$$\varepsilon \leq \mathbb{E}[\|x(\underline{t})\|^2] < \varrho.$$

Define $m(t) = \mathbb{E}[V(t, x(t))]$ for all $t \in [t_0, \underline{t}]$. By Theorem 4.2, we get

$$m(t) < r(t; t_0, a(\mathbb{E}[\|\phi\|_r^2])), \qquad \forall t \in [t_0, \underline{t}],$$

where $r(t; t_0, a(\mathbb{E}[\|\phi\|_r^2]))$ is the maximal solution of the auxiliary comparison system.

Finally, by condition (i), we obtain

$$b(\varepsilon) \leq m(\underline{t}) < r(\underline{t}; t_0, a(\mathbb{E}[\|\phi\|_r^2])) \leq r(\underline{t}; t_0, a(\delta)) < b(\varepsilon),$$

which contradicts with our supposition. Therefore, it must be true that

$$\mathbb{E}[\|x(t)\|^2] < \varepsilon, \qquad \forall t \geq t_0.$$

As for the uniform property, it suffices to choose δ independent of t_0.

To prove the uniform attractivity, we choose $0 < \eta < \varrho_1 < \varrho$. Assume that the comparison system is uniformly attractive; i.e., for a given $b(\eta) > 0$, there exist $\delta > 0$ and a constant $T = T(\eta) > 0$ such that

$$v_0 \le \delta \quad \text{implies} \quad v(t; t_0, v_0) < b(\eta), \qquad \forall t \ge t_0 + T.$$

Following the argument used in proving the stability property, we obtain

$$b(\mathbb{E}[\|x(t)\|^2]) \le v(t; t_0, v_0) < b(\eta), \qquad \forall t \ge t_0 + T,$$

i.e., the system (4.1) is uniformly attractive in the m.s., which leads to the m.s. uniformly asymptotic stability property of $x \equiv 0$. This completes the proof of Theorem 4.3.

Corollary 4.2 *In Theorem 4.3, assume that, for any $(t, \psi(0)) \in \mathbb{R}_+ \times \mathscr{PC}([t - r, \infty); S(\varrho))$,*

$$\alpha_k(V(\tau_k^-, \psi(0^-))) = \alpha(d_k)V(\tau_k^-, \psi(0^-)), \tag{4.7}$$

where d_k is a nonnegative constant such that $d = \sum_{k=1}^{\infty} d_k < \infty$ and $\alpha(d_k) > 1$ for all k. If

(i) $h(t, V(t, \psi(0))) = 0$, (a.s.) provided that $V(t + s, \psi(s)) \le q(V(t, \psi(0)))$ for some $s \in [-r, 0]$ and $q \in \mathscr{K}_3$, then trivial solution $x \equiv 0$ of (4.1) is uniformly stable in the m.s.;

(ii) $h(t, V(t, \psi(0))) = -c(V(t, \psi(0)))$, (a.s.), where q is defined in (i), then the trivial solution $x \equiv 0$ of (4.1) is asymptotically stable in the m.s.

Proof (i). Let $x(t) = x(t; t_0, \phi)$ be the unique solution of system (4.1) and $0 < \varepsilon \le \varrho_1$. Define $\overline{d} = \prod_{k=1}^{\infty} \alpha(d_k)$. Then, $1 \le \overline{d} < \infty$ because $d < \infty$. Choose $\delta = \delta(\varepsilon)$ so that $\delta < \hat{a}^{-1}\big(\hat{b}(\varepsilon)/\overline{d}\big)$ and clearly $0 < \delta < \varepsilon$, where $\hat{a}$ and $\hat{b}$ are defined in the proof of Theorem 4.1.

Let $t_0 \in [\tau_{l-1}, \tau_l)$ for some positive integer l and ϕ for which $\mathbb{E}[\|\phi\|_r^2] \le \delta$. We claim that the trivial solution is uniformly stable in the m.s. If our claim were not true, there would exist a t^* such that, for all $t \in [t_0 - r, t^*)$, we have

$$\mathbb{E}[\|x(t)\|^2] < \varepsilon < \varrho_1$$

and either

$$\mathbb{E}[\|x(t^*)\|^2] = \varepsilon,$$

which implies that

$$\mathbb{E}[\|x(t^*)\|^2] = \mathbb{E}[\|x_{t^*}\|_r^2] = \varepsilon$$

or

$$\varepsilon < \mathbb{E}[\|x(t^*)\|^2], \quad \text{where} \quad t^* = \tau_k \quad \text{for some } k$$

and, by Assumption A1,

$$\varepsilon < \mathbb{E}[\|x(t^*)\|^2] < \varrho$$

since $\mathbb{E}[\|x_{t^{*-}}\|^2] \leq \varepsilon < \varrho_1$. Thus, in either case, $V(t, x(t))$ is defined for $t \in [t_0, t^*]$. Moreover, from (i), we have

$$\mathscr{L}V(t, x_t) \leq 0.$$

Applying the Itô formula to process $V(t, x(t))$ for $t \in [t_0, t^*]$ and taking the mathematical expectation yield

$$\begin{aligned} \mathbb{E}[V(t, x(t))] &\leq \mathbb{E}[V(s, x(s))] + \mathbb{E}\int_s^t \mathscr{L}V(u, x_u)du \qquad \forall\, t_0 \leq s \leq t \leq t^* \\ &\leq \mathbb{E}[V(s, x(s))]. \end{aligned}$$

Define $m(t) = \mathbb{E}[V(t, x(t))]$ for all $t \in [t_0, t^*]$. Then, from the last inequality,

$$D^+ m(t) \leq 0$$

provided that $m(t+s) \leq q(m(t))$; that is, the function $m(t)$ is nonincreasing for all $t \in (t_0, t^*]$ between the impulse moments. Furthermore, from (4.7), we have

$$m(\tau_k) \leq \alpha(d_k) m(\tau_k^-), \qquad \forall\, t \in (t_0, t^*].$$

Consider the following comparison impulsive system

$$\begin{cases} D^+ v(t) = 0 & t \neq \tau_k \\ v(t) = \alpha(d_k) v(t^-) & t = \tau_k \\ v(t_0) = v_0 > m_0 = \mathbb{E}[V(t_0\, x_0)]. \end{cases}$$

This implies that

$$v(t) \leq v_0 < \delta < \varepsilon, \qquad t \in [t_0, t^*)$$

for the same δ and, by Theorem 4.3, one can easily see that $\mathbb{E}[\|x(t^*)\|^2] < \varepsilon$.

On the other hand, let $t^* \in [\tau_k, \tau_{k+1})$ for some $k \geq l$. In this case, we have

$$v(t^*) \leq v(\tau_k), \qquad \text{because} \quad v \quad \text{is nonincreasing} \quad \forall\, t \leq t^*, \tag{4.8}$$

$$v(\tau_k^-) \leq v(t_0) < \hat{a}(\delta), \tag{4.9}$$

$$v(\tau_i^-) \leq v(\tau_{i-1}), \quad i = l+1, l+2, \ldots k \tag{4.10}$$
$$v(\tau_i) \leq \alpha(d_i)v(\tau_i^-) \quad i = l l + 1 l + 2 \ldots, k. \tag{4.11}$$

By (4.11), we have

$$\begin{aligned} v(\tau_i) &\leq \alpha(d_i)v(\tau_i^-) \\ &\leq \alpha(d_i)v(\tau_{i-1}) && \text{by (4.10)} \\ &\leq \alpha(d_i)\alpha(d_{i-1})v(\tau_{i-1}^-) && \text{by(4.11)} \\ &\ \vdots \\ &\leq \prod_{i=1}^{l} \alpha(d_i)v(t_0) \\ v(\tau_i) &\leq \overline{d}\, v(t_0) \leq \overline{d}\, \hat{a}(\delta). && \text{by (4.8)} \end{aligned}$$

Namely, $v(\tau_i) \leq \overline{d}\, \hat{a}(\delta)$ which implies that

$$v(t^*) \leq v(\tau_i) \leq \overline{d}\, \hat{a}(\delta),$$

where the first inequality is from (4.8). With the aid of Theorem 4.3, we have

$$\hat{b}(\varepsilon) < \hat{b}(\mathbb{E}\|x(t^*)\|^2) \leq m(t^*) \leq v(t^*) < \overline{d}\, \hat{a}(\delta) < \hat{b}(\varepsilon).$$

This is a contradiction. It turns out that $x \equiv 0$ is uniformly stable in m.s. This completes the proof of (i).

(ii) The assertion of this part can be proved easily; thus, it is left here as an exercise.

Remark 4.2 Assumptions (i) (or (ii)) in Corollary 4.2 is made to ensure that the Lyapunov function V is nonincreasing (or strictly decreasing) in main, which in turn implies that the continuous system is m.s. uniformly stable (or asymptotically stable). To guarantee that the overall behavior of V decreases for all time (including the impulse moments), we assume that V is nonincreasing at these moments, because, otherwise, the reduction of V may not compensate the jump increases. This condition is summarized in (4.7).

Remark 4.3 Using the efficient comparison method, Theorem 4.3 does not impose any restriction on the stability of continuous system. This fact will be further seen in Corollary 4.3, where the impulsive effects can have a stabilizing role even when the underlying continuous system is unstable. The requirement in this circumstance is that the impulses be small enough to reduce the growth of the continuous part and be applied to the system more frequently.

Corollary 4.3 *In Theorem 4.3, assume that*

(i) there exist a function $p \in \mathscr{PC}(\mathbb{R}_+; \mathbb{R}_+)$ *and* $c \in \mathscr{K}_c$ *such that, for any* $(t, \psi(0)) \in \mathbb{R}_+ \times \mathscr{PC}([t-r, \infty); S(\varrho))$,

$$h(t, V(t, \psi(0))) = p(t)c(V(t, \psi(0))); \tag{4.12}$$

(ii) there exist $\gamma_k \geq 0$ *and* $\varrho_0 > 0$ *such that, for all* $z \in (0, \varrho_0)$ *and any* $k \in \mathbb{N}$,

$$\int_{\tau_{k-1}}^{\tau_k} p(s)ds + \int_z^{\alpha_k(z)} \frac{ds}{c(s)} \leq -\gamma_k. \tag{4.13}$$

Then, the trivial solution $x \equiv 0$ *of (4.1) is uniformly stable in the m.s. If, moreover,* $\sum_{k=1}^{\infty} \gamma_k = \infty$, *then* $x \equiv 0$ *is asymptotically stable in the m.s.*

Proof In the light of Theorem 4.2, defining $m(t) = \mathbb{E}[V(t, x(t))]$ for all $t \geq t_0$ leads to the comparison system

$$\begin{cases} D^+m(t) \leq p(t)c(m(t)), & t \neq \tau_k, \\ m(t) \leq \alpha_k(m(t^-)), & t = \tau_k, \\ m(t_0) = m_0 = \mathbb{E}[V(t_0, x_0)]. \end{cases} \tag{4.14}$$

Consider the auxiliary scalar impulsive comparison system

$$\begin{cases} D^+v(t) = p(t)c(v(t)), & t \neq \tau_k, \\ v(t) = \alpha_k(v(t^-)), & t = \tau_k, \\ v(t_0) = v_0 > m_0. \end{cases} \tag{4.15}$$

Now, we are aiming to prove the stability properties of the comparison system (4.15), which, by Theorem 4.3, imply the corresponding properties of (4.1).

Let $0 < \varepsilon < \varrho_0$ and $t_0 \in [\tau_1, \tau_2)$. Choose $\delta > 0$ for which $\delta < \min\{\varepsilon, \alpha_k(\varepsilon)\}$ and $0 \leq v_0 < \delta$. We claim that $v(t) < \varepsilon$ for all $t \in [t_0, \tau_2)$, where v is any solution of (4.15). If our claim were not true, then there would exist a $t^* \in [t_0, \tau_2)$ such that $v(t^*) \geq \varepsilon$. Integrating the differential inequality in (4.14) over (t_0, t^*) gives

$$\int_{v(t_0)}^{v(t^*)} \frac{ds}{c(s)} \leq \int_{t_0}^{t^*} p(s)ds, \tag{4.16}$$

where a variable substitution is performed. By our choice of t_0 and t^* and the positiveness of p, we have

$$\int_{t_0}^{t^*} p(s)ds \leq \int_{\tau_1}^{\tau_2} p(s)ds$$

and, by the early analysis,

$$\int_{v(t_0)}^{v(t^*)} \frac{ds}{c(s)} > \int_{\alpha_1(\varepsilon)}^{\varepsilon} \frac{ds}{c(s)}.$$

Therefore, (4.16) becomes

$$\int_{\tau_1}^{\tau_2} p(s)ds + \int_{\varepsilon}^{\alpha_1(\varepsilon)} \frac{ds}{c(s)} > 0,$$

which contradicts with (4.13). Thus, it must be true that $v(t) < \varepsilon$ for all $t \in [t_0, \tau_2)$ or $t \in [\tau_1, \tau_2)$.

Suppose that, for all $t \in [t_0, \tau_k)$ (or generally $t \in [\tau_{k-1}, \tau_k)$), $v(t) < \varepsilon$. Then, it follows from (4.14) that, for all $t \in [\tau_k, \tau_{k+1})$,

$$\int_{v(\tau_k)}^{v(t)} \frac{ds}{c(s)} \leq \int_{\tau_k}^{t} p(s)ds \leq \int_{\tau_k}^{\tau_{k+1}} p(s)ds. \tag{4.17}$$

Noting that $v(\tau_k) = \alpha_k(v(\tau_k^-))$, the last inequality becomes

$$\int_{v(\tau_k^-)}^{v(t)} \frac{ds}{c(s)} \leq \int_{\tau_k}^{\tau_{k+1}} p(s)ds + \int_{v(\tau_k^-)}^{\alpha(v(\tau_k^-))} \frac{ds}{c(s)} \leq -\gamma_k. \tag{4.18}$$

Thus, $v(t) \leq v(\tau_k^-) < \varepsilon$ for all $t \in [\tau_k, \tau_{k+1})$ and, by induction, $v(t) < \varepsilon$ for all $t \geq t_0$; that is, the trivial solution $v \equiv 0$ is uniformly stable.

To prove asymptotic stability of $v \equiv 0$, let $\varepsilon = \varrho_0$ and choose $\delta_0 = \delta_0(\varrho) > 0$ such that $v_0 < \delta_0$ implies that $v(t) < \varrho_0$ for all $t \geq t_0$. We will prove under the given assumption and obtained result that $\lim_{k\to\infty} v(\tau_k) = 0$. If this were not the case, there would exist an $\eta > 0$ such that $\lim_{k\to\infty} v(\tau_k) = \eta$. From (4.18), we get

$$\int_{v(\tau_k)}^{v(\tau_{k+1})} \frac{ds}{c(s)} = \frac{v(\tau_{k+1}) - v(\tau_k)}{c(\eta)} \leq -\gamma_k,$$

where

$$\frac{1}{c(\eta)} = \sup\left\{\frac{1}{c(s)} \,\Big|\, \forall s \in [v(\tau_k), v(\tau_{k+1})]\right\},$$

which also implies, by consecutive induction, that

$$v(\tau_k) \leq v(\tau_{k-1}) - c(\eta)\sum_{i=1}^{k} \gamma_k.$$

Letting k go to infinity leads to a contradiction. Therefore, it must be true that $\eta = 0$, which proves the asymptotic stability of $v \equiv 0$. Finally, applying Theorem 4.3 implies that $x \equiv 0$ is asymptotically stable in the m.s. This completes the proof of Corollary 4.3.

Remark 4.4 A similar result can be obtained if $p(t)$ in (4.12) is replaced by $-p(t)$ for all t or, particularly, $p(t) = \pm p$ and the impulsive condition (iii) of Theorem 4.3 is replaced by $\alpha_k V(\tau_k^-, \psi(0))$. In the latter case, the inequality in (4.13) is simplified to

$$\pm p(\tau_k - \tau_{k-1}) + \ln \alpha_k \le -\gamma_k, \qquad \forall k \in \mathbb{N}. \tag{4.19}$$

Example 4.2 Consider the following impulsive system

$$\begin{aligned} dx &= \big(-4x + x(t-1)e^{-|x|}\big)dt - 0.1 \sin x(t-1)dW, && t \neq \tau_k, \\ \Delta x(t) &= \frac{1}{k^2} x_{t^-}, && t = \tau_k. \end{aligned}$$

Define $V(x) = x^2$ as a Lyapunov function candidate. Then, one can easily show that $\mathscr{L}V(x) \le -c(x) < 0$ with $q = 2$, where $c(s) = 3s^2$. At $t = \tau_k$, we have

$$|x(\tau_k)| = |x(\tau_k^-) + \frac{1}{k^2} x_{\tau_k^-}| \le |x(\tau_k^-)| + \frac{\sqrt{2}}{k^2}|x(\tau_k^-)| \le (1 + \frac{\sqrt{2}}{k^2})|x(\tau_k^-)|,$$

from which we have $V(x(\tau_k)) \le \alpha(d_k)V(x(\tau_k^-))$, where $\alpha(d_k)=(1+\sqrt{2}d_k)^2$ and $d_k=\frac{1}{k^2}$. We also have $\varrho_1 < \varrho/(1+\sqrt{2}d_k)$. Choose $a(s) = b(s) = s^2$. Thus, the assumptions of Corollaries 4.2 and 4.3 are satisfied; i.e., the trivial solution $x \equiv 0$ is asymptotically stable in the m.s. The simulation result of this example is shown in Fig. 4.2.

Example 4.3 Consider the following impulsive system

$$\begin{aligned} dx &= \Big(-7x - 0.5y(t-1)e^{-x^2}\Big)dt, && t \neq \tau_k,\ k \in \mathbb{N} \\ dy &= \Big(-5y + \sin x(t-1)\Big)dt + \Big(-\frac{0.1x(t-1)}{1+y^2})\Big)dW_2, && t \neq \tau_k \\ \Delta x(\tau_k) &= -2x(\tau_k^-), \\ \Delta y(\tau_k) &= 0.2y(\tau_k^- - 1). \end{aligned}$$

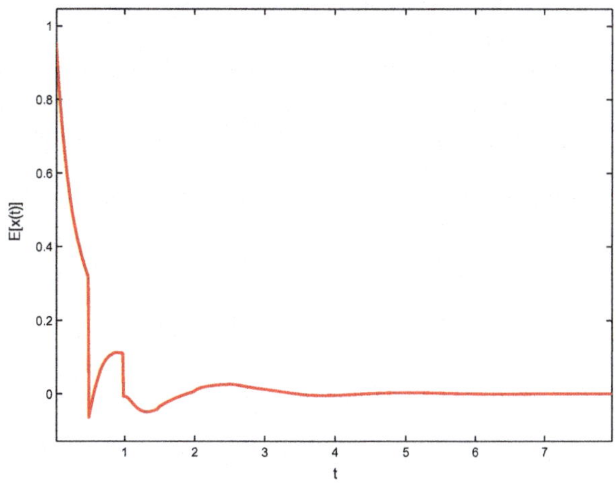

Fig. 4.2 First moment asymptotic stability of $x \equiv 0$

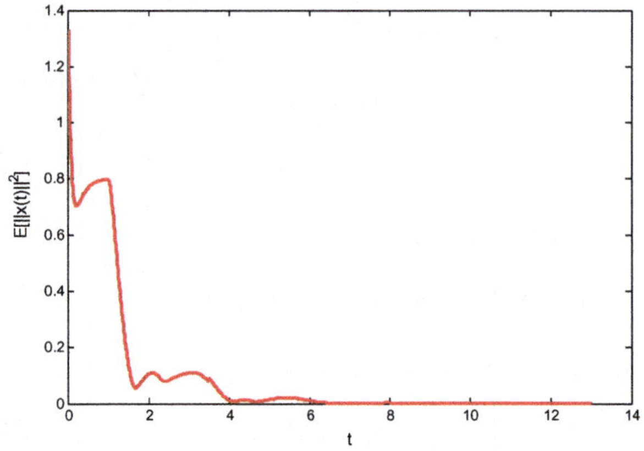

Fig. 4.3 Mean square asymptotic stability of $(x\ y)^T = (0\ 0)$

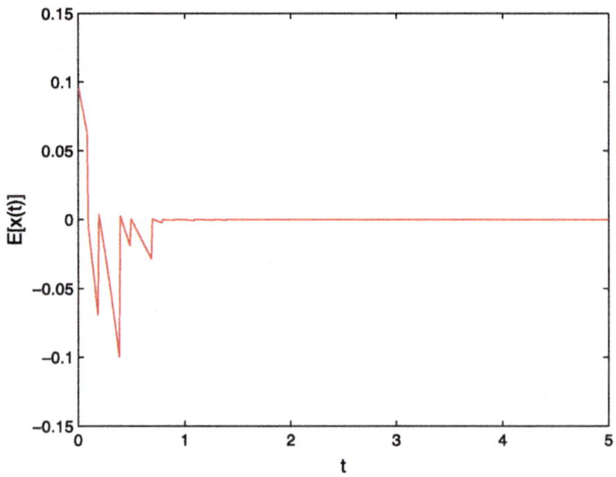

Fig. 4.4 First moment asymptotic stability of $x \equiv 0$

Define $V(x, y) = \frac{1}{2}(x^2 + y^2)$ as a Lyapunov function candidate. Then, after cumbersome calculation, we get $\mathscr{L}V(x, y) \leq -6.98V(x, y)$, where $q = 2$, and, at $t = \tau_k$, we get $V(x(\tau_k), y(\tau_k)) \leq \alpha_k V(x(\tau_k^-), y(\tau_k^-))$, where $\alpha_k = 6$. By (4.19), we find the upper bound on the time between impulses to be $\tau_k - \tau_{k-1} < 0.6$ for all k. Thus, the trivial solution is asymptotically stable in the m.s. Figure 4.3 shows the simulation result.

Example 4.4 Consider the following impulsive system

$$
\begin{aligned}
dx &= \Big(4x - x^2(t-1)\Big)dt + 0.1x\,dW, && t \neq \tau_k \\
\Delta x(t) &= -\frac{k+2}{k+1}x(t^-) && t = \tau_k \quad k \in \mathbb{N}.
\end{aligned}
$$

Considering the Lyapunov function $V(x) = \frac{1}{2}x^2$ leads to $\mathscr{L}V(x) \leq 5.55x^2$; i.e., the underlying continuous system has an unstable trivial solution. At the impulsive effects $t = \tau_k$, $V(x(\tau_k)) \leq \frac{1}{(k+1)^2} V(x(\tau_k^-))$, i.e., $\alpha_k = \frac{1}{(k+1)^2} < 1$. From (4.19), we get $\tau_k - \tau_{k-1} \leq 0.2$ for all k. The simulation result is shown in Fig. 4.4, which shows the stabilizing effects of impulses.

4.3 Notes and Comments

Throughout this chapter, the focus has been on establishing m.s. uniform stability and uniform asymptotic stability for impulsive stochastic differential equations with time delay, where we have used two different approaches, the classical Lyapunov method (Sect. 4.1) and comparison method (Sect. 4.2). The material of this chapter is taken from [3]. Also, the stability theory of the nonlinear deterministic impulsive systems with time delay can be read, for instance, in [1, 4]. In both sections, the method of Lyapunov–Razumikhim in which we use Lyapunov function is efficient to examine qualitative properties of delay systems, because it provides results that are independent of time delay. In contrast, one may use Lyapunov functionals to address the same qualitative properties; however, the obtained result in this case will be delay dependent.

Particularly, in Sect. 4.1, the underlying continuous systems are stable or unstable that are perturbed by impulsive actions, which are not necessarily bounded. It has been shown that the stable continuous system can preserve its stability property if the impulses are relatively small and infrequently applied to the system and, if the continuous system is unstable, the impulses have to be applied frequently in order to reduce the growth of continuous states. In Sect. 4.2, it has been shown that systems can maintain their stability properties even if they are disturbed by unbounded impulses (Corollary 4.2). Moreover, it is evident that impulses can help in stabilizing systems which are originally unstable (Corollary 4.3).

References

1. Liu XZ, Ballinger GH (2001) Uniform asymptotic stability of impulsive delay differential equations. Comput Math Appl 41:903–915
2. Lakshmiknatham V, Bainov DD, Simeonov PS (1989) Theory of impulsive differential equations. World Scientific, Singapore
3. Alwan MS, Liu XZ, Xie W-C. (2013) Stability properties of nonlinear stochastic impulsive systems with time delay. J Nonlinear Syst Appl:86–91
4. Ballinger GH (1999) Qualitative theory of impulsive delay differential equations. PhD thesis, University of Waterloo, On, Canada

Chapter 5
Large-Scale Stochastic Impulsive Systems with Time Delay

In this chapter, we consider large-scale nonlinear stochastic systems with time delay and subject to impulsive effects. The random noise is described by Wiener process, the time delay is finite and the impulsive actions are applied at fixed times. The focus is on establishing uniform asymptotic stability property of the system in the mean square. In fact, stability property of large-scale systems can be achieved in different ways. As presented in Sect. 2.11, an efficient approach to deal with such a complex system is to decompose the composite (or interconnected) systems into simpler, more manageable isolated (also called uncoupled or unperturbed) subsystems at different hierarchical levels. Analyze each individual subsystem by initially ignoring the interconnection between the subsystems, then combine the available results together with interconnection, which is usually regarded as a perturbation, to draw a conclusion on the qualitative property of the composite system.

We use Razumikhin technique and comparison method to develop Lyapunov-like sufficient conditions. We also consider two cases of continuous systems. In the first case, the isolated subsystems are assumed to be stable in the m.s. and the rest (i.e., the interconnection) will be viewed as perturbation, which is required to have magnitude be smaller than the degree of stability of each isolated subsystem. This type of relation between isolated subsystems and their interconnection is usually represented in a special type of matrices called *test matrices*. In the second case, the isolated continuous subsystems are assumed to be unstable and are stabilized to be impulsive effects, which also help to stabilize the entire composite system.

5.1 Problem Formulation

Typically, an interconnected or composite system with decomposition $\mathbb{D}_i$ may have the form

M. S. Alwan and X. Liu, *Theory of Hybrid Systems: Deterministic and Stochastic*, Nonlinear Physical Science, https://doi.org/10.1007/978-981-10-8046-3_5

$$\mathbb{D}_i : \begin{cases} dw^i(t) = f_i(t, w_t^i)dt + g_i(t, w_t^1, w_t^2, \ldots, w_t^l)dt \\ \qquad\qquad + \sum_{j=1}^{l} \sigma_{ij}(t, w_t^j)dW_j(t), \ t \neq \tau_k, \\ \Delta w^i(t) = \mathscr{I}_i(t, w_{t^-}^i), \qquad\qquad t = \tau_k, \\ w_{t_0}^i = \phi_i(s), \qquad s \in [-r, 0], \end{cases} \tag{5.1}$$

where $k \in \mathbb{N}$ and $i = 1, 2, \ldots l$ for some $l \in \mathbb{N}$. Here, we have w^i (or w_t^i) $\in \mathbb{R}^{n_i}$, which is an n_i-dimensional vector state (or delayed state, respectively) and $n = \sum_i^l n_i$ for some $n_i \in \mathbb{N}$. $f_i : \mathbb{R}_+ \times \mathbb{R}^{n_i} \to \mathbb{R}^{n_i}$, $g_i : \mathbb{R}_+ \times \mathbb{R}^n \to \mathbb{R}^{n_i}$, $\sigma_{ij} : \mathbb{R}_+ \times \mathbb{R}^{n_j} \to \mathbb{R}^{n_i \times m_j}$, $m = \sum_i^l m_i$ for some $m_i \in \mathbb{N}$, $\phi_i : [-r, 0] \to \mathbb{R}^{n_i}$, $\mathscr{I}_i : \mathbb{T} \times \mathbb{R}^{n_i} \to \mathbb{R}^{n_i}$ with $\mathbb{T} = \{\tau_k \mid k \in \mathbb{N}\}$ where τ_k represents constant impulsive moments and satisfies $0 < \tau_1 < \tau_2 < \cdots$ and $\lim_{k\to\infty} \tau_k = \infty$.

Define the isolated subsystems $\mathbb{S}_i$ as follows

$$\mathbb{S}_i : \begin{cases} dw^i(t) = f_i(t, w_t^i)dt + \sigma_{ii}(t, w_t^i)dW_i(t), & t \neq \tau_k, \\ \Delta w^i(t) = \mathscr{I}_i(t, w_{t^-}^i), & t = \tau_k, \\ w_{t_0}^i = \phi_i(s), \qquad s \in [-r, 0]. \end{cases} \tag{5.2}$$

For $x \in \mathbb{R}^n$, let $x^T = [(w^1)^T \ (w^2)^T \ \cdots \ (w^l)^T]$ and $x_t^T = [(w_t^1)^T \ (w_t^2)^T \ \cdots \ (w_t^l)^T]$ denote the composite system state and delayed state, respectively. Define the composite system vector field $f : \mathbb{R}_+ \times \mathbb{R}^n \to \mathbb{R}^n$ by

$$f^T(t, x_t) = [f_1^T(t, w_t^1) \ f_2^T(t, w_t^2) \ \cdots \ f_l^T(t, w_t^l)],$$

the interconnection $g : \mathbb{R}_+ \times \mathbb{R}^n \to \mathbb{R}^n$ by

$$\begin{aligned} g^T(t, x_t) &= [g_1^T(t, x_t) \ g_2^T(t, x_t) \ \cdots \ g_l^T(t, x_t)] \\ &= [g_1^T(t, w_t^1, w_t^2, \ldots, w_t^l) \ g_2^T(t, w_t^1, w_t^2, \ldots, w_t^l) \ \cdots \ g_l^T(t, w_t^1, w_t^2, \ldots, w_t^l)], \end{aligned}$$

the diffusion matrix function $\sigma : \mathbb{R}_+ \times \mathbb{R}^n \to \mathbb{R}^{n \times m}$ by

$$\sigma(t, x_t) = [\sigma_{ij}(t, w_t^j)],$$

with $\sigma_{ij} : \mathbb{R}_+ \times \mathbb{R}^n \to \mathbb{R}^{n_i \times n_i}$ being representing the noise function perturbing the ith isolated subsystem, and Wiener process vector $W : \mathbb{R}_+ \to \mathbb{R}^m$ by

$$W^T = [W_1^T \ W_2^T \ \cdots \ W_l^T],$$

where, for any $i = 1, 2, \ldots, l$, $W_i : \mathbb{R}_+ \to \mathbb{R}^{m_i}$. We also define the impulsive functional vector of the composite system $\mathscr{I} : \mathbb{T} \times \mathbb{R}^n \to \mathbb{R}^n$ by

$$\mathscr{I}^T(t, x_{t^-}) = [\mathscr{I}_1^T(t, w_{t^-}^1) \ \mathscr{I}_2^T(t, w_{t^-}^2) \ \cdots \ \mathscr{I}_l^T(t, w_{t^-}^l)]$$

and the initial vector state of the composite system $\Phi : [-r, 0] \to \mathbb{R}^n$ by

$$\Phi^T = [\phi_1^T \ \phi_2^T \ \cdots \ \phi_l^T].$$

By adopting these notations, the impulsive composite system with decomposition $\mathbb{D}_i$ can be written in the form $\mathbb{S}$

$$\mathbb{S} : \begin{cases} dx(t) = F(t, x_t)dt + \sigma(t, x_t)dW(t), & t \neq \tau_k, \\ \Delta x(t) = \mathscr{I}(t, x_{t^-}), & t = \tau_k, \\ x_{t_0} = \Phi(s), \qquad s \in [-r, 0], \end{cases} \tag{5.3}$$

where $F(t, x_t) = f(t, x_t) + g(t, x_t)$ is an $\mathscr{L}_{ad}(\Omega; L[t_0, t_0 + \alpha])$ function for some $\alpha > 0$, $\sigma \in \mathscr{L}_{ad}(\Omega; L^2[t_0, t_0 + \alpha])$ and the initial function of the composite system $\Phi \in \mathscr{L}^2_{\mathscr{F}_0}([-r, 0]; \mathbb{R}^n)$.

Integrating the differential equation and making use of the initial condition yield

$$x(t) = \Phi(0) + \int_{t_0}^{t} F(s, x_s)ds + \int_{t_0}^{t} \sigma(s, x_s)\, dW(s) + \sum_{\{k \mid \tau_k \in (t_0, t]\}} \mathscr{I}(t, x_{\tau_k^-}),$$

for $t \neq \tau_k$, where the first integral is a Riemann integral almost surely (a.s.) and the second one is an Itô integral satisfying

$$\mathbb{E}\left[\int_{t_0}^{t} \sigma(s, x_s)\, dW(s)\right] = 0, \text{ and } \mathbb{E}\left[\left\| \int_{t_0}^{t} \sigma(s, x_s)\, dW(s)\right\|^2\right] = \int_{t_0}^{t} \mathbb{E}\|\sigma(s, x_s)\|^2\, ds.$$

Definition 5.1 The trivial solution $x \equiv 0$ of (5.3) is said to be

(i) *stable in the m.s.* if for every $\varepsilon > 0$ and $t_0 \in \mathbb{R}_+$, there exists a $\delta = \delta(t_0, \varepsilon) > 0$ such that

$$\mathbb{E}[\|\Phi\|_r^2] \leq \delta \quad \text{implies} \quad \mathbb{E}[\|x(t)\|^2] < \varepsilon, \quad \forall t \geq t_0,$$

where $x(t) = x(t; t_0, \Phi)$ is any solution of (5.3), with $x \in \mathscr{PC}([t_0 - r, t_0 + \alpha]; \mathscr{D})$ for some $\alpha > 0$ and $\Phi \in \mathscr{L}^2_{\mathscr{F}_0}([-r, 0], \mathscr{D})$;

(ii) *uniformly stable in the m.s.* if δ in (i) is independent of t_0;

(iii) *asymptotically stable in the m.s.* if it is stable and for any $t_0 \in \mathbb{R}_+$, there exists $\eta = \eta(t_0) > 0$ such that

$$\mathbb{E}[\|\Phi\|_r^2] \leq \eta \quad \text{implies} \quad \lim_{t\to\infty} x(t) = 0;$$

(iv) *uniformly asymptotically stable in the m.s.* if it is uniformly stable in the m.s. and there exists some $\eta > 0$ such that, for every $\gamma > 0$, there exists $T = T(\eta, \gamma) > 0$ for which

$$\mathbb{E}[\|\Phi\|_r^2] \le \eta \quad \text{implies} \quad \mathbb{E}[\|x(t)\|^2] < \gamma, \quad \forall t \ge t_0 + T;$$

(v) *exponentially stable in the m.s.* if there exist positive constants K and λ such that

$$\|x(t)\| \le K\mathbb{E}[\|\Phi\|_r^2]e^{-\lambda(t-t_0)}, \qquad \forall t \ge t_0;$$

(vi) *unstable in m.s.* if (i) fails to hold.

For convenient presentation, the following properties (or definitions) will be used in the theorem statements of this chapter.

Definition 5.2 For $i = 1, 2, \ldots, l$, the isolated subsystem $\mathbb{S}_i$ in (5.2) is said to possess **Property A** if Assumptions A1 and A2 are satisfied, there exist functions $a_i \in \mathcal{K}_c, b_i, c_i \in \mathcal{K}_v$, a constant $\sigma_i < 0$ and $V^i \in \mathscr{C}^{1,2}\big([-r, \infty) \times S(\varrho)); \mathbb{R}_+\big)$ such that the following hold:

(i) for all $(t, \psi^i(0)) \in [-r, \infty) \times S(\varrho)$,

$$b_i(\|\psi^i(0)\|^2) \le V^i(t, \psi^i(0)) \le a_i(\|\psi^i(0)\|^2), \qquad \text{(a.s.)};$$

(ii) for all $t \ne \tau_k \in \mathbb{R}_+$ and $\psi^i \in \mathscr{PC}([-r, \infty); S(\rho))$,

$$\mathscr{L}_i V^i(t, \psi) \le \sigma_i c_i(V^i(t, \psi^i(0))), \qquad \text{(a.s.)}$$

provided that $V^i(t + s, \psi^i(s)) \le g^i(V^i(t, \psi^i(0)))$ for some $s \in [-r, 0]$, where $g^i \in \mathcal{K}_3$;

(iii) at any impulsive moment $\tau_k \in \mathbb{T}$ and $\psi^i \in \mathscr{PC}([-r, \infty); S(\rho))$,

$$V^i(\tau_k, \psi^i(0) + \mathscr{I}^i(\tau_k, \psi^i(\tau_k^-))) \le g^i(V^i(\tau_k^-, \psi^i(0))), \qquad \text{(a.s.)}$$

with $\psi^i(0^-) = \psi^i(0)$, where $(\tau_k, \psi^i(\tau_k^-)) \in \mathbb{R}_+ \times \mathscr{PC}([-r, 0]; S(\rho_1))$ and $g^i \in \mathcal{K}_3$; and

(iv) $M_1^i = \sup_{q \ge 0} \int_q^{g^i(q)} \frac{ds}{c_i(s)} < \infty$ and $-\sigma_i\mu > M_1^i$, with $\mu = \inf_{k\in\mathbb{N}}\{\tau_k - \tau_{k-1}\} > 0$.

Definition 5.3 For $i = 1, 2, \ldots, l$, the isolated subsystem $\mathbb{S}_i$ in (5.2) is said to possess **Property B** if Assumptions A1 and A2 are satisfied, there exist functions $a_i, c_i \in \mathcal{K}_c, b_i \in \mathcal{K}_v, \sigma_i > 0$ and $V^i \in \mathscr{C}^{1,2}([-r, \infty) \times S(\rho); \mathbb{R}_+)$ such that the following hold:

(i) condition (i) in Definition 5.2 is satisfied;

(ii) for all $t \ne \tau_k \in \mathbb{R}_+$ and $\psi^i \in \mathscr{PC}([-r, \infty); S(\rho))$,

$$\mathscr{L}_i V^i(t, \psi^i) \le \sigma_i c_i(V(t, \psi^i(0))), \qquad \text{(a.s.)},$$

provided that $g^i(V^i(t + s, \psi^i(s))) \le V^i(t, \psi^i(0))$ for some $s \in [-r, 0]$, where $g^i \in \mathcal{K}_3$;

(iii) at any impulsive moment $\tau_k \in \mathbb{T}$ and $\psi^i \in \mathscr{PC}([-r, \infty); S(\rho))$,

$$V^i(\tau_k, \psi^i(0) + \mathscr{I}^i(\tau_k, \psi^i(\tau_k^-))) \leq g^i(V^i(\tau_k^-, \psi^i(0))), \qquad \text{(a.s.)}$$

with $\psi^i(0^-) = \psi^i(0)$, where $(\tau_k, \psi^i(\tau_k^-)) \in \mathbb{R}_+ \times \mathscr{PC}([-r, 0]; S(\rho_1))$; and

(iv) $\inf_{q>0} \int_{g^i(q)}^{q} ds/c_i(s) > \tau\sigma_i$ with $\tau = \sup_{k\in\mathbb{N}}\{\tau_k - \tau_{k-1}\} < \infty$.

Remark 5.1 Properties A and B, which are, respectively, extracted from Theorem 4.1 and Corollary 4.1, state that every isolated subsystem $\mathbb{S}_i$ (for $i = 1, 2, \ldots l$) is uniformly asymptotically stable in the m.s. Also, as can be seen, we assume that all states of isolated subsystems have impulsive jump discontinuity occurring simultaneously.

Throughout this chapter, we prove some m.s. stability properties of (5.3) using the classical Lyapunov theorems and comparison method. Also, in both cases, we use Razumikhin methodology in which we define a Lyapunov function $V(t, \psi(0))$ for all $t \geq 0$, but not functional $V(s, \psi(s))$ for all $s \in [-r, 0]$.

5.2 Analysis by Lyapunov Method

The focus here is on the classical Lyapunov technique to write some sufficient conditions to guarantee m.s. asymptotic stability of trivial solution, $x \equiv 0$, of composite SISD (5.3). We should also remark that impulses applied to the systems do not have to be bounded or vanishing. In Theorem 5.1, the impulsive effects are considered to be a perturbation to a stable system. While in Theorem 5.2, the underlying continuous system is unstable that is stabilized by an impulsive controller.

Theorem 5.1 *Suppose that composite system* (5.3) *satisfies the following conditions:*

(i) *for $i = 1, 2, \ldots l$, the isolated subsystem $\mathbb{S}_i$ possesses Property A;*

(ii) *for $i, j = 1, 2, \ldots, l$, there exists a positive constant b_{ij} such that*

$$g_i^T(t, \psi^i) V^i_{\psi^i(0)}(t, \psi^i(0)) \leq c_i^{1/2}(\|\psi^i(0)\|^2) \sum_{j=1}^{l} \bar{q} b_{ij} c_j^{1/2}(\|\psi^j(0)\|^2),$$

where $\bar{q} \geq 1$ and c_i is defined in (i) in Definition 5.1;

(iii) *for $i = 1, 2, \ldots, l$, there exists $e_i > 0$ such that*

$$(y^i)^T V^i_{\psi^i(0)\psi^i(0)}(t, \psi^i(0)) y^i \leq \bar{q} e_i \|y^i(0)\|^2,$$

where $y^i = \sigma_{ij}(t, \psi_t^j)$, i.e., the ith row of matrix σ;

(iv) *for any* $\sigma_{ij}(t, \psi_t^j)$, $i, j = 1, 2, \ldots, l$, *there exists* $d_{ij} \geq 0$ *such that*

$$\|\sigma_{ij}(t, \psi^j)\|^2 \leq \bar{q} d_{ij} c_i(\|\psi^j(0)\|^2);$$

(v) *matrix* $S = [s_{ij}]_{l\times l}$ *is negative definite where*

$$s_{ij} = \begin{cases} \alpha_i(\sigma_i + \bar{q} b_{ii}) + \frac{1}{2}\sum_{k=1,k\neq i} \bar{q}\alpha_k e_k d_{ki}, & i = j, \\ \frac{1}{2}\bar{q}(\alpha_i b_{ij} + \alpha_j b_{ji}), & i \neq j, \end{cases}$$

for some positive constant α_i *for any* i*; and*

(vi) *there exist functions* $g \in \mathscr{K}_3$, $c \in \mathscr{K}_c$ *and a constant* $\sigma < 0$ *such that* $\sup_{q>0} \int_q^{g(q)} ds/c(s) < -\sigma\mu$ *where* μ *is defined in Definition 5.1.*

Then, the trivial solution $x \equiv 0$ *of composite system* (5.3) *is uniformly asymptotically stable in the m.s.*

Proof For all $t \geq t_0$ with $t_0 \in \mathbb{R}_+$, let $x = x(t; t_0, \Phi)$ be the solution of the composite system (5.3). Define the composite Lyapunov function candidate

$$V(t, x) = \sum_{i=1}^{l} \alpha_i V^i(t, w^i),$$

where V^i is the Lyapunov function related to the ith isolated subsystem and $\alpha_i > 0$. This also implies that

$$\begin{aligned}
\mathscr{L}V(t, x) &= \sum_{i=1}^{l} \alpha_i \Big\{\mathscr{L}_i V^i(t, w^i) + g_i^T(t, x_t) V^i_{w^i}(t, w^i) \\
&\quad + \frac{1}{2}\sum_{j=1,i\neq j}^{l} \operatorname{tr}\big[\sigma_{ij}^T(t, w_t^j) V^i_{w^i w^i}(t, w^i)\sigma_{ij}(t, w_t^j)\big]\Big\} \\
&\leq \sum_{i=1}^{l} \alpha_i \Big\{\sigma_i c_i(\|w^i\|^2) + c_i^{1/2}(\|w^i\|^2) \sum_{j=1}^{l} \bar{q} b_{ij} c_j^{1/2}(\|w^j\|^2) \\
&\quad + \frac{1}{2}\sum_{j=1,i\neq j}^{l} \bar{q} e_i \|\sigma_{ij}(t, w_t^j)\|^2\Big\} \\
&\leq \sum_{i=1}^{l} \alpha_i \Big\{\sigma_i c_i(\|w^i\|^2) \\
&\quad + c_i^{1/2}(\|w^i\|^2) \sum_{j=1}^{l} \bar{q} b_{ij} c_j^{1/2}(\|w^j\|^2) + \frac{1}{2}\sum_{j=1,i\neq j}^{l} \bar{q} e_i d_{ij} c_i(\|w^j\|^2)\Big\} \\
&= z^T S z,
\end{aligned}$$

where $z^T = \left(c_1^{1/2}(\|w^1\|^2)\ c_2^{1/2}(\|w^2\|^2)\ \cdots\ c_l^{1/2}(\|w^l\|^2)\right) \in \mathbb{R}^l$ and S is the $l \times l$ negative-definite matrix defined in (v). It follows that the eigenvalues of S are strictly negative (i.e., $\lambda_M(S) < 0$). Therefore,

$$\mathscr{L}V(t,x) \le \lambda_M(S) \sum_{i=1}^{l} c_i(\|w^i\|^2),$$

i.e., $\mathscr{L}V(t,x)$ is negative definite, which implies that

$$\mathscr{L}V(t,x) \le \sigma c(\|x(t)\|^2),$$

where $\sigma < 0$ and $c \in \mathscr{K}_c$ satisfying the condition in (vi). Finally, at the impulsive moments $t = \tau_k$, we have

$$V(\tau_k, x(\tau_k)) = \sum_{i=1}^{l} \alpha_i V^i(\tau_k, w^i(\tau_k))$$

and, from the isolated subsystems,

$$V^i(\tau_k, \psi^i(0) + \mathscr{I}^i(\tau_k, \psi^i(\tau_k^-))) \le g^i(V^i(\tau_k^-, \psi^i(0))).$$

Hence,

$$\begin{aligned} V(\tau_k, \psi(0) + \mathscr{I}(\tau_k, \psi(\tau_k))) &= \sum_{i=1}^{l} \alpha_i V^i(\tau_k, \psi^i(0) + \mathscr{I}^i(\tau_k, \psi^i(\tau_k^-))) \\ &\le \sum_{i=1}^{l} \alpha_i g^i(V^i(\tau_k^-, \psi^i(0))) \\ &=: g(V(\tau_k^-, \psi(0))), \end{aligned}$$

where $g \in \mathscr{K}_3$. Thus, all the conditions of Corollary 4.1 are satisfied; therefore, the trivial solution, $x \equiv 0$, of composite system (5.3) is uniformly asymptotically stable in the m.s. This completes the proof.

In the following theorem, the continuous isolated subsystems and composite systems are assumed to be unstable and can be stabilized by an impulsive controller.

Theorem 5.2 *Suppose that composite system* (5.3) *satisfies the following conditions:*

(i) for $i = 1, 2, \ldots l$, the isolated subsystem $\mathbb{S}_i$ in (5.2) *possesses Property B;*
(ii) assumptions (ii)–(iv) of Theorem 5.1 are satisfied;
(iii) the matrix $S = [s_{ij}]_{l \times l}$ is positive definite, where

$$s_{ij} = \begin{cases} \alpha_i(\sigma_i + \bar{q}b_{ii}) + \frac{1}{2}\sum_{k=1,k\neq i} \bar{q}\alpha_k e_k d_{ki}, & i = j, \\ \frac{1}{2}\bar{q}(\alpha_i b_{ij} + \alpha_j b_{ji}), & i \neq j, \end{cases}$$

for some positive constant α_i for any i; and

(iv) there exist functions $g \in \mathcal{K}_3$, $c \in \mathcal{K}_v$ and a constant $\sigma > 0$ such that $\inf_{q>0} \int_{g(q)}^{q} ds/c(s) > \sigma\tau$, where $\tau = \inf_{k\in\mathbb{N}}\{\tau_k - \tau_{k-1}\} > 0$.

Then, the trivial solution, $x \equiv 0$, of composite system (5.3) is uniformly asymptotically stable in the m.s.

Proof For all $t \geq t_0$ with $t_0 \in \mathbb{R}_+$, let $x = x(t; t_0, \Phi)$ be the solution of composite system (5.3). Define the composite Lyapunov function candidate by

$$V(t, x) = \sum_{i=1}^{l} \alpha_i V^i(t, w^i),$$

where V^i is the Lyapunov function related to the ith isolated subsystem and $\alpha_i > 0$. Then,

$$\mathcal{L}V(t, x) \leq z^T S z,$$

where $z^T = \left(c_1^{1/2}(\|w^1\|^2)\ c_2^{1/2}(\|w^2\|^2)\ \cdots\ c_l^{1/2}(\|w^l\|^2)\right)$, and S is the positive-definite matrix defined in (iii). It follows that the eigenvalues of S are strictly positive (i.e., $\lambda_M(S) > 0$). Therefore,

$$\mathcal{L}V(t, x) \leq \lambda_M(S) \sum_{i=1}^{l} c_i(\|w^i\|^2)$$

which implies that

$$\mathcal{L}V(t, x) \leq \sigma c(\|x(t)\|^2),$$

where $\sigma > 0$ and $c \in \mathcal{K}_v$ are defined in (iv). As achieved in Theorem 5.1, we have, at the impulsive moments $t = \tau_k$,

$$V(\tau_k, \psi(0) + \mathcal{I}(\tau_k, \psi(\tau_k))) \leq g(V(\tau_k^-, \psi(0))),$$

for some function $g \in \mathcal{K}_3$. Thus, all the conditions of Theorem 5.1 are satisfied; therefore, the trivial solution $x \equiv 0$ of composite system (5.3) is uniformly asymptotically stable in the m.s. This completes the proof.

Remark 5.2 In Theorems 5.1 and 5.2, we have assumed that the individual isolated subsystems $\mathbb{S}_i$ possess Properties A and B, respectively, so as to guarantee their m.s. uniformly asymptotic stability. Assumptions (ii) (and (iii) and (iv)) in Theorem 5.1

describe the upper bound on the deterministic (and stochastic) interconnections of the system. While assumption (v) describes the relationship between the degree of stability of each subsystem and their interconnections magnitude, which is formed in the *test matrix* S. The negative definiteness of the matrix, which is required to guarantee the stability of the composite system, ensures that the stability margin of each individual is stronger than the interconnection magnitude. In Theorem 5.2, the isolated continuous systems $\mathbb{S}_i$ are unstable and stabilized by impulsive effects, which also have the role of stabilizing the entire composite system.

5.3 Comparison Method

In this section, depending on the type of composite Lyapunov function candidate used, we adopt two approaches to analyze the stability property using the comparison method. In Sect. 5.3.1, a *scalar* Lyapunov function is considered, while in Sect. 5.3.2, we use a *vector* of Lyapunov functions.

5.3.1 Method of Lyapunov Function

In the following theorem, we establish m.s. stability properties of (5.3) after being compared with an auxiliary scalar comparison system, which enjoys the same stability properties. In fact, we use Theorem 4.3 in proving the stability properties in this subsection.

Theorem 5.3 *Assume that the assumptions of Theorem 5.1 hold except that, whenever* $V^i(t+s, \psi^i(s)) \leq \bar{q} V(t, \psi^i(0))$ *for some* $\bar{q} > 1$ *and* $s \in [-r, 0]$,

$$\mathscr{L}_i V^i(t, \psi^i) \leq h_{1_i}(t, V^i(t, \psi^i(0))), \qquad (a.s.)$$

and

$$g_i^T(t, \psi) V^i_{\psi^i(0)}(t, \psi^i(0)) + \frac{1}{2} \sum_{j=1, i \neq j} tr\big[\sigma_{ij}^T(t, \psi^j) V^i_{\psi^i(0)\psi^i(0)}(t, \psi^i(0)) \sigma_{ij}(t, \psi^j)\big]$$
$$< h_{2_i}(t, V(t, \psi(0))),$$

where $\bar{h} \in \mathscr{C}\big([\tau_{k-1}, \tau_k) \times \mathbb{R}_+; \mathbb{R}\big)$, $\bar{h}(t, u)$ *is concave in* u *for all* $t \in \mathbb{R}_+$ *and*

$$\lim_{(t,y) \to (\tau_k^-, x)} \bar{h}(t, y) = \bar{h}(\tau_k^-, x),$$

where $\bar{h}$ refers to both h_{1_i} and h_{2_i}. Then, the stability properties of composite system (5.3) *are implied by those of the following auxiliary impulsive comparison system*

$$\begin{cases} D^+v = h(t, v), & t \neq \tau_k, \\ v(t) = \alpha(v(t^-)), & t = \tau_k, \\ v(t_0) = v_0 \geq 0, \end{cases} \tag{5.4}$$

where h is a scalar function defined later and $\alpha \in \mathscr{K}_3$.

Proof Let $x^T = \left((w^1)^T\ (w^2)^T\ \cdots\ (w^l)^T\right)$ be the solution of composite system (5.3). Define the composite Lyapunov function candidate by

$$V(t, x) = \sum_{i=1}^{l} \alpha_i V^i(t, w^i).$$

Then, for all $t \neq \tau_k$ with $k \in \mathbb{N}$, whenever $V(t, x_t) \leq \bar{q}V(t, x)$,

$$\begin{aligned} \mathscr{L}V(t, x) =& \sum_{i=1}^{l} \alpha_i \Big\{ \mathscr{L}_i V^i(t, w^i) + g_i(t, x_t)^T V^i_{w^i}(t, w^i) \\ &+ \frac{1}{2} \sum_{j=1, i\neq j}^{l} \operatorname{tr}\big[\sigma_{ij}(t, w^j_t) V^i_{w^i w^i}(t, w^i) \sigma_{ij}(t, w^j_t)\big]\Big\} \\ \leq& \sum_{i=1}^{l} \alpha_i \Big\{ h_{1_i}(t, V^i(t, w^i)) + h_{2_i}(t, V^i(t, w^i))\Big\} \\ =:& h(t, V(t, x)). \end{aligned}$$

It follows that, after applying Itô formula to process V and taking the mathematical expectation,

$$D^+m(t) \leq h(t, m(t)),$$

where $m(t) = \mathbb{E}[V(t, x(t))]$ for all $t \neq \tau_k$. At the impulsive moments, $t = \tau_k$, we have

$$m(\tau_k) \leq \alpha(m(\tau_k^-)).$$

In summary, we have obtained

$$\begin{cases} D^+m \leq h(t, m(t)), & t \neq \tau_k, \\ m(t) \leq \alpha(m(t^-)), & t = \tau_k, \\ m(t_0) \leq v_0, \end{cases}$$

which is compared with the auxiliary comparison system (5.4). To conclude the desired result, it suffices to apply Theorem 4.3. This completes the proof.

The following corollary is analogous to Corollary 4.4; thus, we state it without a proof.

Corollary 5.1 *In Theorem 5.3, let* $p \in \mathscr{PC}(\mathbb{R}_+; \mathbb{R}_+)$ *and* $c \in \mathscr{K}_v$ *such that*

$$h(t, V(t, x)) = p(t)c(V(t, x))$$

and

$$\int_{\tau_{k-1}}^{\tau_k} p(s)ds + \ln \alpha_M(d_k) \leq -\gamma_k, \qquad k \in \mathbb{N}, \tag{5.5}$$

where $\alpha_M(d_k) = \max\{\alpha^i(d_k) \mid i = 1, 2, \ldots, l\}$ *with* $\alpha^i(d_k)$ *being a constant for which*

$$V^i\big(\tau_k, \psi^i(0^-) + \mathscr{I}^i(\tau_k, \psi^i(\tau_k^-))\big) \leq \alpha^i(d_k)V^i(\tau_k^-, \psi^i(0^-))$$

and satisfying $\alpha^i(d_k) > 1$, $\prod_{k=1}^{\infty} \alpha^i(d_k) < \infty$ *and* $\sum_{k=1}^{\infty} d_k < \infty$. *Then, if* $\gamma_k \geq 0$, *the composite system in* (5.3) *is uniformly stable in m.s. and, if* $\sum_{k=1}^{\infty} \gamma_k = +\infty$, *the system is asymptotically stable in the m.s.*

5.3.2 Method of Vector Lyapunov Functions

In this subsection, we continue to use a comparison method to prove the stability properties for composite large-scale SISD (5.3), where we use a *vector* of Lyapunov functions having components which are Lyapunov functions related to the isolated subsystems and, in this case, the finding of Theorem 5.5 will be carried over to every individual subsystem. In other words, the comparison occurs between a vector of differential inequalities and a vector of differential equations whose solutions are known and enjoy some stability properties. As done early in this chapter and for convenient theorem statement, we define Property C.

Definition 5.4 For $i = 1, 2, \ldots, l$, the isolated subsystem $\mathbb{S}_i$ in (5.2) is said to possess **Property C** if Assumptions A1 and A2 hold, there exist functions $c_i \in \mathscr{K}_c$, a_i which satisfies the conditions of $\bar{h}$ in Theorem 5.5 and $V^i \in \mathscr{C}^{1,2}\big([-r, \infty) \times S(\varrho)\big); \mathbb{R}_+\big)$ which is decreasing and satisfies

(i) for all $(t, \psi^i(0)) \in [-r, \infty) \times S(\rho)$,

$$c_i(\|\psi^i(0)\|^2) \leq V^i(t, \psi^i(0)), \qquad \text{(a.s.)}$$

and, for all $t \neq \tau_k$ in $\mathbb{R}_+$ and $\psi^i \in \mathscr{PC}\big([-r, 0]; S(\rho)\big)$,

$$\mathscr{L}_i V^i(t, \psi^i) \le a_i(t, V^i(t, \psi^i(0))), \qquad \text{(a.s.)}$$

provided that $V^i(t+s, \psi^i(s)) \le \bar{q} V^i(t, \psi^i(0))$ for some $\bar{q} > 1$ and $s \in [-r, 0]$; and

(ii) for any $\tau_k \in \mathbb{T}$ and $\psi^i \in \mathscr{PC}\big([t_0 - r, \infty); S(\rho)\big)$,

$$V^i\big(\tau_k, \psi^i(0) + \mathscr{I}_i(\tau_k, \psi^i(\tau_k^-))\big) \le \alpha^i(d_k) V^i(\tau_k^-, \psi^i(0)), \qquad \text{(a.s.)}$$

where $\psi^i(0^-) = \psi^i(0)$ and $\prod_{k=1}^{\infty} \alpha^i(d_k) < \infty$ with $\alpha^i(d_k) > 1$ for all k.

Definition 5.5 A function $g(t, u)$ (or $g : \mathbb{R}_+ \times \mathbb{R}^n \to \mathbb{R}$) is said to be *quasi-monotone nondecreasing in* u if, for any $u, v \in \mathbb{R}^n$ such that $0 \le u_j < v_j$ for all $i \ne j$ and $0 \le u_i = v_i$, we have $g(t, u) < g(t, v)$ for any fixed t in $\mathbb{R}_+$.

Theorem 5.4 *(Comparison theorem) Assume that the following assumptions hold:*

(i) for $i = 1, 2, \ldots, l$, the isolated subsystem $\mathbb{S}_i$ in (5.2) has Property C;

(ii) for $i = 1, 2, \ldots, l$, there exists a function $\bar{b}_i(t, u) \in \mathscr{C}\big([\tau_{k-1}, \tau_k) \times \mathbb{R}_+; \mathbb{R}\big)$ that is quasi-monotone nondecreasing in u such that

$$g_i^T(t, \psi) V^i_{\psi^i(0)}(t, \psi^i(0)) + \frac{1}{2} \sum_{j=1, i \ne j}^{l} tr\big[\sigma_{ij}^T(t, \psi^j) V^i_{\psi^i(0)\psi^i(0)}(t, \psi^i(0)) \sigma_{ij}(t, \psi^j)\big]$$
$$< \bar{b}_i(t, V(t, \psi(0))),$$

where $V^T(t, x) = \big(V^1(t, w^1)\ V^2(t, w^2)\ \cdots\ V^l(t, w^l)\big)$;

(iii) let $a^T(\cdot) = \big(a_1(\cdot)\ a_2(\cdot)\ \cdots\ a_l(\cdot)\big) \in \mathscr{L}_{ad}\big(\Omega, L[t_0, t_0 + \alpha]\big)$ and $\bar{b}^T(\cdot) = \big(\bar{b}_1(\cdot)\ \bar{b}_2(\cdot)\ \cdots\ \bar{b}_l(\cdot)\big) \in \mathscr{L}_{ad}\big(\Omega, L^2[t_0, t_0 + \alpha]\big)$, where $a_i(\cdot)$ and $\bar{b}_i(\cdot)$ are defined in assumptions (i) and (ii), respectively, and assume that the following inequalities hold

$$|a(t, v') + \bar{b}(t, v')|^2 \le h_1(t) + h_2(t)\kappa(\|v'\|^2),$$
$$|a(t, v') + \bar{b}(t, v') - a(t, v'') - \bar{b}(t, v'')| \le K\|v' - v''\|,$$

for all $t \in \mathbb{R}_+$, where h_1, h_2 are Borel measurable functions (or $\mathscr{PC}(\mathbb{R}_+; \mathbb{R}_+)$ functions), $\kappa : \mathbb{R}_+ \to \mathbb{R}_+$ is continuous, increasing, concave function, v', $v'' \in \mathbb{R}^l_+$ and $K > 0$; and

(iv) there exists an adapted function $p : \mathbb{R}^l \times \mathbb{R}_+ \to \mathbb{R}$ such that

$$\sup_{V(t,x) \le v} \sum_{i,j=1}^{l} \|\sigma_{ij}^T(t, \psi^j) V_{\psi^i(0)^i(t, \psi^i(0))}\|^2 \le p(t, v),$$

where

$$p(t, v) \le h_1(t) + h_2(t)\kappa(\|v\|^2).$$

Then, provided that $V(t_0, x_0) < v_0$, *we have* $V(t, x(t)) < v(t)$, *for all* $t \geq t_0$, *where* $v = (v^1\ v^2\ \cdots\ v^l)^T$ *(i.e.,* $v \in \mathbb{R}^l$*) is a solution of the vector stochastic impulsive differential equation*

$$\begin{cases} dv = \big(a(t, v) + \bar{b}(t, v)\big)dt + \mathbb{V}dW(t), & t \neq \tau_k, \\ \Delta v(t) = \alpha_M(d_k)v(t^-), & t = \tau_k, \end{cases} \tag{5.6}$$

with $\mathbb{V} = [v_{ij}]_{l\times l}$ *being a matrix random process such that*

$$\|\mathbb{V}\|^2 \leq p(t, v),$$

and $\alpha_M(\cdot) = \max\{\alpha^i(\cdot) : i = 1, 2, \ldots, l\}$.

Proof For all $t \geq t_0$ with $t_0 \in \mathbb{R}_+$, let $x = x(t; t_0, \Phi)$ be the solution of composite impulsive system (5.3). Define

$$V^T(t, x(t)) = \big(V^1(t, w^1)\ V^2(t, w^2)\ \cdots\ V^l(t, w^l)\big)$$

as the vector Lyapunov function candidate for the composite system with V^i being the Lyapunov function related to the ith isolated subsystem $\mathbb{S}_i$. Then, by the vector form of Itô formula, we have

$$dV^T(t, x(t)) = \big(dV^1(t, w^1)\ dV^2(t, w^2)\ \cdots\ dV^l(t, w^l)\big),$$

where, for $i = 1, 2, \ldots, l$,

$$dV^i(t, w^i) < \Big(a_i(t, V^i(t, w^i)) + \bar{b}_i(t, V^i(t, w^i))\Big)dt + \sum_{i,j=1}^{l} v_{ij} dW_i(t),$$

with $v_{ij} = V_{w^i}^{i^T}(t, w^i)\sigma_{ij}(t, w_t^j)$. It follows that the vector differential inequality is

$$dV(t, x(t)) < \Big(a(t, V(t, x(t))) + \bar{b}(t, V(t, x(t)))\Big)dt + \mathbb{V}dW(t),$$

for all $t \in [\tau_{k-1}, \tau_k)$ and $k \in \mathbb{N}$.

At the impulsive moments $t = \tau_k$, we have

$$\begin{aligned} V^T&(\tau_k, x(\tau_k)) \\ &= \big(V^1(\tau_k, w^1(\tau_k))\ V^2(t, w^2(\tau_k))\ \cdots\ V^l(t, w^l(\tau_k))\big) \\ &\leq \Big(\alpha^1(d_k)V^1(\tau_k^-, w^1(\tau_k^-))\ \alpha^2(d_k)V^2(\tau_k^-, w^2(\tau_k^-))\ \cdots\ \alpha^l(d_k)V^l(\tau_k^-, w^l(\tau_k^-))\Big) \\ &\leq \alpha_M(d_k)\big(V^1(\tau_k^-, w^1(\tau_k^-))\ V^2(\tau_k^-, w^2(\tau_k^-))\ \cdots\ V^l(\tau_k^-, w^l(\tau_k^-))\big) \\ &= \alpha_M(d_k)V^T(\tau_k^-, x(\tau_k^-)). \end{aligned}$$

Particularly, for all $t \in [\tau_0, \tau_1)$ and $i = 1, 2, \cdots, l$, we have $V^i(t_0, w^i(t_0)) < v_0$ and

$$dV^i(t, w^i) - dv_i < \Big\{[a_i(t, V^i(t, w^i)) - a_i(t, v_i)] + [\bar{b}_i(t, V(t, x(t))) - \bar{b}_i(t, v(t))]\Big\}dt.$$

Since the composite system satisfies the existence-uniqueness conditions, $V(t, x(t))$ is a continuous process (a.s.) for all $[\tau_0, \tau_1)$. Similar conclusion can be drawn for the process $v(t)$. Therefore, to ensure that, given $V(t_0, x_0) < v_0$, $V(t, x(t)) < v(t)$ (a.s.) for all $[\tau_0, \tau_1)$, it suffices to show that $dV^i(t, w^i) - dv^i(t) < 0$ whenever $V^i(t, w^i) = y^i(t)$. But this inequality is true because $\bar{b}_i$ is quasi-monotone nondecreasing. Thus, we obtain that $V^i(t, w^i(t)) < v_i(t)$ for all $t \in [\tau_0, \tau_1)$ and, at the impulsive moment τ_1, we have

$$V^i\big(\tau_1, w^i(\tau_1)\big) - v_i(\tau_1) < \alpha_M(d_k)\Big[V^i(\tau_1^-, w^i(\tau_1^-)) - v_i(\tau_1^-)\Big] < 0,$$

i.e.,

$$V^i(\tau_1, w^i(\tau_1)) < v_i(\tau_1).$$

Similarly, for $k = 2, 3, \ldots$ and $t \in [\tau_{k-1}, \tau_k)$, we get $V^i(t, w^i(t)) < v_i(t)$ and at $t = \tau_k$, $V^i(\tau_k, w^i(\tau_k)) < v_i(\tau_k)$. Therefore, for all $t \ge t_0$ and $i = 1, 2, \ldots, l$, $V_i(t, w^i(t)) < v_i(t)$, from which we get the vector inequality

$$V(t, x(t)) < v(t), \qquad \forall t \ge t_0.$$

This completes the proof.

Theorem 5.5 *Suppose that the assumptions of Theorem 5.4 hold, there exist class-$\mathcal{K}_c$ functions α_1 and c, a function $\bar{h} \in \mathscr{C}\big([\tau_k, \tau_{k-1}) \times \mathbb{R}^l; \mathbb{R}_+\big)$, $z \in \mathbb{R}^l$ and $U \in \mathscr{C}^{1,2}\big([\tau_{k-1}, \tau_k) \times \mathbb{R}^l; \mathbb{R}_+\big)$ which is decreasing, $U(t, 0) = 0$ and satisfies*

(i) for all $t \in \mathbb{R}_+$ and $v \in \mathscr{PC}(\mathbb{R}_+; \mathbb{R}^l)$,

$$\alpha_1(\|v\|^2) \le U(t, v), \qquad (a.s.)$$
$$z^T U_{vv}(t, v)z \le \bar{h}(t, v)\|z\|^2, \qquad (a.s.)$$

and

$$U_t(t, v) + U_v(t, v)\big[a(t, v) + \bar{b}(t, v)\big] + \frac{1}{2}h(t, v)p(t, v) \le -c(\|v\|), \qquad (a.s.);$$

(ii) for any $\tau_k \in \mathbb{T}$ and $v \in \mathscr{PC}(\mathbb{R}_+; \mathbb{R}^l)$,

$$U(\tau_k, v(\tau_k)) = \alpha(d_k)U(\tau_k^-, v(\tau_k^-)), \qquad (a.s.).$$

Then, comparison system (5.6) *and, hence, composite SISD* (5.3) *have asymptotically stable trivial solutions in the m.s.*

Proof Let $v \geq 0$ be the solution vector of the comparison system (5.6). Apply Itô formula to process U to get

$$\mathscr{L}U(t, v) \leq -c(\|v\|),$$

which shows that, by the previous analysis, (5.6) has the desired stability property.

As for composite system (5.3), we have shown in Theorem 5.4 that the vector inequality $V(t, x(t)) < v(t)$ holds for all $t \geq t_0$. It follows that

$$\alpha_1(\|x(t)\|^2) \leq \Big[\sum_{i=1}^{l} c_i^2(\|w^i\|^2)\Big]^{1/2} \leq \|V(t, x(t))\| < \|v(t)\|,$$

where $c \in \mathscr{K}_c$. Taking the mathematical expectation and then applying α_1^{-1} to both sides imply the desired result, i.e., $\mathbb{E}[\|x(t)\|^2] \leq \alpha_1^{-1}(\mathbb{E}[\|v(t)\|^2])$ for all $t \geq t_0$. This completes the proof.

Corollary 5.2 *In Theorem 5.5, assume that there exists a positive constant c such that $c(s) = c\,s$ for all $s > 0$ and*

$$\beta^T\big(a(t, v) + \bar{b}(t, v)\big) \leq -c\|v\|,$$

for some positive vector $\beta \in \mathbb{R}^l$. Then, system (5.6) *possesses the same stability property.*

Proof Let $U(t, v) = \beta^T v > 0$ be a Lyapunov function candidate. Then, $U_v = \beta^T$ and $U_{vv} = 0 \in \mathbb{R}^{l\times l}$, from which $\mathscr{L}U(t, v) \leq -c\|v\|$. Applying the impulsive effects yields the desired result.

5.4 Examples

As an application of the proposed results, we consider an *indirect control system* in automatic control, which describes the longitudinal motion of an aircraft.

Example 5.1 Consider the control SISD

$$\begin{cases} dx = Axdt + bf(y)dt + \sigma_{11}(x(t-1))dW_1 + \sigma_{12}(y)dW_2,\ t \neq \tau_k, \\ dy = \big(-\zeta y - \xi f(y)\big)dt + a^T xdt + \sigma_{21}(x)dW_1 \\ \quad +\sigma_{22}(y(t-1))dW_2, \qquad\qquad t \neq \tau_k, \end{cases} \tag{5.7}$$

where $x^T = (x_1\ x_2\ x_3\ x_4)$ is the system state, $y \in \mathbb{R}$ is the controller (i.e., $n_1 = 4$, $n_2 = 1$), $A \in \mathbb{R}^{4\times 4}$, $b \in \mathbb{R}^4$, $\zeta, \xi \in \mathbb{R}$, $f \in \mathbb{R}$ is continuous for all $y \in \mathbb{R}$, $f(y) =$

0 if and only if $y = 0$, and $0 < yf(y) < k|y|^2$ for all $y \neq 0$ and $k > 0$, $a \in \mathbb{R}^4$, $\sigma_{11} \in \mathbb{R}^{4\times 4}$, $\sigma_{12} \in \mathbb{R}^{1\times 1}$, $\sigma_{21} \in \mathbb{R}^{4\times 1}$, $\sigma_{22} \in \mathbb{R}^{1\times 1}$, $W_1 \in \mathbb{R}^4$ and $W_2 \in \mathbb{R}$.

The isolated subsystems are

$$\mathbb{S}_i : \begin{cases} dx = Axdt + \sigma_{11}(x(t-1))dW_1, & t \neq \tau_k, \\ dy = \big(-\zeta y - \xi f(y)\big)dt + \sigma_{22}(y(t-1))dW_2, & t \neq \tau_k. \end{cases} \tag{5.8}$$

The impulses are given by the following difference equations

$$\begin{cases} \Delta x(\tau_k) = \mathscr{I}_1(\tau_k, x(\tau_k^-)) = \frac{1}{k^2}\big(-2x_1(\tau_k^-), -2x_2(\tau_k^-), 2x_3(\tau_k^-), 0\big)^T, \\ \Delta y(\tau_k) = \mathscr{I}_2(\tau_k, y(\tau_k^-)) = -\frac{1}{1+k^2}y(\tau_k^-). \end{cases} \tag{5.9}$$

Let $A = \begin{pmatrix} -5 & 0 & 0 & 0 \\ 0 & -6 & 0 & 0 \\ 0 & 0 & -8 & 0 \\ 0 & 0 & 0 & -10 \end{pmatrix}$, $\sigma_{11} = 0.01 \begin{pmatrix} \sin x_1(t-1) & 0 & \frac{x_2(t-1)}{1+x_4^2} & 0 \\ 0 & \frac{x_2(t-1)}{1+x_1^2} & 0 & -x_3^2(t-1) \\ 0 & 0 & x_3(t-1) & 0 \\ 0 & 0 & 0 & -x_4(t-1) \end{pmatrix}$,

$b^T = (1\ 1\ 1\ 1)$, $a^T = (1\ 1\ 1\ 1)$, $\zeta = 5$, $\xi = 2$, $\sigma_{12} = 0.01\frac{y}{1+y^2}$, $\sigma_{21}^T = 0.01(x_2\ x_1\ x_4\ x_3)$ and $\sigma_{22} = 0.01 \sin y(t-1)$.

Let $V^1(x) = \|x\|^2$ and $V^2(y) = y^2$ be the Lyapunov function candidates for the isolated subsystems in (5.8). After cumbersome calculations, one may get $\mathscr{L}_1 V^1(x) \leq (-10 + 0.0002\bar{q})\|x\|^2$ and $\mathscr{L}_2 V^2(y) \leq (-2\zeta + 0.0001\bar{q})y^2 = (-10 + 0.0001\bar{q})y^2$ (i.e., $\sigma_1 = -10 + 0.0002\bar{q}$ and $\sigma_2 = -10 + 0.0001\bar{q}$). For the stability of the continuous isolated subsystems, we take $\bar{q} = 2$. As for the interconnections, we have $V_x^{1^T}(x)g_1(x, y) = 2x^T\xi f(y) \leq 4k\|x\|\,|y|$ (i.e., $b_{12} = 4k$), $V_y^2(y)g_2(x, y) = 2ya^T x \leq 4\|x\|\,|y|$ (i.e., $b_{21} = 4$). The (noisy) interconnections are: $\sigma_{12}^T(y)V_{xx}^1\sigma_{12}(y) = 2\|\sigma_{12}(y)\|^2 \leq 0.0002y^2$ and $\sigma_{21}^T(x)V_{yy}^2\sigma_{21}(x) = 2\|\sigma_{21}(x)\|^2 \leq 0.0002\|x\|^2$ (i.e., $e_1 = e_2 = 2$ and $d_{12} = d_{21} = 0.0001$).

Let $V(x, y) = \alpha_1 V^1(x) + \alpha_2 V^2(y) = \|x\|^2 + y^2$ (i.e., $\alpha_1 = \alpha_2 = 1$) be the composite Lyapunov function candidate for composite system (5.7). Then, the matrix

$$S = \begin{pmatrix} -9.9997 & 2\bar{k} + 2 \\ 2\bar{k} + 2 & -7.9997 \end{pmatrix}$$

is negative definite if $\bar{k} < 3.9998$. Let $f(y) = \frac{2y}{1+y^2}$. Clearly, if we choose $\bar{k} = 2$, the required conditions are satisfied. Therefore, the condition $\mathscr{L}V(x, y) \leq z^T Sz < 0$ is also satisfied, where $z^T = (\|x\|\ |y|)$.

At the impulsive moments τ_k, we have

$$\begin{aligned} V(x(\tau_k), y(\tau_k)) &= \|x(\tau_k)\|^2 + y^2(\tau_k) \\ &\leq (1 + \frac{2}{k^2})\|x(\tau_k^-)\|^2 + (1 - \frac{5}{1+k^2})y^2(\tau_k^-) \\ &\leq \alpha_M(d_k)V(x(\tau_k^-), y(\tau_k^-)), \end{aligned}$$

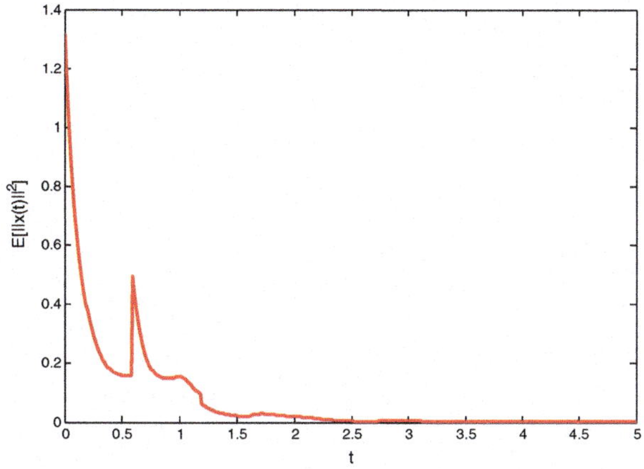

Fig. 5.1 Mean square asymptotic stability of $(x\ y)^T \equiv (0\ 0)$

where $\alpha_M(d_k) = 1 + \frac{2}{k^2}$. As for the impulsive moments, we have, after choosing $\bar{k} = 2$, $\mu > 0.04$. For $i = 1, 2$ and $s > 0$, choose $a_i(s) = b_i(s) = s^2$ to ensure the asymptotic stability in the m.s. of isolated subsystems. The eigenvalues of matrix S are -15.082, -2.917. Choose $\sigma = 2.917$ to obtain $\mu > 0.14$. Also, the trivial solution $(x\ y)^T = (0\ 0) \in \mathbb{R}^5$ (with $x \in \mathbb{R}^4$ and $y \in \mathbb{R}$) of composite SISD system given in (5.7)–(5.9) is exponential stable in the m.s. if $a(s) = b(s) = s^2$ and $c(s) = s$ for all $s > 0$. The simulation result is shown in Fig. 5.1, where we have taken $\mu = 0.5$.

Example 5.2 Consider again the continuous control composite system given in (5.7) and same composite Lyapunov scalar function $V(x, y) = \|x\|^2 + y^2$. By the previous analysis, we have found

$$
\begin{aligned}
&V_x^1(x)g_1^T(x, y) \le 2k(V^1(x) + V^2(y)) = 2kV(x, y),\\
&V_x^2(y)g_2^T(x, y) \le 2(V^1(x) + V^2(y)) = 2V(x, y),\\
&\sigma_{12}^T(y)V_{xx}^1\sigma_{12}(y) \le 0.0002V^2(y),\\
&\sigma_{21}^T(x)V_{yy}^2\sigma_{21}(y) \le 0.0002V^1(x),
\end{aligned}
$$

that is $h_{1_1}(V^1(x)) = \sigma_1 V^1(x), h_{1_2}(V^2(y)) = \sigma_2 V^2(y), h_{2_1}(V^1(x)) = (2k + 2.0001)$ $V^1(x)$ and $h_{2_2}(V^2(y)) = (2k + 2.0001)V^2(y)$. Therefore,

$$
\begin{aligned}
h(V(x, y)) &= \sum_{i=1}^{l} \alpha_i \left\{ h_{1_i}(V^i(t, w^i)) + h_{2_i}(V^i(t, w^i)) \right\}\\
&= (\sigma_1 + 2k + 2.0001)V^1(x) + (\sigma_2 + 2k + 2.0002)V^2(y)\\
&\le pV(x, y),
\end{aligned}
$$

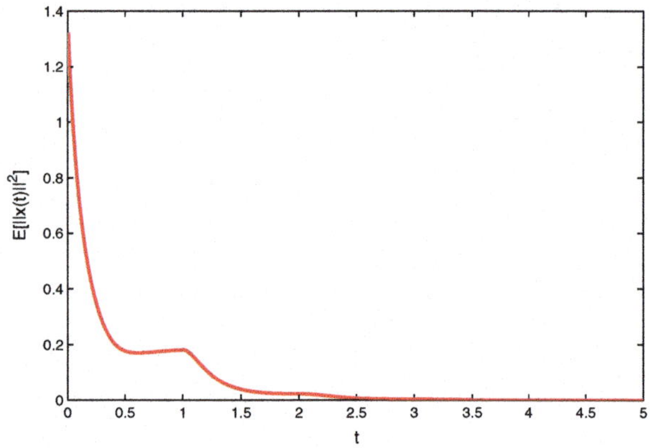

Fig. 5.2 Mean square asymptotic stability of $(x\ y)^T \equiv (0\ 0)$

where $p = \sigma_1 + 2k + 2.0001 = -3.9997$, from which one has

$$\mathscr{L}V(x, y) \leq pV(x, y).$$

Consider now the following impulsive difference equations

$$\begin{cases} \Delta x(\tau_k) = -\frac{5}{4}x(\tau_k^-), \\ \Delta y(\tau_k) = -\frac{5}{4}y(\tau_k^-). \end{cases} \tag{5.10}$$

It follows that $V(x(\tau_k), y(\tau_k)) \leq \alpha_k V(x(\tau_k^-), y(\tau_k^-))$ where $\alpha_k = \frac{1}{16}$. Making use of condition (5.5), one obtains $\tau_k - \tau_{k-1} > 0.69$ for any k. Therefore, the trivial solution, $x \equiv 0$, of the composite SISD given in (5.7) and (5.10) is exponentially stable in the m.s. The simulation result is shown in Fig. 5.2.

Reconsider the control composite continuous system in (5.7) with unstable state subsystem where

$$A = \begin{pmatrix} 5 & 0 & 0 & 0 \\ 0 & -6 & 0 & 0 \\ 0 & 0 & -8 & 0 \\ 0 & 0 & 0 & -10 \end{pmatrix}.$$

Following the same analysis, we obtain $\mathscr{L}_1 V^1(x) \leq (10 + 0.0001\bar{q})V^1(x)$; that is, the state of the isolated subsystem is unstable, while $\mathscr{L}_2 V^2(y) \leq -9.9998V^2(y)$. It follows that the composite system is unstable where $h(V(x, y), u) = 6.0005V(x, y) > 0$. Considering the stabilizing impulsive effects in (5.10) gives $\tau_k - \tau_{k-1} \leq 0.3$. Figure 5.3 shows the simulation result.

Example 5.3 Consider the composite system in (5.7) and the same Lyapunov functions. We have found $\mathscr{L}_1 V^1(x) \leq \sigma_1 V^1(x)$ and $\mathscr{L}_2 V^2(x) \leq \sigma_2 V^2(x)$, from which

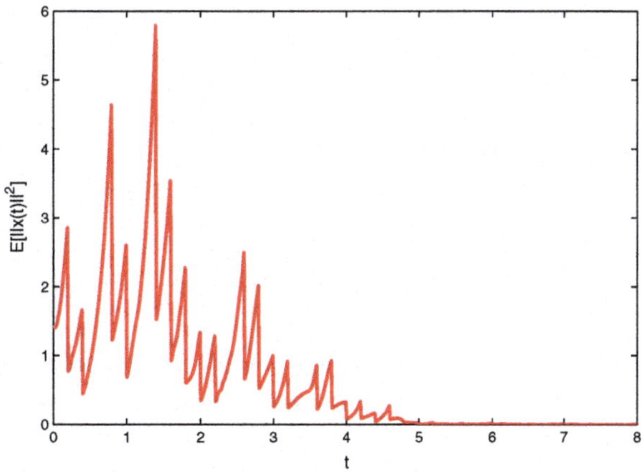

Fig. 5.3 Mean square asymptotic stability of $(x\ y)^T \equiv (0\ 0)$

we get $a(V(x,y))^T = (a_1(V^1(x))\ a_2(V^2(y))) = (\sigma_1 V^1(x)\ \sigma_2 V^2(y))$. From the interconnection, we have found $\bar{b}(V(x,y))^T = \big((2k+0.0001)V(x,y)\ 2.0001V(x,y)\big)$. Clearly, the functions a and b satisfy the conditions in (iii) of Theorem 5.4. As for condition (iv), we have

$$\begin{aligned}
&\sup_{V\le v}\sum_{i,j=1}^{l}\|\sigma_{ij}^T(w^i)V_{w^i}(w^i)^i\|^2\\
&= \sup_{V\le v}\Big[\|\sigma_{11}^T(x(t-1))V_x^1(x)\|^2+\|\sigma_{12}^T(y))V_x^1(x)\|^2\|\sigma_{21}^T(x)V_y^2(x)\|^2\\
&+\|\sigma_{22}^T(y(t-1)))V_y^2(x)\|^2\Big]\\
&\le 4\sup_{V\le v}\Big[\xi_1(V^1(x))^2+0.0004V^1(x)V^2(y)+\xi_2(V^2(y))^2\Big]\\
&\le 4\sup_{V\le v}\Big[\xi_1 v_1^2+0.0004v_1v_2+\xi_2 v_2^2\Big]\\
&\le 8\bar{\xi}\|v\|^2,
\end{aligned}$$

i.e., $p(v)\le 8\bar{\xi}\|v\|^2$, where $\bar{\xi}=\max\{\xi_1,\xi_2\}$, ξ_1=1.0004 and ξ_2=1.0002 with $\bar{q}$=2.

Making use of the impulsive effect given in Example 5.1, we get

$$\begin{aligned}
V^T(x(\tau_k),y(\tau_k)) &= (V^1x(\tau_k)),V^2(y(\tau_k))\\
&\le (1+\frac{1}{k^2})(V^1(x(\tau_k^-)),V^2(y(\tau_k^-)))=(1+\frac{1}{k^2})V^T(x(\tau_k^-),y(\tau_k^-))\\
&\le (1+\frac{1}{k^2})v^T(\tau_k^-)=v^T(\tau_k).
\end{aligned}$$

Thus, by Theorem 5.4, $V(x(t), y(t)) \leq v(t)$, for all $t \geq t_0$. As for the stability result, choose $U(v) = v_1 + v_2$, i.e., $\beta^T = (1\ 1)$. It is easy to show that $\mathscr{L}U(v) \leq -5.9997U(v)$, where we have chosen $k = 2$. Also, $U(v(\tau_k)) = \alpha_M(d_k)U(v(\tau_k^-))$, where $\alpha_M(d_k) = 1 + \frac{1}{k^2}$. Therefore, the trivial solution of composite system in (5.7) is asymptotically stable in the m.s.

5.5 Notes and Comments

In this chapter, nonlinear large-scale SISD with fixed impulses has been considered. The interest has been to demonstrate some qualitative properties by decomposing the interconnected system into smaller isolated subsystems, and the rest has been treated as system perturbation. The material of this chapter is taken from [1]. Assuming that the isolated subsystems have asymptotic stable trivial solutions in the m.s. and the perturbation, the connection among the subsystems is estimated by an upper bound, which is smaller than the stability margin of the individual subsystems, and we have been able to conclude that the interconnected SISD has a trivial solution that is asymptotically stable in the m.s. Also, it has been shown that if the continuous system is unstable, helpful impulses can contribute to stabilize such a system. In the stability analysis, we have used the classical Lyapunov theorems and comparison method using scalar Lyapunov function and vector Lyapunov functions. In fact, the stability results obtained by the first two approaches are extension to the results developed in Chap. 4. Moreover, for further reading about the qualitative notions analyzed by decomposing the system states of large-scale systems, one may refer to [2–6]. Finally, to demonstrate the effectiveness of the theoretical results of this chapter, we have presented the stability and stabilization problems of an automated indirect control system, which is a modification of Example 4.6.1 in [2], where we have involved time delay and impulsive effects.

References

1. Alwan MS, Liu XZ, Xie W-C, Stability results for large-scale stochastic impulsive systems with time delay. (Submitted)
2. Michel AN, Miller RK (1977) Qualitative analysis of large scale dynamical systems. Academic Press Inc, New York
3. Khalil HK (2002) Nonlinear systems. Prentice-Hall Inc, NJ
4. Šilijak DD (1991) Decentralized control of complex systems. Academic Press, Inc
5. Šiljak DD (2007) Large-Scale dynamic systems: stability and structure. Dover Publications, Inc
6. Zecevic A, Silijak DD (2010) Control of complex systems: structural constraints and uncertainty. Springer (2010)

Chapter 6
Input-to-State Stability for Stochastic Switched Systems

This chapter is mainly concerned with the input-to-state stability concept of nonlinear stochastic switched systems with bounded disturbance input. The primary objective is to develop Lyapunov-like sufficient conditions guaranteeing the stability property in the pth moment. To control the switching among the system modes, we adopt two switching rules, an initial-state-dependent dwell-time switching signal and Markovian switching. We consider systems consisting of a set of all stable modes and a set of stable and unstable modes. Also, implications of these results are stated with enhancing examples.

6.1 Problem Formulation

Consider now the following stochastic switched system

$$dx(t) = f_{\sigma(t)}\big(t, x(t), u(t)\big)dt + g_{\sigma(t)}\big(t, x(t), u(t)\big)dW(t), \tag{6.1a}$$

$$x(t_0) = x_0, \tag{6.1b}$$

where the state vector $x \in \mathbb{R}^n$ is assumed to be a right-continuous stochastic process, the input $u : [t_0, \infty) \to \mathbb{R}^l$ is an essentially bounded function with $\|u(t)\|_\infty \leq 1$, where $\|u(t)\|_\infty := \text{ess.}\sup_{t\geq t_0} \|u(t)\|$, and the switching signal $\sigma(t) : [t_0, \infty) \to \mathscr{S}$ is a piecewise constant function taking values in a finite compact set $\mathscr{S} = \{1, 2, \dots, N\}$.

If the switching among the elements of $\mathscr{S}$ occurs randomly, we assume that the switching signal $\sigma(\cdot)$ is a right-continuous Markov chain taking values in $\mathscr{S}$ with the generator $\Gamma = [\gamma_{ij}]_{N\times N}$ and its evolution is governed by the following probability transitions

M. S. Alwan and X. Liu, *Theory of Hybrid Systems: Deterministic and Stochastic*, Nonlinear Physical Science, https://doi.org/10.1007/978-981-10-8046-3_6

$$\mathbb{P}\{\sigma(t+h)=j|\sigma(t)=i\}=\begin{cases}\gamma_{ij}h+o(h), & \text{if } i\neq j,\\ 1+\gamma_{ii}h+o(h), & \text{if } i=j,\end{cases}$$

where $h>0$, γ_{ij} is the transition rate from mode i to mode j with $\gamma_{ij}\geq 0$, when $i\neq j$, and $\gamma_{ii}=-\sum_{j\neq i}^{N}\gamma_{ij}$ and $o(h)$ is such that $\lim_{h\to 0}o(h)/h=0$. The switching signal $\sigma(\cdot)$ is assumed to be independent of $W(\cdot)$.

The switching times $\{t_k\}_{k\in\mathbb{N}}$ (with $t_k\in\mathbb{R}_+$) form a strictly increasing sequence such that $\lim_{k\to\infty}t_k=\infty$. For any $i\in\mathscr{S}$ and $k\in\mathbb{N}$, the functions $f_i:[t_{k-1},t_k)\times\mathbb{R}^n\times\mathbb{R}^l\to\mathbb{R}^n$, $g_i:[t_{k-1},t_k)\times\mathbb{R}^n\times\mathbb{R}^l\to\mathbb{R}^{n\times m}$, which belong to $\mathscr{L}_{ad}(\Omega, L^p[t_{k-1},t_k))$ with $p=1$ and $p=2$, respectively, are assumed to be smooth enough to guarantee a unique solution, and $f_i(t,0,0)=0$ and $g_i(t,0,0)=0$; that is, system (6.1) admits a trivial solution, $x\equiv 0$. We also assume the initial state x_0 to be $\mathscr{F}_0$-measurable with finite pth moment (i.e., $\mathbb{E}[\|x_0\|^p]<\infty$).

Definition 6.1 (*Itô formula*) For all $t\geq t_0$ with $t_0\in\mathbb{R}_+$, let $x(t)$ be an n-dimensional continuous adapted process satisfying

$$dx(t)=f(t,x(t),u(t))dt+g(t,x(t),u(t))dW(t), \qquad \text{(a.s.)}$$

where f and g are as defined before. Assume that $V\in\mathscr{C}^{1,2}(\mathbb{R}_+\times\mathbb{R}^n;\mathbb{R}_+)$. Then, $V(t,x(t))$ is a scalar-valued stochastic process satisfying

$$dV(t,x(t))=\mathscr{L}V(t,x(t),u(t))dt+V_x(t,x(t))g(t,x(t),u(t))dW(t), \qquad \text{(a.s.)}$$

where the infinitesimal operator $\mathscr{L}$, associating $x(t)$ to $V(t,x(t))$ is defined by

$$\begin{aligned}\mathscr{L}V(t,x(t),u(t))=&V_t(t,x(t))+V_x(t,x(t))f(t,x(t),u(t))\\&+\frac{1}{2}\mathrm{tr}[g^T(t,x(t),u(t))V_{xx}(t,x(t))g(t,x(t),u(t))]\end{aligned} \tag{6.2}$$

with $V_x(t,x(t))$ and $V_{xx}(t,x(t))$ being the gradient and Hessian matrix of process $V(t,x(t))$, respectively.

The following lemma ensures the global boundedness of the mean value of process V when the operator $\mathscr{L}V$ (as a single operator) is estimated by a positive bound. The lemma is also interesting on its own because it guarantees a global unique solution even if a *local* Lipschitz condition holds.

Lemma 6.1 *Assume that a unique solution $x(t)=x(t;t_0,x_0)$ of the initial-value problem*

$$dx(t)=f(t,x(t),u(t))dt+g(t,x(t),u(t))dW(t), \qquad x(t_0)=x_0, \qquad \text{(a.s.)},$$

exists for all $t\in[t_0,\tau_\infty)$ with $t_0\in\mathbb{R}_+$ and τ_∞ being the explosion time. Let $V\in\mathscr{C}^{1,2}(\mathbb{R}_+\times\mathbb{R}^n;\mathbb{R}_+)$ such that it is radially unbounded (i.e., for all $(t,x)\in\mathbb{R}_+\times\mathbb{R}^n$, the limit $\lim_{\|x\|\to\infty}\left[\inf_{t\geq t_0}V(t,x)\right]=\infty$) and

$$\mathscr{L}V(t,x,u) \leq \alpha(V(t,x)), \qquad (a.s.)$$

where $\alpha \in \mathscr{K}_c$. *Then*

$$\mathbb{E}[V(t,x(t))] = G^{-1}\Big[G\big(\mathbb{E}[V(t_0,x_0)]\big) + (T-t_0)\Big] < \infty, \qquad \forall T \geq t_0,$$

where $G(s) = \int_1^s \frac{dt}{\alpha(t)}$, G^{-1} *is the inverse function of* G *and* $G\big(\mathbb{E}[V(t_0,x_0)]\big) + (T - t_0) \in Domain(G^{-1})$. *Moreover, the solution* $x(t)$ *is unique and defined for all* $t \geq t_0$.

Proof For all $t \geq t_0$ with $t_0 \in \mathbb{R}_+$, let $x(t) = x(t; t_0, x_0)$ be a local solution of the system. We claim that $\tau_\infty = \infty$. If our claim were not true, there would exist positive constants ε and T such that

$$\mathbb{P}\{\tau_\infty \leq T\} > \varepsilon.$$

Define a sequence of stopping times τ_l (for $l \geq 1$) of the process x from the ball $\|x\| > l$, i.e.,

$$\tau_l = \inf\{t \geq t_0 \mid \|x(t)\| > l\}$$

such that $\tau_l \to \tau_\infty$ (a.s.). This implies that, for sufficiently large l^*,

$$\mathbb{P}\{\tau_l \leq T\} > \varepsilon', \qquad \text{for some } \varepsilon' < \varepsilon, \quad l \geq l^*.$$

For all $t \in [t_0, T]$ and $l \geq l^*$, let $\tau_l(t) = \min\{\tau_l, t\}$. Apply the generalized Itô formula for the process $V(\tau_l(t), x(\tau_l(t)))$ and then take the mathematical expectation to get

$$\begin{aligned}
\mathbb{E}[V(\tau_l(t), x(\tau_l(t)))] &= \mathbb{E}[V(t_0,x_0)] + \mathbb{E}\int_{t_0}^{\tau_l(t)} \mathscr{L}V(s,x(s),u(s))ds \\
&\leq \mathbb{E}[V(t_0,x_0)] + \mathbb{E}\int_{t_0}^{t} \mathscr{L}V(\tau_l(s),x(\tau_l(s)),u(\tau_l(s)))ds \\
&\leq \mathbb{E}[V(t_0,x_0)] + \int_{t_0}^{t} \alpha\big(\mathbb{E}[V(\tau_l(s),x(\tau_l(s)))]\big)ds.
\end{aligned}$$

By Bihari's inequality [1, 2], we get

$$\begin{aligned}
\mathbb{E}[V(\tau_l(t), x(\tau_l(t)))] &= G^{-1}\Big[G\big(\mathbb{E}[V(t_0,x_0)]\big) + (t-t_0)\Big] \\
&\leq G^{-1}\Big[G\big(\mathbb{E}[V(t_0,x_0)]\big) + (T-t_0)\Big] < \infty,
\end{aligned}$$

where $G(s)=\int_1^s \frac{dt}{\alpha(t)}$, G^{-1} is the inverse function of G and $G\big(\mathbb{E}[V(t_0,x_0)]\big)+(T-t_0)\in \text{Domain}(G^{-1})$. From the above inequality, we see $\mathbb{E}[V(t,x(t)]<\infty$ for any $t\in[t_0,T]$.

On the other hand,

$$\mathbb{E}\big[1_{\{\tau_l\le T\}}V(\tau_l,x(\tau_l))\big]\le G^{-1}\Big[G\big(\mathbb{E}[V(t_0,x_0)]\big)+(T-t_0)\Big],$$

where 1_A is the indicator function of a set A. Define

$$\eta_l=\inf\{V(t,x)\mid \|x\|\ge l,\ t\ge t_0\}.$$

Thus

$$G^{-1}\Big[G\big(\mathbb{E}[V(t_0,x_0)]\big)+(T-t_0)\Big]\ge \eta_l\mathbb{P}\{\tau_l\le T\}\ge \varepsilon'\eta_l.$$

Letting $l\to\infty$ implies contradiction because V is radially unbounded; therefore, it must be true that

$$\mathbb{P}\{\tau_l\ge T\}=1.$$

The uniqueness follows from the definition of x up to equivalence, i.e., if y is another solution, then

$$\mathbb{P}\{\|x(t)-y(t)\|=0,\quad t_0\le t\le \sigma_\infty\}=1.$$

This completes the proof.

Definition 6.2 System (6.1) is said to be uniformly asymptotically ISS (aISS) in the pth moment if there exist functions $\beta\in\mathcal{KL}$ and $\gamma\in\mathcal{K}$ such that, for any u and $p>1$, the solution satisfies

$$\mathbb{E}[\|x(t)\|^p]\le\beta\big(\mathbb{E}[\|x_0\|^p],t-t_0\big)+\gamma(\|u(t)\|_\infty),\quad \forall t>t_0 \text{ with } t_0\in\mathbb{R}_+$$

where $\mathbb{E}[\|x_0\|^p]<\infty$ and $x(t)=x(t;t_0,x_0)$ is any solution of system (6.1). It is said to be exponentially ISS (eISS) in the pth moment if, in addition, $\beta\big(\mathbb{E}[\|x_0\|^p],t-t_0\big)\le K\mathbb{E}[\|x_0\|^p]e^{-\lambda(t-t_0)}$, for some positive constants K and λ.

Remark 6.1 Immediate implications of the above definition are, for instance, if $u\equiv 0$, it reduces to the uniform pth moment asymptotic (or exponential) stability of the trivial solution of unforced system. If $u\not\equiv 0$ and $g\equiv 0$, it reduces to the standard definitions of uniform ISS for deterministic systems.

6.2 Initial-State-Dependent Dwell-Time

In this section, we present the stability property of (6.1), where we use the initial-state-dependent dwell-time condition (denoted by τ_{isd}) to organize the switching among the system modes. In Theorem 6.1, the switching occurs among all aISS modes, while in Theorem 6.2, the switching occurs among stable and unstable modes. In both cases, we show that the solution converges to a ball of radius depending on the input magnitude.

Theorem 6.1 *Let $p \geq 1$. For any $i \in \mathscr{S}$, all $t \in [t_{k-1}, t_k)$ and $x \in \mathbb{R}^n$, let $V_i \in \mathscr{C}^{1,2}\big([t_{k-1}, t_k) \times \mathbb{R}^n; \mathbb{R}_+\big)$ with $V_i(t, 0) = 0$ satisfy the following assumptions:*

(i) there exist a concave function $\alpha_{1_i} \in \mathscr{K}_\infty$ and a convex function $\alpha_{2_i} \in \mathscr{K}_\infty$ such that

$$\alpha_{2_i}(\|x\|^p) \leq V_i(t, x) \leq \alpha_{1_i}(\|x\|^p), \qquad (a.s.); \tag{6.3}$$

(ii) there exist $\alpha_{3_i} \in \mathscr{K}_\upsilon$ and a function $\gamma \in \mathscr{K}$ such that

$$\mathscr{L}V_i(t, x(t), u(t)) \leq -\alpha_{3_i}(\|x\|^p), \qquad (a.s.) \tag{6.4}$$

provided that $\|x\|^p > \Big[\alpha_{3_i}^{*^{-1}}\Big(\frac{1}{\nu}\gamma(\|u\|_\infty)\Big)\Big] =: \rho_i(\|u\|_\infty)$ (a.s.) where $0 < \nu < 1$ and $\alpha_{3_i}^*(\cdot) = \frac{1}{(1-\nu)}\alpha_{3_i}(\cdot)$; and

(iii) for all $k \in \mathbb{N}$, the τ_{isd} condition

$$\tau_{isd} \geq \ln \frac{\theta_{2_i}\big(a_{k-1}\mathbb{E}[\|x_0\|^p]\big)}{\theta_{1_i}\big(a_k\mathbb{E}[\|x_0\|^p]\big)} \tag{6.5}$$

holds, where a_k are positive real numbers with $a_0 = 1$, $a_k < a_{k-1}$ and $\lim_{k\to\infty} a_k = 0$, and θ_{i_1} and θ_{2_i} are some class$-\mathscr{K}_\infty$ functions.

Then, system (6.1) is pth moment aISS with ISS-gain

$$\rho_M(\cdot) = \max\{\rho_i(\cdot) = \alpha_{3_i}^{*^{-1}}\big(\gamma^*(\cdot)\big) \mid i \in \mathscr{S}\},$$

where $\gamma^* = \frac{1}{\nu}\gamma$.

Proof For all $t \geq t_0$ with $t_0 \in \mathbb{R}_+$, let $x(t) = x(t; t_0, x_0)$ be the solution of (6.1) and, for all $t \in [t_{k-1}, t_k)$, let $V_i(t, x(t))$ be a Lyapunov function candidate related to the ith mode. By (ii), we can define the time-varying function $m_i(t) = \mathbb{E}[V_i(t, x(t))]$ for all $t \in [t_{k-1}, t_k)$. Applying Itô formula to $m_i(t)$, taking the mathematical expectation, and using Fubini's Theorem and the property of α_i function, we get

$$\begin{aligned} m_i(t) &= m_i(t_{k-1}) + \mathbb{E}\int_{t_{k-1}}^{t} \mathscr{L}V_i(s, x(s), u(s))ds \\ &\le m_i(t_{k-1}) - \int_{t_{k-1}}^{t} \alpha_i(m(s))ds, \end{aligned}$$

which implies

$$D^+ m_i(t) \le -\alpha_i(m_i(t)),$$

where $\alpha_i(\cdot) = \alpha_{3_i}\left(\alpha_{1_i}^{-1}(\cdot)\right)$. Then, by the classical stability result (see e.g. [3]), there exists a class$-\mathscr{K}\mathscr{L}$ function β_i^* such that

$$m_i(t) \le \beta_i^*(m_i(t_{k-1}), t - t_{k-1}).$$

Then, there exist class$-\mathscr{K}_\infty$ functions $\theta_{1_i}^*$ and $\theta_{2_i}^*$ [4, 5] such that, by (i),

$$\mathbb{E}[\|x(t)\|^p] \le \theta_{1_i}^{-1}\left[\theta_{2_i}(\mathbb{E}[\|x(t_{k-1})\|^p])e^{-(t-t_{k-1})}\right], \tag{6.6}$$

where $\theta_{1_i}(\cdot) := \theta_{1_i}^{*^{-1}}(\alpha_{2_i}(\cdot))$, $\theta_{2_i}(\cdot) := \theta_{2_i}^*(\cdot)$.

To show the solution convergence, we run the first mode on $[t_0, t_1)$, then last inequality reduces to

$$\mathbb{E}[\|x(t)\|^p] \le \theta_{1_1}^{-1}\left[\theta_{2_1}(\mathbb{E}[\|x(t_0)\|^p])e^{-(t-t_0)}\right]$$

and, by the τ_{isd} condition, we obtain at the switching time $t = t_1$

$$\mathbb{E}[\|x(t_1)\|^p] \le a_1\mathbb{E}[\|x_0\|^p],$$

which also implies after operating the second mode on $[t_1, t_2)$,

$$\begin{aligned} \mathbb{E}[\|x(t)\|^p] &\le \theta_{1_2}^{-1}\left[\theta_{2_2}\mathbb{E}[\|x(t_1)\|^p])e^{-(t-t_1)}\right] \\ &\le \theta_{1_2}^{-1}\left[\theta_{2_2}(a_1\mathbb{E}[\|x_0\|^p])e^{-(t-t_1)}\right]. \end{aligned}$$

By the same argument, we get

$$\mathbb{E}[\|x(t)\|^p] \le \theta_{1_i}^{-1}\left[\theta_{2_i}(a_{k-1}\mathbb{E}[\|x_0\|^p])e^{-(t-t_{k-1})}\right], \qquad t \in [t_{k-1}, t_k),$$

whenever $\|x(t)\| > [\rho_i(\|u\|_\infty)]^{1/p}$ (a.s.) and at the switching time $t = t_k$, $\mathbb{E}[\|x(t_k)\|^p] \le a_k\mathbb{E}[\|x_0\|^p]$. Since $\lim_{k\to\infty} a_k = 0$, the system states will eventually approach (in the pth moment) the ultimate bound $[\rho(\|u\|_\infty)]^{1/p}$ where $\rho = \max_i\{\rho_i\}$; that is, the solution of (6.1) is aISS in the pth moment. This completes the proof.

Remark 6.2 Assumptions (i) and (ii) are made to ensure the aISS property in the pth moment of each subsystem. In this case, the function V_i satisfying these assumptions is called *stochastic* ISS Lyapunov function related to the ith subsystem.

Remark 6.3 The idea behind the dwell-time-based condition, τ_{isd}, in (iii) is to generate a sequence of solution trajectories at the switching times that converges (in the pth moment) to a limit set with a radius depending on the maximum ISS gain of the system modes. That is to say, the switching among all stable modes ensures the stability of the switched system. Furthermore, compared with the existing state-dependent dwell-time condition which requires the knowledge of the state at the switching times, the proposed criterion is easier to work with because it depends on the system initial state only.

Implications of this result are stated in the following corollary whose proofs are straightforward; thus, they are left here as an exercise.

Corollary 6.1 *In Theorem 6.1,*

(i) if $\alpha_{1_i}(r) = \alpha_{1_i} r$, $\alpha_{2_i}(r) = \alpha_{2_i} r$ *and* $\alpha^*_{3_i}(r) = \alpha^*_{3_i} r$ *for all* $r > 0$, *where* α_{1_i}, α_{2_i} *and* $\alpha^*_{3_i}$ *are positive constants, then the above aISS properties reduce to eISS, respectively;*

(ii) if $u(t) \equiv 0$ *for all* $t \in \mathbb{R}_+$, *then aISS reduces to the* p*th moment global uniform asymptotic stability (g.u.a.s.) of the trivial solution of the nonlinear stochastic switched system*

$$dx(t) = f_{\sigma(t)}(t, x(t))dt + g_{\sigma(t)}(t, x(t))dW(t), \qquad x(t_0) = x_0;$$

(iii) if $g(t, x(t), u(t)) \equiv 0$ *and* $u(t) \not\equiv 0$ *for all* t, *then the aISS property reduces to the standard aISS of the nonlinear switched system*

$$\dot{x}(t) = f_{\sigma(t)}(t, x(t), u(t)), \qquad x(t_0) = x_0,$$

where $\mathscr{L}V(t, x(t), u(t)) = \dot{V}(t, x(t)) = V_t(t, x(t)) + V_x f(t, x(t), u(t))$; *and*

(iv) if $u(t) \equiv 0$ *and* $g(t, x(t)) \equiv 0$ *for all* t, *then the aISS property reduces to g.u.a.s. of the nonlinear switched system*

$$\dot{x}(t) = f_{\sigma(t)}(t, x(t)), \qquad x(t_0) = x_0,$$

where $\mathscr{L}V(t, x(t)) = \dot{V}(t, x(t)) = V_t(t, x(t)) + V_x f(t, x(t))$.

In the following example, we illustrate these results.

Example 6.1 Consider the following switched system

$$dx = (-a_i x + u(t))dt + u(t) \sin x \, dW(t),$$

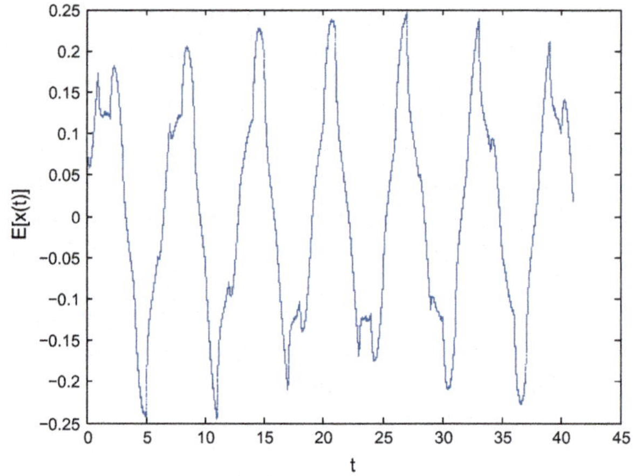

Fig. 6.1 First moment aISS with $u(t) = \sin(t)$

where $i \in \mathscr{S} = \{1, 2\}$ and a_i is a positive real number. Let $V_i(x) = \frac{1}{4}x^4$ be a stochastic ISS Lyapunov function candidate related to the ith subsystem. We also choose $\alpha_{1_i}(\cdot) = \alpha_{2_i}(\cdot) = V_i(\cdot)$. Then, with little effort, one may get

$$\begin{aligned}\mathscr{L}V_i(x, u) &\leq -a_i x^4 + |x|^3|u| + \frac{3}{2}x^4 \\ &\leq -a_i x^4 + a_i\theta x^4 - a_i\theta x^4 + |x|^3|u| + \frac{3}{2}x^4, \qquad 0 < \theta < 1, \\ &\leq -\alpha_{3_i} V(x), \qquad \text{provided that } |x| \geq |u|/(a_i\theta - 3/2),\end{aligned}$$

with $a_i\theta > 3/2$, where $\alpha_{3_i} = 4a_i(1-\theta) > 0$. Thus, both subsystems are aISS in the fourth moment. Taking $a_1 = 4$, $a_2 = 8$ and $\theta = 1/2$ gives $\alpha_{3_1} = 8$ and $\alpha_{3_2} = 14$. By Theorem 6.1, we have $m_i(t) \leq m_i(t_{k-1})e^{-\alpha_{3_i}(t-t_{k-1})} \leq e^{-(t-t_{k-1})}$. This also implies that $\theta_{1_i}(r) = \theta_{2_i}(r) = r$ and hence $\mathbb{E}[x^4] \leq \mathbb{E}[x^4(t_{k-1})] \leq e^{-(t-t_{k-1})}$. Therefore, the τ_{isd} becomes $t_k - t_{k-1} \geq \ln(\frac{a_{k-1}}{a_k})$, where we choose $a_k = \frac{1}{k+1}, k = 0, 1, \ldots$. Figure 6.1 shows that the solution is aISS in the first moment where $u(t) = \sin(t)$.

The standard aISS property of the deterministic switched system is shown in Fig. 6.2.

The classical asymptotic stability property of $x \equiv 0$ of the unforced stochastic switched system is shown in Fig. 6.3.

In the following theorem, we establish the pth moment aISS property of the switched system (6.1) with stable and unstable subsystems. For convenience of notation, we denote by $\mathscr{S}_s = \{1, 2, \ldots, N_s\}$ ($\mathscr{S}_u = \{1, 2, \ldots, N_u\}$), with $N_s + N_u = N$, the index set of stable (unstable, respectively) subsystems and $\mathscr{S} = \mathscr{S}_s \cup \mathscr{S}_u$.

Theorem 6.2 *Consider system* (6.1) *with* $\mathscr{S} = \mathscr{S}_s \cup \mathscr{S}_u$. *Let* $V_i \in \mathscr{C}^{1,2}([t_{k-1}, t_k) \times \mathbb{R}^n; \mathbb{R}_+)$ *with* $V_i(t, 0) = 0$ *satisfy the following assumptions:*

(i) for each $i \in \mathscr{S}$, *there exist a concave function* $\alpha_{1_i} \in \mathscr{K}_\infty$ *and convex function* $\alpha_{2_i} \in \mathscr{K}_\infty$ *such that*

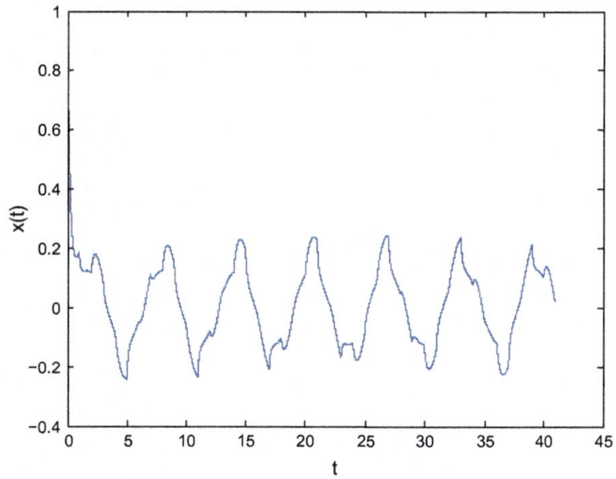

Fig. 6.2 aISS property with $u(t) = \sin(t)$ and $g(t, x(t), u(t)) \equiv 0$

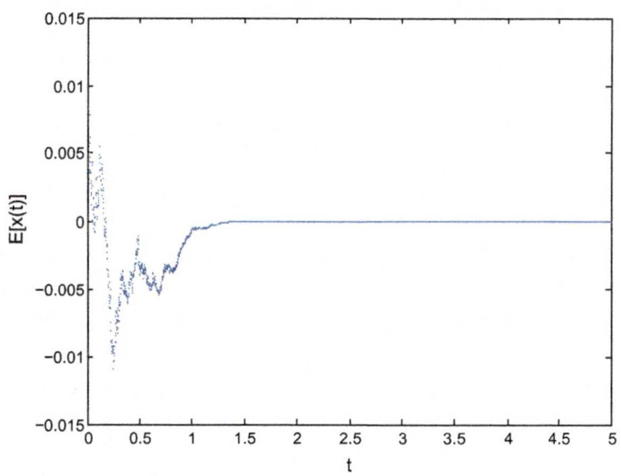

Fig. 6.3 First moment asymptotic stability of $x \equiv 0$

$$\alpha_{2_i}(\|x\|^p) \leq V_i(t, x) \leq \alpha_{1_i}(\|x\|^p), \qquad (a.s.);$$

(ii) (1) for each $i \in \mathscr{S}_s$, there exist $\alpha_{3_i} \in \mathscr{K}_v$ and $\rho_i \in \mathscr{K}_\infty$ such that

$$\mathscr{L}V_i(t, x, u) \leq -\alpha_{3_i}(\|x\|^p), \qquad (a.s.), \quad \text{whenever } \|x\|^p > \rho_i(\|u\|_\infty);$$

(ii) (2) for each $i \in \mathscr{S}_u$, there exist $\alpha_{3_i} \in \mathscr{K}_c$ such that

$$\mathscr{L}V_i(t, x, u) \leq \alpha_{3_i}(\|x\|^p), \qquad (a.s.);$$

(iii) the τ_{isd} condition satisfies

(1) for each $i \in \mathscr{S}_s = \{1, 2, 3, \ldots, N_s\}$ and $k = 1, 3, 5, \ldots$

$$t_1 - t_0 \geq \ln \frac{\theta_{2_1}(\mathbb{E}[\|x_0\|^p])}{\theta_{1_1}(a_1\mathbb{E}[\|x_0\|^p])} > 0,$$
$$t_3 - t_2 \geq \ln \frac{\theta_{2_3}(a_1 A_1\mathbb{E}[\|x_0\|^p])}{\theta_{1_3}(a_2\mathbb{E}[\|x_0\|^p])} > 0;$$
$$\ldots$$

(2) for each $i \in \mathscr{S}_u = \{1, 2, 3, \ldots, N_u\}$ *and* $k = 2, 4, 6, \ldots$

$$0 < t_2 - t_1 \leq G_2\Big[\alpha_{2_2}(a_1 A_1\mathbb{E}[\|x_0\|^p])\Big] - G_2\Big[\alpha_{1_2}(a_1\mathbb{E}[\|x_0\|^p])\Big],$$
$$0 < t_4 - t_3 \leq G_4\Big[\alpha_{2_4}(a_2 A_2\mathbb{E}[\|x_0\|^p])\Big] - G_4\Big[\alpha_{1_4}(a_2\mathbb{E}[\|x_0\|^p])\Big],$$
$$\ldots,$$

where $0 < a_k < a_k A_k \leq a_{k-1}$ *with* $a_0 = 1$, $\theta_{1_i}(\cdot) := \theta^{*^{-1}}_{1_i}\big(\alpha_{1_i}(\cdot)\big)$ *and* $\theta_{2_i}(\cdot) := \theta^*_{2_i}(\cdot)$ *are functions of class* $\mathscr{K}_\infty$, *and* $G_2, G_4, \cdots$ *are functions defined in Lemma 6.1.*

Then, the solution of (6.1) *is pth moment aISS stable with the ISS gain* $\rho_M(\cdot) := \max_{i \in \mathscr{S}} \rho_i(\cdot)$.

Proof For all $t \geq t_0$ with $t_0 \in \mathbb{R}_+$, let $x(t) = x(t; t_0, x_0)$ be the solution of (6.1). For all $t \in [t_{k-1}, t_k)$ and $i \in \mathscr{S}$, define $V_i(t, x(t)) \in \mathscr{C}^{1,2}([t_{k-1}, t_k) \times \mathbb{R}^n; \mathbb{R}_+)$ as a Lyapunov function candidate related to the ith subsystem.

For convenience, we adopt the case where the switching among the stable and unstable modes occurs alternatively.

For $i = 1$ and $t \in [t_0, t_1)$, we have by Theorem 6.1

$$\mathbb{E}[\|x(t)\|^p] \leq \theta^{-1}_{1_1}\Big[\theta_{2_1}(\mathbb{E}[\|x_0\|^p])e^{-(t-t_0)}\Big],$$

and, by the stable τ_{isd} condition in (ii)(1), we have at the switching time $t = t_1$,

$$\mathbb{E}[\|x(t_1)\|^p] \leq a_1\mathbb{E}[\|x_0\|^p].$$

If the system switches to an unstable mode on $[t_1, t_2)$, then by Lemma 6.1, assumption (i) and the last inequality, we have

$$\mathbb{E}[\|x(t)\|^p] \leq \alpha^{-1}_{2_2}\Big\{G^{-1}_2\Big(G_2(\alpha_{1_2}(a_1\mathbb{E}\|x_0\|^p])) + (t - t_1)\Big)\Big\},$$

which implies that, with the aid of the unstable τ_{isd} condition in (ii)(2),

$$\mathbb{E}[\|x(t_2)\|^p] \leq a_1 A_1\mathbb{E}[\|x_0\|^p].$$

By the same manner, one generates a sequence of states at the switching times

$$\mathbb{E}[\|x(t_k)\|^p] \leq a_k \mathbb{E}[\|x_0\|^p] \quad \text{and} \quad \mathbb{E}[\|x(t_{k+1})\|^p] \leq a_k A_k \mathbb{E}[\|x_0\|^p].$$

Since $a_k < a_k A_k \leq a_{k-1}$, $\lim_{k\to\infty} a_k = 0$ ($\forall k \in \mathbb{N}$,) and $\mathbb{E}[\|x_0\|^p] < \infty$, then $\lim_{k\to\infty} \mathbb{E}[\|x(t_k)\|^p] = 0$, which means that, when $t \to \infty$, the solution will eventually linger on at the ultimate bound of the system input. This completes the proof.

Remark 6.4 As can be seen in the assumptions (ii) and (iii) of Theorem 6.2, if the system switches between stable and unstable modes, the stable modes have to dominate over the unstable ones. This in turn implies that dwell times of stable modes are larger than the corresponding times of unstable modes.

The following example illustrates this result.

Example 6.2 Consider the switched system with the following unstable mode

$$dx = (-ax^3 + xu(t))dt + \sqrt{2a}x^2 dW(t)$$

and the stable mode

$$dx = (-ax^3 - bx + u(t))dt + \sqrt{2a}x^2 dW(t),$$

where a and b are positive constants. Here, $\mathscr{S}_u = \{1\}$ and $\mathscr{S}_s = \{2\}$. For any $i \in \mathscr{S} = \{1, 2\}$, define $V_i(x) = \frac{1}{2}x^2$ as a Laypunov function candidate related to the ith subsystem. Then, for $i = 1$, we have $\mathscr{L}V_1(x, u) = x^2$, where $u(t) = 1$ for all $t \in [t_{k-1}, t_k)$, i.e., the subsystem is unstable. This also implies that $D^+\mathbb{E}[V_1(x(t))] = 2\mathbb{E}[V_1(x(t))]$ and, by Lemma 6.1,

$$\mathbb{E}[V_1(x(t))] = \mathbb{E}[V_1(x(t_{k-1}))]e^{2(t-t_{k-1})},$$

i.e., $G(r) = \ln(r)$ and $G^{-1}(r) = e^r$. If we choose $\alpha_{1_1}(x) = \alpha_{2_1}(x) = V_1(x)$, we obtain

$$\mathbb{E}[x^2(t)] = \mathbb{E}[x^2(t_{k-1})]e^{2(t-t_{k-1})}.$$

Similarly, for $i = 2$, we have $\mathscr{L}V_2(x, u) \leq -\alpha_{3_2} V_2(x)$ provided that $|x| \geq |u|/b\theta$, where $\alpha_{3_2} = 2b(1-\theta) > 0$ and $0 < \theta < 1$, which implies

$$\mathbb{E}[V_2(x)] \leq \mathbb{E}[V_2(x(t_{k-1}))]e^{-\alpha_{3_2}(t-t_{k-1})} \leq \mathbb{E}[V_2(x(t_{k-1}))]e^{-2(t-t_{k-1})}$$

if we choose $b = 2$ and $\theta = 1/2$ (i.e., $\theta_{1_2}(r) = \theta_{2_2}(r) = r$). Choose $\alpha_{1_2}(x) = \alpha_{2_2}(x) = V_2(x)$. Then

$$\mathbb{E}[x^2(t)] \leq \mathbb{E}[x^2(t_{k-1})]e^{-(t-t_{k-1})}, \quad t \in [t_{k-1}, t_k).$$

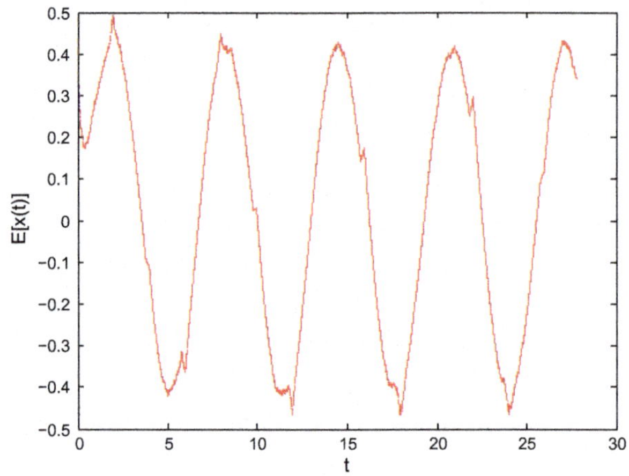

Fig. 6.4 First moment aISS with $u(t) = \sin(t)$

As for the dwell time, let us first run a stable mode (i.e., $k = 1$). Then, from (iii)(1), if $a_1 = 1/2$, we get $t_1 - t_0 \geq \ln 2 = 0.7$ and, for $k = 2$, we run the unstable mode with $A_1 = 1.5 > 1$ which gives $t_2 - t_1 \leq \frac{1}{2} \ln 1.5 = 0.2$. By the same argument, for $k = 3$, we get $t_3 - t_2 \geq \ln \frac{a_1 A_1}{a_2}$, where $a_1 A_1 > a_2$ which implies $a_2 < 3/4$; so that, taking $a_2 = 1/4$ gives $t_3 - t_2 \geq 1.1$. For $k = 4$, we get $1 < A_2 \leq 2$, so that taking $A_2 = 1.5$ gives $t_4 - t_3 \leq 0.2$. Figure 6.4 shows the first moment aISS of the switched system where $u(t) = \sin(t)$.

6.3 Markovian Switching

In this section, we consider a more general approach than the dwell-time condition called Markovian switching to control the mode switchings. In this case, the switching signal is represented by a Markov chain which takes values in a finite set. An interesting issue in adopting this type of switching arises from involving the transition rates of the Markov chain in the calculation of dwell times. One can recognize that the stability requirement of individual modes is neither sufficient nor necessary for guaranteeing a stability-like property of a switched system.

Consider the nonlinear system with Markovian switching

$$\begin{cases} dx(t) = f(t, x(t), u(t), \sigma(t))dt + g(t, x(t), u(t), \sigma(t))dW(t), \\ x(t_0) = x_0, \qquad \sigma(t_0) = \sigma_0 \in \mathscr{S}, \end{cases} \tag{6.7}$$

where, for all $t \geq t_0$ with $t_o \in \mathbb{R}_+$, the switching signal $\sigma(t)$ is a Markov process taking values in a finite state space $\mathscr{S} = \{1, 2, \ldots, N\}$ and σ_0 is an initial state. In this case, the operator $\mathscr{L}$ associating $(x(t), i)$ to $V(t, x(t), i) \in \mathscr{C}^{1,2}\big(\mathbb{R}_+ \times \mathbb{R}^n \times \mathscr{S}; \mathbb{R}_+\big)$ for any $i \in \mathscr{S}$, is defined by

$$\begin{aligned}\mathscr{L}V(t,x(t),u(t),i) =& V_t(t,x(t),i) + V_x(t,x(t),i)f(t,x(t),u(t),i)\\ &+ \frac{1}{2}\mathrm{tr}[g^T(t,x(t),u(t),i)V_{xx}(t,x(t),i)g(t,x(t),u(t),i)]\\ &+ \sum_{j=1}^{N}\gamma_{ij}V(t,x(t),j),\end{aligned} \tag{6.8}$$

with γ_{ij} being the transition rate defined earlier.

In the following theorem, we state and prove the pth moment eISS of the forced system (6.7).

Theorem 6.3 *For any $i \in \mathscr{S}$ and all $t \in [t_{k-1}, t_k)$, assume that the following assumptions hold:*

(i) there exist constants $K > 0$, $\alpha_i > 0$, $\rho_i \geq 0$, and $\sigma_i \geq 0$ such that

$$\begin{aligned}&\|f(t,x,0,i)\| \leq K\|x\|,\ \|x^T f(t,x,0,i)\| \leq \alpha_i\|x\|^2, \quad (a.s.)\\ &\|g(t,x,0,i)\| \leq \rho_i\|x\|,\ \|x^T g(t,x,0,i)\| \leq \sigma_i\|x\|^2, \quad (a.s.);\end{aligned}$$

(ii) there exist positive constants λ, c_1, and c_2 such that

$$c_1\|x\|^p \leq V(t,x,i) \leq c_2\|x\|^p, \quad (a.s.) \tag{6.9}$$

$$\mathscr{L}V(t,x,u,i) \leq -\lambda\|x\|^p, \quad (a.s.) \tag{6.10}$$

whenever $\|x\| > \rho(\|u\|_\infty)$, where $V \in \mathscr{C}^{1,2}([t_0,\infty)\times\mathbb{R}^n\times\mathscr{S};\mathbb{R}_+)$ and ρ is a class-$\mathscr{K}$ function; and

(iii) the functions f and g are locally Lipschitz in u, uniformly in t and x, i.e., there exist positive constants c_3 and c_4 such that

$$\begin{aligned}&\|f(t,x,u,i) - f(t,x,0,i)\| \leq c_3\|u\|, \quad (a.s.)\\ &\|g(t,x,u,i) - g(t,x,0,i)\| \leq c_4\|u\|, \quad (a.s.).\end{aligned}$$

Then, the solution of (6.7) is pth moment eISS for $0 < p < \min\{2, (3c_4 + 4\sigma_i)/(c_4 + 2\sigma_i)\}$ with Lyapunov exponent being not larger than $-\lambda/c_2$.

Proof For all $t \geq t_0$ with $t_0 \in \mathbb{R}_+$, let $x(t) = x(t; t_0, x_0)$ be the solution of (6.7). For any $i \in \mathscr{S}$ and $\beta_i > 0$, define $V(t,x(t),i) = \beta_i\|x(t)\|^p$ as a stochastic ISS Lyapunov function candidate related to the ith mode. Then, by (6.8) we have

$$\begin{aligned}\mathscr{L}V(t,x,u,i) =& p\beta_i\|x\|^{p-2}x^T f(t,x,u,i) + \frac{1}{2}p\beta_i\|x\|^{p-2}\|g(t,x,u,i)\|^2\\ &- \frac{1}{2}p(2-p)\beta_i\|x\|^{p-4}\|x^T g(t,x,u,i)\|^2 + \sum_{j=1}^{N}\gamma_{ij}\beta_j\|x\|^p\end{aligned}$$

$$
\begin{aligned}
&= p\beta_i \|x\|^{p-2}x^T\Big[f(t,x,u,i) - f(t,x,0,i) + f(t,x,0,i)\Big] \\
&\quad + \frac{1}{2}p\beta_i \|x\|^{p-2}\|g(t,x,u,i) - g(t,x,0,i) + g(t,x,0,i)\|^2 \\
&\quad - \frac{1}{2}p(2-p)\beta_i \|x\|^{p-4}\|x^T[g(t,x,u,i) - g(t,x,0,i) + g(t,x,0,i)]\|^2 \\
&\quad + \sum_{j=1}^{N}\gamma_{ij}\beta_i \|x\|^p \\
&\le p\beta_i \|x\|^{p-2}\Big\{\|x^T[f(t,x,u,i) - f(t,x,0,i)]\| + \|x^T f(t,x,0,i)\|\Big\} \\
&\quad + \frac{1}{2}p\beta_i \|x\|^{p-2}\Big\{\|g(t,x,u,i) - g(t,x,0,i)\|^2 + \|g(t,x,0,i)\|^2 \\
&\quad + 2\|g(t,x,u,i) - g(t,x,0,i)\|\,\|g(t,x,0,i)\|\Big\} \\
&\quad + \frac{1}{2}p(2-p)\beta_i \|x\|^{p-4}\Big\{\|x^T[g(t,x,u,i) - g(t,x,0,i)]\|^2 + \|x^T g(t,x,0,i)\|^2 \\
&\quad + 2\|x^T[g(t,x,u,i) - g(t,x,0,i)]\|\,\|x^T g(t,x,0,i)\|\Big\} + \sum_{j=1}^{N}\gamma_{ij}\beta_j \|x\|^p \\
&\le p\beta_i \|x\|^{p-2}\|x^T[f(t,x,u,i) - f(t,x,0,i)]\| + p\beta_i \|x\|^{p-2}\|x^T f(t,x,0,i)\| \\
&\quad + \frac{1}{2}p\beta_i \|x\|^{p-2}\|g(t,x,u,i) - g(t,x,0,i)\|^2 + \frac{1}{2}p\beta_i \|x\|^{p-2}\|g(t,x,0,i)\|^2 \\
&\quad + 2\frac{1}{2}p\beta_i \|x\|^{p-2}\|g(t,x,u,i) - g(t,x,0,i)\|\,\|g(t,x,0,i)\|\Big\} \\
&\quad + \frac{1}{2}p(2-p)\beta_i \|x\|^{p-4}\|x^T[g(t,x,u,i) - g(t,x,0,i)]\|^2 \\
&\quad + \frac{1}{2}p(2-p)\beta_i \|x\|^{p-4}\|x^T g(t,x,0,i)\|^2 \\
&\quad + 2\frac{1}{2}p(2-p)\beta_i \|x\|^{p-4}\|x^T[g(t,x,u,i) - g(t,x,0,i)]\|\,\|x^T g(t,x,0,i)\| \\
&\quad + \sum_{j=1}^{N}\gamma_{ij}\beta_j \|x\|^p \\
&\le p\beta_i c_3 \|x\|^{p-1}\|u\| + p\beta_i|\alpha_i|\,\|x\|^p + \frac{1}{2}p\beta_i c_4^2\|x\|^{p-2}\|u\|^2 + \frac{1}{2}p\beta_i\rho_i^2\|x\|^p \\
&\quad + p\beta_i\rho_i c_4\|x\|^{p-1}\|u\|_\infty + \frac{1}{2}p(2-p)\beta_i c_4^2\|x\|^{p-2}\|u\|_\infty^2 + \frac{1}{2}p(2-p)\beta_i\sigma_i^2\|x\|^p \\
&\quad + p(2-p)\beta_i\sigma_i c_4\|x\|^{p-2}\|u\|_\infty + \sum_{j=1}^{N}\gamma_{ij}\beta_i \|x\|^p \\
&\le \Big\{p\beta_i|\alpha_i| + \frac{1}{2}p\beta_i\rho_i^2 + \frac{1}{2}p(2-p)\beta_i\sigma_i^2 + \sum_{j=1}^{N}\gamma_{ij}\beta_j\Big\}\|x\|^p \\
&\quad + \Big\{\beta_i c_3\|x\|^{p-1} + \frac{1}{2}p\beta_i c_4^2\|x\|^{p-2} + p\beta_i\rho_i c_4\|x\|^{p-1} + \frac{1}{2}p(2-p)\beta_i c_4^2\|x\|^{p-2} \\
&\quad + p(2-p)\beta_i\sigma_i c_4\|x\|^{p-2}\Big\}\|u\|_\infty
\end{aligned}
$$

$$
\begin{aligned}
&= \Big\{p|\alpha_i| + \frac{1}{2}p\rho_i^2 + \frac{1}{2}p(2-p)\sigma_i^2\Big\}\beta_i + \sum_{j=1}^{N}\gamma_{ij}\beta_j\|x\|^p \\
&\quad + \Big\{\Big[c_3 + p\rho_i c_4\Big]\|x\|^{p-1} + \Big[\frac{1}{2}pc_4^2 + \frac{1}{2}p(2-p)c_4^2 + p(2-p)\sigma_i c_4\Big]\|x\|^{p-2}\Big\}\beta_i\|u\|_\infty \\
&= \left\{\Big[p|\alpha_i| + \frac{1}{2}p\rho_i^2 + \frac{1}{2}p(2-p)\sigma_i^2\Big]\beta_i + \sum_{j=1}^{N}\gamma_{ij}\beta_j\right\}\|x\|^p \\
&\quad + \Big\{\Big[c_3 + p\rho_i c_4\Big]\|x\|^{p-1} + \Big[-(0.5c_4 + \sigma_i)p + (1.5c_4 + 2\sigma_i)\Big]c_4 p\|x\|^{p-2}\Big\}\beta_i\|u\|_\infty \\
&= -\beta_i^*\|x\|^p + 2M(\|x\|)\|u\|_\infty \\
&\le -\lambda^*\|x\|^p + 2M(\|x\|)\|u\|_\infty,
\end{aligned} \tag{6.11}
$$

where $\lambda^* = \min\{-\beta_i^* \mid i \in \mathscr{S}\}$ with

$$
\beta_i^* = -\beta_i\left[p|\alpha_i| + \frac{1}{2}p\rho_i^2 + \frac{1}{2}p(2-p)\sigma_i^2\right] + \sum_{j=1}^{N}\gamma_{ij}\beta_j < 0,
$$

and

$$
\begin{aligned}
M(\|x\|) = \max\Big\{&\beta_i\Big[c_3 + p\rho_i c_4\Big]\|x\|^{p-1}, \\
&c_4 p\beta_i\Big[-(0.5c_4 + \sigma_i)p + (1.5c_4 + 2\sigma_i)\Big]\|x\|^{p-2}\Big\}.
\end{aligned}
$$

To use $\lambda^*\|p\|^p$ to dominate $2M(\|x\|)\|u\|_\infty$, we write the last inequality in (6.11) as follows

$$
\begin{aligned}
\mathscr{L}V(t,x,u,i) &\le -(\lambda^* - \nu)\|x\|^p - \nu\|x\|^p + 2M(\|x\|)\|u\|_\infty, \qquad 0 < \nu < \lambda^*, \\
&\le -(\lambda^* - \nu)\|x\|^p = -\lambda\|x\|^p,
\end{aligned}
$$

where $\lambda := \lambda^* - \nu > 0$ provided that $\nu\|x\|^p > 2M(\|x\|)\|u\|_\infty$ or

$$
\begin{cases}
\|x\| > 2\beta_i/\nu\cdot[c_3 + p\rho_i c_4]\,\|u\|_\infty, \quad \text{if} \quad M(\|x\|) = \beta_i\,[c_3 + p\rho_i c_4]\,\|x\|^{p-1}, \\
\|x\| > \{2\beta_i p c_4/\nu\cdot[-(0.5c_4 + \sigma_i)p + (1.5c_4 + 2\sigma_i)]\,\|u\|_\infty\}^{1/2}, \quad \text{if} \\
\qquad M(\|x\|) = c_4 p\beta_i\,[-(0.5c_4 + \sigma_i) + (1.5c_4 + 2\sigma_i)]\,\|x\|^{p-2}.
\end{cases} \tag{6.12}
$$

Applying the generalized Itô formula to $e^{\lambda t/c_2}V(t,x,i)$ and taking the mathematical expectation yield

$$
\begin{aligned}
\mathbb{E}[e^{\frac{\lambda}{c_2}t}V(t,x,i)] &= \mathbb{E}[V(t_0,x_0,\sigma_0)]e^{\frac{\lambda}{c_2}t_0} + \mathbb{E}\Big[\int_{t_0}^{t} e^{\frac{\lambda}{c_2}s}[\frac{\lambda}{c_2}V(s,x,i) + \mathscr{L}V(s,x,i)]ds\Big] \\
&\le \mathbb{E}[V(t_0,x_0,\sigma_0)]e^{\frac{\lambda}{c_2}t_0} + \mathbb{E}\Big[\int_{t_0}^{t} e^{\frac{\lambda}{c_2}s}[\frac{\lambda}{c_2}V(s,x,i) - \frac{\lambda}{c_2}V(s,x,i)]ds\Big] \\
&= \mathbb{E}[V(t_0,x_0,\sigma_0)]e^{\frac{\lambda}{c_2}t_0}.
\end{aligned}
$$

With the aid of (ii) and, after some algebraic manipulations, one may obtain

$$\mathbb{E}[\|x(t)\|^p] \le K\mathbb{E}[\|x_0\|^p]e^{-\frac{\lambda}{c_2}(t-t_0)}, \qquad K = c_2/c_1, \quad \forall t \ge t_0.$$

This result shows that system (6.7) is pth moment eISS with the ultimate bound given in (6.12) and Lyapunov exponent $-\lambda/c_2$. This completes the proof.

Example 6.3 Consider the switched system in (6.7) with

$$f(t,x,u,1) = \frac{a}{1+t}(x+u(t)), \qquad g(t,x,u,1) = b(\sin x + u(t)),$$
$$f(t,x,u,2) = c(xe^{-|x|} + u(t)), \qquad g(t,x,u,2) = b(x + u(t)\ln|1+x|),$$

where a, b, c, and d are some constants to be chosen later. The probability transition matrix is

$$\Gamma = \begin{pmatrix} -1 & 1 \\ 1 & -1 \end{pmatrix}.$$

Clearly, the vector field functions satisfy the conditions in (i) and (ii) of the theorem, where $K_1 = \alpha_1 = c_{3_1} = |a|$, $K_2 = \alpha_2 = c_{3_2} = |c|$ with $c \in \{-1, 1\}$ or $\rho_1 = \sigma_1 = c_{4_1} = |b|, \rho_2 = \sigma_2 = c_{4_2} = |d|, c_3 = \max\{|a|, |c|\}$, and $c_4 = \max\{|b|, |d|\}$. For $i = 1, 2$, let $V(x,i) = \beta_i|x|^p$ with $0 < p < \min\{2, (\frac{3}{2}c_4 + 2\sigma_i)/(\frac{1}{2}c_4 + \sigma_i)\}$. Taking $|a| = |b| = |c| = |d| = p = 1$ yields $\beta_1^* = -3\beta_1 + \beta_2$ and $\beta_2^* = \beta_1 - 3\beta_2$ and by choosing $\beta_1 = \beta_2 = 1$, we get $\lambda^* = \min\{\beta_1, \beta_2\} = -2$. Therefore, if $\nu = 1 \le -\lambda^*$, $\mathscr{L}V(x,u,i) \le -|x| < 0$ provided that $|x| > 4|u|$. By our choice of the probability transition matrix $\Gamma = [\gamma_{ij}]_{2\times 2}$, we get $\pi_1 = \pi_2 = 0.5$, which represent the time spent in the first and second modes. Figures 6.5 and 6.6 illustrate the first moment aISS property with $u(t) = \sin(t)$ and $u(t) = e^{-t}$, respectively. In both cases, the switching occurs between two stable and unstable modes.

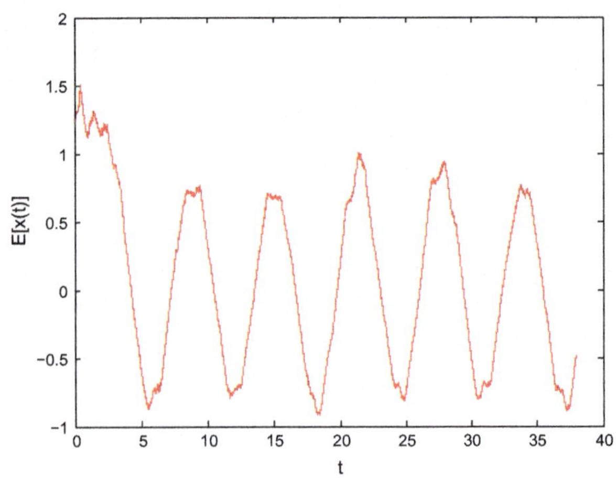

Fig. 6.5 First moment *a*ISS with $a = c = -1$ and $u(t) = \sin(t)$

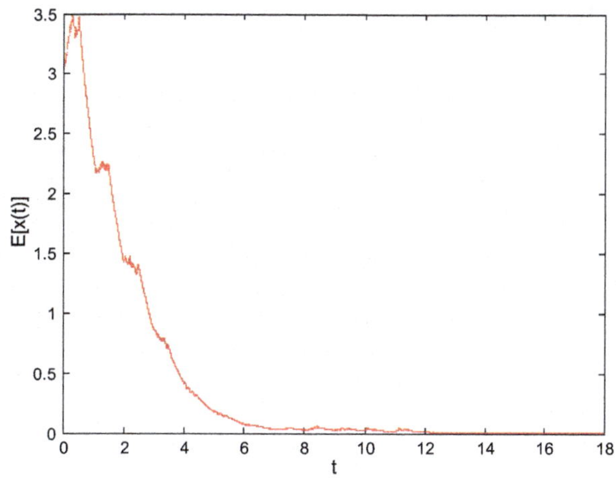

Fig. 6.6 First moment aISS with $a = c = -1$ and $u(t) = e^{-t}$

6.4 Notes and Comments

In this chapter, the pth moment ISS property of nonlinear stochastic switched systems has been presented. The Lyapunov-like sufficient conditions have been written and switching rules to guarantee the system stability properties has been designed. The material of this chapter is taken from [6]. In Sect. 6.2, we have used the τ_{isd} condition to control the switching among the system modes. Two cases were discussed; namely, systems with all stable modes, and systems with stable and unstable modes. The latter case required developing Lemma 6.1, in which Bihari's lemma, but not Bellman-Gronwall lemma, plays an important role. We have shown that the result of Theorem 6.1 has some implications that can be applied to some special cases, such as deterministic or unforced systems or systems without these types of perturbations. In fact, one can also derive some analogous implications from Theorems 6.2 and 6.3.

We should remark that, in [7], the pth moment asymptotic ISS was developed for a stochastic retarded systems using Markovian switching. In Sect. 6.3, we have discussed the pth moment exponential ISS property, which necessitates the vector fields to satisfy Lipschitz condition in the input variable and grow linearly for all time.

We have also showed that in Theorems 6.2 and 6.3, the stability of each individual subsystem is not necessary to achieve the stability of the switched system. In such a case, ISS is guaranteed if stable modes are activated longer than unstable ones. We should also remark that throughout this chapter the switched system is subject to the same input. Therefore, one may consider a more general case in which each subsystem is subject to a different input.

References

1. Bihari I (1956) A generalization of a lemma of Bellman and its application to uniqueness problems of differential equations. Acta Math. Hung. 7:81–94
2. Mao X (1994) Exponential stability of stochastic differential equations. Marcel Dekker
3. H.K. Khalil HK (2002) Nonlinear systems. Prentice-Hall, Inc., NJ
4. Sontag ED (1998) Comments on integral variants of ISS. Syst Control Lett 34:93–100
5. De Persis C, De Santis R, Morse AS (2003) Switched nonlinear systems with state-dependent dwell-time. Syst Control Lett 50:291–302
6. Alwan MS, Liu XZ, Xie W-C (2012) On input-to-state stability of nonlinear stochastic hybrid systems. Dyn Contin Discrete Impuls Syst Series A 19, 513–533
7. Huang L, Mao X (2009) On input-to-state stability of stochastic retarded systems with Markovian switching. IEEE Trans Autom Control 54(8):1898–1902

Chapter 7
Reliable Control for Stochastic Switched Systems with State Delay

In the feedback control design of real systems, an unavoidable, undesirable control performance may occur due to the failure(s) in control components, such as actuator or/and sensor failures. Therefore, it is necessary to design controllers to achieve the desired plant performance, not only when the control components are properly operational, but also in the presence of failures. Control systems that tolerate sensor/actuator outages are called *reliable control systems*.

This chapter deals with the design problem of a switching reliable control for a class of stochastic switched systems. The stochastic differential equation is of Itô type with constant time delay, the nonlinear lumped disturbances have linear growth bounds and the random noise is approximated by Wiener process. Two sets of actuators are considered, a set of operational actuators that never fail and a set of actuators that are susceptible to failure. Primarily, the focus here is to design a state feedback sub-controller for each system mode such that, for all admissible nonlinear uncertainties and actuator failures occurring in a pre-specified subset of actuators, the closed-loop modes are exponential stable in the mean square (m.s.). Moreover, to maintain the stability property for the closed-loop switched system, the initial state dwell-time switching rule and the technique of multiple Lyapunov function together with the Razumikhin methodology are used. This approach leads to solving a set of algebraic Riccati-like matrix equations.

7.1 Problem Formulation

Consider the following stochastic switched control system with time delay

$$dx(t) = \big(A_{\kappa(t)}x + B_{\kappa(t)}u + f_{\kappa(t)}(x_t)\big)\,dt + g_{\kappa(t)}(x_t)\,dW(t), \tag{7.1a}$$

$$x_{t_0} = \phi(s),\ s \in [-r, 0], \tag{7.1b}$$

M. S. Alwan and X. Liu, *Theory of Hybrid Systems: Deterministic and Stochastic*, Nonlinear Physical Science, https://doi.org/10.1007/978-981-10-8046-3_7

where for all $t \geq t_0$ with $t_0 \in \mathbb{R}_+$, $x(t) \in \mathbb{R}^n$ is the system state vector and $u(t) \in \mathbb{R}^q$ is the control system input. $\kappa : [t_0, \infty) \to \mathscr{S}$ where $\mathscr{S} = \{1, 2, \cdots, N\}$ (with $N \in \mathbb{N}$) is a piecewise constant function representing the switching signal. For every $i \in \mathscr{S}$, we assume that the functionals $f_i \in \mathbb{R}^n$ and $g_i \in \mathbb{R}^{n\times m}$, which represent lumped uncertainties, are bounded above by a linear growth bound and u is a state feedback controller of the form $K_i x$, where $K_i \in \mathbb{R}^{n\times q}$ is a control matrix gain. $\phi \in \mathbb{R}^n$ is the initial state function which is assumed to be in $\mathscr{L}^2_{\mathscr{F}_0}([-r, 0]; \mathbb{R}^n)$. A_i and B_i are real known constant matrices of appropriate dimensions. To guarantee that (7.1) has a unique regular solution, we assume that $f_i \in \mathscr{L}_{ad}(\Omega, L[a, b])$ and $g_i \in \mathscr{L}_{ad}(\Omega, L^2[a, b])$ and they satisfy Lipschitz condition in their argument. We also assume that $f_i(0) = 0 \in \mathbb{R}^n$ and $g_i(0) = 0 \in \mathbb{R}^{n\times m}$ to ensure that the system admits a trivial solution, $x \equiv 0$.

To analyse the reliable stabilization with respect to actuator failures, for every $i \in \mathscr{S}$ consider the decomposition of the control matrix $B^i = B^i_\Sigma + B^i_{\bar{\Sigma}}$ where $\Sigma \subseteq \{1, 2, \cdots, q\}$ the set of actuators that are susceptible to failure (i.e., they may occasionally fail) and the other set of actuators which are assumed to be robust to failures $\bar{\Sigma} \subseteq \{1, 2, \cdots, q\} - \Sigma$ and essential to stabilize the given system. Moreover, B^i_Σ and $B^i_{\bar{\Sigma}}$ are the control matrices associated with Σ and $\bar{\Sigma}$, respectively, and B^i_Σ and $B^i_{\bar{\Sigma}}$ are generated by zeroing out the columns corresponding to $\bar{\Sigma}$ and Σ, respectively. The elements of Σ are redundant in terms of the stabilization, though they are necessary to improve the system performance. On the other hand, the elements of $\bar{\Sigma}$ are required to stabilize the system and assumed that they never fail, i.e., the pair $(A^i, B^i_{\bar{\Sigma}})$ is assumed to be stabilizable. For a fixed $i \in \mathscr{S}$, let $\sigma \subseteq \Sigma$ corresponds to some of the actuators that experience failure and assume that the output of faulty actuators is zero, i.e., outage case. Then, the decomposition becomes $B^i = B^i_\sigma + B^i_{\bar{\sigma}}$, where B^i_σ and $B^i_{\bar{\sigma}}$ have the same definition of B^i_Σ and $B^i_{\bar{\Sigma}}$, respectively. We should emphasize that the pre-specified subset $\sigma \in \Sigma$ of faulty actuators of the ith sub-controller may differ from $\sigma \in \Sigma$ of the jth sub-controller, for any $i, j \in \mathscr{S}$.

Applying the control input u of the form

$$u(t) = K_i x(t), \qquad \forall i \in \mathscr{S} \text{ and } t \in [t_k, t_{k+1}) \tag{7.2}$$

to the system plant through the normal actuators and, since we assumed that the outputs of the faulty actuators are zeros, the closed-loop system of (7.1) becomes

$$dx(t) = \big((A_i + B_{i\bar{\sigma}} K_i)x + f_i(x_t)\big)dt + g_i(x_t)\, dW(t), \tag{7.3a}$$

$$x_{t_0} = \phi(s), \qquad s \in [-r, 0], \tag{7.3b}$$

for every $i \in \mathscr{S}$ and all $t \in [t_k, t_{k+1})$.

7.2 Stability Analysis

In this section, we prove the m.s. global asymptotic stability for the closed-loop system (7.3), where all actuators are operational (Theorem 7.1). This result will be carried over in Theorem 7.2 to achieve stabilization property for the system in the presence of possible actuator failures in all sub-controllers.

Theorem 7.1 *For any $i \in \mathscr{S}$, let the controller gain K_i be given. Assume there exist positive constants ε_i and positive-definite matrix P_i such that the following algebraic Riccati-like matrix equation*

$$\Big(A_i + B_i K_i + \frac{1}{2}\gamma_i I\Big)^T P_i + P_i\Big(A_i + B_i K_i + \frac{1}{2}\gamma_i I\Big) + \frac{1}{\varepsilon_i} P_i^2 + \varepsilon_i \bar{q}_i \|U_i\|^2 I + \alpha_i P_i = 0,$$

holds, where $\alpha_i > 0$, $\bar{q}_i > 1$ and $\gamma_i > 0$ such that, for any $\psi \in \mathscr{C}([-r, 0]; \mathbb{R}^n)$,

$$tr[g_i^T(\psi)P_i g_i(\psi)] \leq 2\gamma_i \bar{q}_i \psi(0)^T P_i \psi(0)$$

and U_i is an $n \times n$ matrix such that $\|f_i(\psi)\|^2 \leq \bar{q}_i \|U_i\|^2 \|\psi(0)\|^2$. Suppose further that, for any $k \in \mathbb{N}$ and $i \in \mathscr{S}$, the following dwell-time-type condition

$$t_k - t_{k-1} \geq \frac{1}{\alpha_i} \ln\Big(\frac{a_i d_{i-1} e^{r\alpha_i}}{b_i d_i}\Big) > 0 \tag{7.4}$$

holds, where $a_i = \lambda_{\max}(P_i)$, $b_i = \lambda_{\min}(P_i)$ and $d_i < d_{i-1} < 1$ (for $i = 2, 3, \cdots$) such that $\lim_{i\to\infty} d_{i-1} = 0$. Then, the closed-loop modes in (7.3) are m.s. globally exponentially stabilized by the control law given in (7.2) and switched system (7.3) is asymptotically stable in the m.s.

Proof For all $t \geq t_0$ with $t_0 \in \mathbb{R}_+$, let $x(t) = x(t; t_0, \phi)$ be the solution of system (7.3) and, for any $i \in \mathscr{S}$, define $V_i(x) = x^T P_i x$ as a Lyapunov function candidate for the ith mode. Then, from Itô formula we have

$$\begin{aligned}\mathscr{L}V_i(x) =& \big((A_i x + B_i K_i)x + f_i(x_t)\big)^T P_i x + x^T P_i\big(A_i + B_i K_i x + f_i(x_t)\big) \\ &+ \frac{1}{2}\mathrm{tr}\big(g_i^T(x_t)P_i g_i(x_t)\big) \\ \leq& x^T\big(A_i^T P_i + P_i A_i + 2K_i^T B_i^T P_i\big)x + f_i^T(x_t)P_i x + x^T P_i f_i(x_t) \\ &+ \gamma_i \bar{q}_i x^T x.\end{aligned}$$

By the fact that [1]

$$2x^T P_i f_i(x_t) \leq x^T\big(\varepsilon_i \bar{q}_i \|U_i\|^2 I + \frac{1}{\varepsilon_i} P^2\big)x, \tag{7.5}$$

we obtain

$$\begin{aligned}\mathscr{L}V_i(x) \leq & x^T\Big((A_i + B_iK_i)^T P_i + P_i(A_i + B_iK_i) \\ & + \varepsilon_i\bar{q}_i\|U_i\|^2 I + \frac{1}{\varepsilon_i}P_i^2 + \gamma_i\bar{q}_i I\Big)x \\ = & -\alpha_i x^T P_i x = -\alpha_i V_i(x) < 0.\end{aligned}$$

Applying the Itô Lemma to process $V_i(x)$ and taking the mathematical expectation yield

$$D^+m_i(t) \leq -\alpha_i m_i(t), \quad \forall i \in \mathscr{S} \text{ and } t \in [t_k, t_{k+1}),$$

where $m_i(t) = \mathbb{E}[V_i(x(t))]$ for any $i \in \mathscr{S}$ and all $t \in [t_k, t_{k+1})$, and D^+m is the Dini derivative of m defined by

$$D^+m(t) = \lim_{h\to 0^+} \sup \frac{1}{h}[m(t+h) - m(t)].$$

It follows that, for any $i \in \mathscr{S}$ and all $t \in [t_k, t_{k+1})$,

$$\mathbb{E}[\|x(t)\|^2] \leq \frac{a_i}{b_i}\mathbb{E}[\|x_{t_k}\|_r^2]e^{-\alpha_i(t-t_k)},$$

that is every closed-loop mode in (7.3) is exponential stability in the m.s. Invoking the dwell-time condition in (7.4) results in

$$\mathbb{E}[\|x(t)\|^2] \leq d_i\,\mathbb{E}[\|\phi\|_r^2] \tag{7.6}$$

and, at the switching moments $t = t_k$, $\mathbb{E}[\|x(t_k)\|^2] \leq d_i\,\mathbb{E}[\|\phi\|_r^2]$. Since $\lim_{i\to\infty} d_i = 0$, $\lim_{t_k\to\infty}\mathbb{E}[\|x(t_k)\|^2] = 0$ asymptotically, which proves the desired result. This completes the proof.

Remark 7.1 The solvability condition of the algebraic Riccati-like equation is to guarantee the existence of the positive-definite matrix P_i for any $i \in \mathscr{S}$, which in turn implies that the system modes are exponentially stabilized by the state feedback control law defined in (7.2). Moreover, the dwell-time-type condition in (7.4) is made to ensure that the solution of the closed-loop switched system (7.3) converges in m.s. to the trivial solution by the rate d_i during the finite time $t_k - t_{k-1}$ for any $i \in \mathscr{S}$ and $k \in \mathbb{N}$; this is can be easily seen in (7.6). We should mention that the dwell-time condition in (7.4) is a special case of the initial-state-dependent dwell-time condition presented in Sect. 6.2. The positive tuning parameter ε_i (for any $i \in \mathscr{S}$) is presented to reduce the conservativeness of the matrix inequality (7.5). We should also remark that the assumptions of Theorem 7.1 do not impose any restriction on the time delay, which makes the proposed stability approach efficiently applicable to systems with state delay.

Having proved the key-role stability theorem, we are in a position to address the robust reliable control design.

Theorem 7.2 *For any $i \in \mathscr{S}$, assume that*

(i) there exist positive constants ε_i and ε_i^*, and positive-definite matrix P_i such that the following algebraic Riccati-like matrix equation

$$\Big(A_i + \frac{1}{2}\gamma_i I\Big)^T P_i + P_i\Big(A_i + \frac{1}{2}\gamma_i I\Big) + P_i(\frac{1}{\varepsilon_i}I - \varepsilon_i^* B_{\bar{\sigma}} B_{\bar{\sigma}}^T)P_i + \varepsilon_i \bar{q}_i \|U_i\|^2 I + \alpha_i P_i = 0$$

holds, where $\alpha_i > 0$, $\bar{q}_i > 1$, $\gamma_i > 0$ and U_i are defined in Theorem 7.1; and

(ii) the dwell-time condition in Theorem 7.1 holds.

Then, system modes in (7.3) are m.s. globally exponentially stabilized by the control law $u = K_i x$, where the control gain $K_i = -\frac{1}{2}\varepsilon_i^* B_{i\bar{\sigma}}^T P_i$, for any nonlinear uncertainties and actuator failures in the pre-specified set $\sigma \in \Sigma$ associated with the ith mode. Moreover, the entire switched system in (7.3) is globally asymptotically stable in the m.s.

Proof Since the control input u is applied to the system plant only through the normal actuators, we have $B_i K_i = -\frac{1}{2}\varepsilon_i^* B_{i\bar{\sigma}} B_{i\bar{\sigma}}^T P_i$, for all $i \in \mathscr{S}$.

For all $t \geq t_0$ with $t_0 \in \mathbb{R}_+$, let $x(t) = x(t; t_0, \phi)$ be the solution of (7.1) and define $V_i(x) = x^T P_i x$ (for $i \in \mathscr{S}$) as Lyapunov function candidate related to the ith mode. Then, as achieved in Theorem 7.1, we have

$$\begin{aligned}
&\Big(A_i + B_i K_i + \frac{1}{2}\gamma_i I\Big)^T P_i + P_i\Big(A_i + B_i K_i + \frac{1}{2}\gamma_i\Big) \\
&\quad + \frac{1}{\varepsilon_i}P_i^2 + \varepsilon_i \bar{q}_i \|U_i\|^2 I \\
&\leq \Big(A_i + \frac{1}{2}\gamma_i I\Big)^T P_i + P_i\Big(A_i + \frac{1}{2}\gamma_i I\Big) \\
&\quad + P_i(\frac{1}{\varepsilon_i}I - \varepsilon_i^* B_{\bar{\sigma}} B_{\bar{\sigma}}^T)P_i + \varepsilon_i \bar{q}_i \|U_i\|^2 I \\
&= -\alpha P < 0,
\end{aligned}$$

where we have used the inequality $B_{i\bar{\Sigma}} B_{i\bar{\Sigma}}^T \leq B_{i\bar{\sigma}} B_{i\bar{\sigma}}^T$, which follows from the fact that $B_{i\bar{\Sigma}} B_{i\bar{\Sigma}}^T = B_{i\bar{\sigma}} B_{i\bar{\sigma}}^T - B_{i\Sigma - i\sigma} B_{i\Sigma - i\sigma}^T$ [2]. That is, the system modes are all m.s. globally exponentially stabilized by the mentioned feedback control law. Thus, as achieved in Theorem 7.1, applying the dwell-time condition results in that $x \equiv 0$ is m.s. globally asymptotically stabilized. This completes the proof of Theorem 7.2.

7.3 Numerical Example

In the following example, we consider a system switching between two modes with two cases, normal and faulty control actuators. In Case 1, the actuators are operational in both modes and, in Case 2, the first mode experiences a failure in the second actuator and the second mode has a failure in the first actuator. In both cases, the normal and reliable state feedback controllers guarantee the stabilization requirement.

Example 7.1 Consider system (7.1) where $x^T = (x_1\ x_2)$, $\mathcal{S} = \{1, 2\}$, $\phi(s) = 1 - s$ for all $s \in [-1, 0]$ (i.e., the time delay $r = 1$), $\varepsilon_i = 1$, $\varepsilon_i^* = 1$, $\bar{q}_i = 2$, $d_i = 1/2$, $\alpha_1 = 2$, $\alpha_2 = 3$, $\gamma_1 = 0.01$, $\gamma_2 = 0.02$,

$$A_1 = \begin{bmatrix} 0.2 & 0.1 \\ 0 & -10 \end{bmatrix},\ B_1 = \begin{bmatrix} 1 & -0.2 \\ 0 & 1 \end{bmatrix},\ K_1 = \begin{bmatrix} -5 & 0.1 \\ 0 & 1 \end{bmatrix},$$

$$f_1(x(t-1)) = \begin{bmatrix} 0.01x_1(t-1) \\ -0.02x_2(t-1) \end{bmatrix},\quad g_1(x(t-1)) = 0.01\begin{bmatrix} x_1(t-1) & 0 \\ 0 & x_2(t-1) \end{bmatrix},$$

and

$$A_2 = \begin{bmatrix} -11 & 0 \\ 0.2 & 0.1 \end{bmatrix},\ B_2 = \begin{bmatrix} 1 & 0 \\ 00.1 & 1 \end{bmatrix},\ K_2 = \begin{bmatrix} 1 & 0 \\ 0.3 & -6 \end{bmatrix},$$

$$f_2(x(t-1)) = \begin{bmatrix} 0.01x_1(t-1) \\ 0.1x_2(t-1) \end{bmatrix},\quad g_2(x(t-1)) = \begin{bmatrix} 0.01x_1(t-1) & 0 \\ 0 & 0.2x_2(t-1) \end{bmatrix}.$$

Case 1: When all actuators are operational, we have from the algebraic Riccati-like equations

$$P_1 = \begin{bmatrix} 0.002 & 0 \\ 0 & 0.0011 \end{bmatrix} \text{ and } P_2 = \begin{bmatrix} 0.001 & 0 \\ 0 & 0.0017 \end{bmatrix};$$

That is, $a_1 = 0.0021$, $a_2 = 0.0017$, $b_1 = 0.0011$ and $b_2 = 0.001$. We also have found $A_1 + B_1K_1 = \begin{bmatrix} -4.8 & 0 \\ 0 & -9 \end{bmatrix}$ and $A_2 + B_2K_2 = \begin{bmatrix} -10 & 0 \\ 0.4 & -5.9 \end{bmatrix}$, which are Hurwitz matrices. The simulation result is shown in Fig. 7.1, where the dwell times of the first and second modes are, respectively, $t_k - t_{k-1} \geq 2.01$ and $t_k - t_{k-1} \geq 2.17$, for any $k \in \mathbb{N}$.

Case 2: When there is a failure in actuator 2 in the first mode and a failure in the first actuator in the second mode, i.e.,

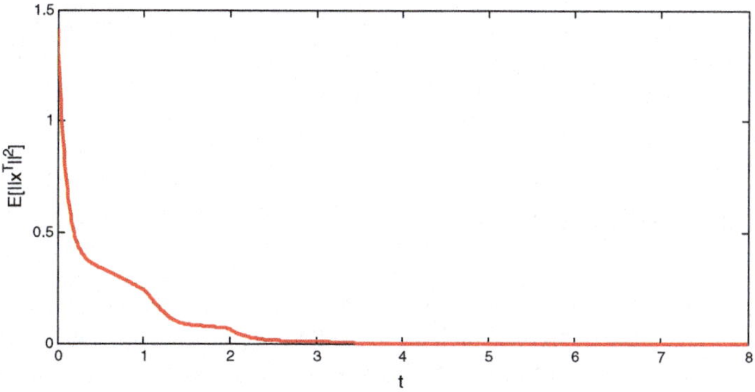

Fig. 7.1 Mean square asymptotic stability: normal actuators in both modes

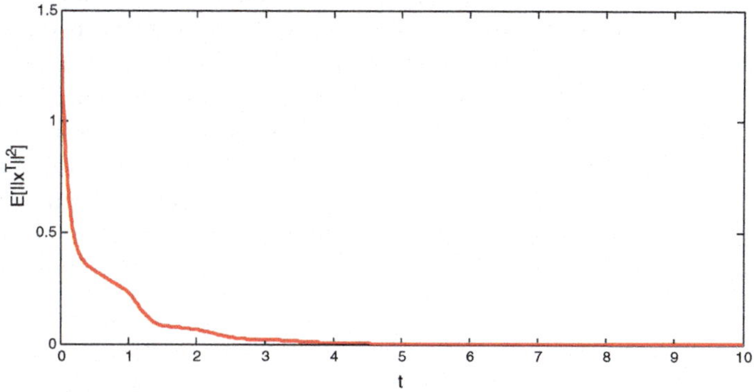

Fig. 7.2 Mean square asymptotic stability with failures in the actuator components

$$B_{1\bar{\sigma}} = \begin{bmatrix} 1 & 0 \\ 0 & 0 \end{bmatrix} \text{ and } B_{2\bar{\sigma}} = \begin{bmatrix} 0 & 0 \\ 0 & 1 \end{bmatrix}$$

we have from the algebraic Riccati-like equations

$$P_1 = \begin{bmatrix} 0.4815 & 0.0047 \\ 0.0047 & 0.001 \end{bmatrix} \text{ and } K_1 = \begin{bmatrix} -0.2407 & -0.0024 \\ 0 & 0 \end{bmatrix} \text{ and}$$

$$P_2 = \begin{bmatrix} 0.001 & 0.0067 \\ 0.0067 & 0.3734 \end{bmatrix} \text{ and } K_2 = \begin{bmatrix} 0 & 0 \\ -0.0033 & -0.1867 \end{bmatrix},$$

which give $a_1 = 0.4815$, $a_2 = 0.3736$, $b_1 = 0.001$ and $b_2 = 0.0009$. We also have found $A_1 + B_{1\bar{\sigma}} K_1 = \begin{bmatrix} -0.0407 & 0.0976 \\ 0 & -10 \end{bmatrix}$ and $A_2 + B_{2\bar{\sigma}} K_2 = \begin{bmatrix} -11 & 0 \\ 0.1967 & -0.0867 \end{bmatrix}$, which are Hurwitz matrices. The simulation result is shown in Fig. 7.2, where the dwell times of the first and second modes are, respectively, $t_k - t_{k-1} \geq 4.78$ and $t_k - t_{k-1} \geq 3.24$, for any $k \in \mathbb{N}$.

Remark 7.2 As can be seen in Figs. 7.1 and 7.2, the convergence of the system state to the equilibrium state in systems with operational control actuators is faster than the convergence in systems with faulty actuators.

7.4 Notes and Comments

The problem of reliable control for stochastic switched systems has been addressed. The focus was on the design of such a controller to guarantee m.s. global exponential stability of each mode, not only when all control actuators are operational, but also when there is a failure in some pre-specified subset of actuators. The material of this chapter is taken from [3]. This result has led to establish the asymptotic stability of the switched system under the prescribed dwell-time condition. The outputs of the faulty actuators are assumed to be zero; therefore, for further investigation, one may extend this result to consider sensor outages or faulty actuators with nonzero output signals, which can be viewed a disturbance inputs. The Lyapunov–Razumikhin approach is efficiently applicable to systems with delayed states, because these results do not impose any restriction on the time delay. Also, employing Lyapunov functions has led to solving Riccati-like matrix equations for positive-definite matrices P_i.

In fact, there have been many studies devoted to design different reliable controllers for systems with various levels of complexities; see, for instance, [4–15].

References

1. Alwan MS, Liu XZ, Xie W-C (2013) Comparison principle and stability results for nonlinear differential equations with piecewise constant arguments. J Frankl Inst 350:211–230
2. Wang Z, Shu H, Liu X (2005) Reliable stabilization of stochastic time-delay systems with nonlinear disturbance. Int J Gen Syst 34(5):523–535
3. Alwan MS, Liu XZ, Xie, W-C (2013) Stochastic switching reliable control for a class of stochastic hybrid systems with state delay. J Nonlinear Syst Appl 86–91
4. Ackremann J (1985) Sampled-data control systems. Springer, New York
5. Alwan MS, Liu XZ, Xie W-C (2011) Robust reliable control for uncertain stochastic impulsive systems with time delay. In: Proceedings of the international conference on dynamical systems and applications, vol 6, pp 1–8
6. Cheng X-M, Gui W-H, Gan Z-J (2005) Robust reliable control for a class of time-varying uncertain impulsive systems. J Cent South Univ Technol 12(1):199–202
7. Gao J, Huang B, Wang Z, Fisher DG (2001) Robust reliable control for a class of uncertain nonlinear systems with time-varying multistate time delays. Int J Syst Sci 32(7):817–824
8. Luo X, Yang L, Yang H, Guan X (2006) Robust reliable control for neutral delay systems. In: Proceeding of the 6th world congress on intelligent control and automation, June 21–23, Dalian, China, pp 144–148
9. Seo C-J, Kim BK (1996) Robust reliable H_∞ control for linear systems with parameter uncertainty and actuator failure. Automatica 32:465–467
10. Veillette RJ (1992) Reliable state feedback and reliable observers. In: Proceedings of the 31st conference on decision and control, pp. 2898–2903

11. Veillette RJ, Medanic JV, Perkins WR (1992) Design of reliable control systems. IEEE Trans Autom Control 37(3):290–304
12. Wang Z, Huang B, Unbehauen H (1999) Robust reliable control for a class of uncertain nonlinear state-delayed systems. Automatica 35:955–963
13. Wang Z, Huang B, Burnham KJ (2001) Stochastic reliable control of a class of uncertain time-delay systems with unknown nonlinearities. IEEE Trans Circuits Syst I Fundam Theory Appl 48(5):646–650
14. Wang J, Shao H (2000) Delay-dependent robust and reliable H_∞ control for uncertain time-delay systems with actuator failures. J Frankl Inst 337:781–791
15. Yang G-H, Wang JL, Soh YC (2001) Reliable H_∞ control design for linear systems. Automatica 37:717–725

Chapter 8
Robust Reliable Control for Impulsive Large-Scale Systems

This chapter deals with the problem of designing a robust reliable decentralized control for impulsive large-scale systems (ILSS) with admissible uncertainties in the system states. Then, the same idea is carried over to design reliable observers to estimate the states of the above systems. The faulty actuator/sensor outputs are assumed to be zero. The reliability analysis is achieved by using a scalar Lyapunov function.

8.1 Problem Formulation

Consider the interconnected system

$$\begin{cases} \dot{w}^i = (A^i + \Delta A^i)w^i + B^i u^i + f_i(w^i) \\ \qquad +g^i(w^1, w^2, \cdots, w^i, \cdots, w^l), \quad t \neq t_k, \\ \Delta w^i(t) = \mathscr{I}_k(w^i(t^-)) = C_{ik} w^i(t^-), \quad t = t_k, \ k \in \mathbb{N}, \\ w^i(t_0) = w_0^i, \end{cases} \tag{8.1}$$

where for $i = 1, 2, \ldots, l$, $w^i \in \mathbb{R}^{n_i}$ is the ith subsystem state such that $\Sigma_{i=1}^{l} n_i = n$, $A^i \in \mathbb{R}^{n_i \times n_i}$ is a non-Hurwitz matrix, the impulsive times t_k satisfy $t_0 < t_1 < t_2 < \cdots < t_k < \cdots$ with $\lim_{k\to\infty} t_k = \infty$, $\Delta w^i(t_k) = w^i(t_k^+) - w^i(t_k^-)$ where $w(t_k^+)$(or $w(t_k^-)$) is the state just after (or before) the impulse at t_k, $\mathscr{I}_k : \mathbb{R}^{n_i} \to \mathbb{R}^{n_i}$ is the impulsive function, $u^i = K^i w^i \in \mathbb{R}^q$ is the control input for the ith subsystem, where $K^i \in \mathbb{R}^{q \times n_i}$ is the control gain matrix, $f_i : \mathbb{R}^{n_i} \to \mathbb{R}^{n_i}$, is some nonlinearity and $g^i : \mathbb{R}^{n_1} \times \mathbb{R}^{n_2} \times \cdots \times \mathbb{R}^{n_l} \to \mathbb{R}^n$ is the interconnection. The functions f_i and g^i satisfy Lipschitz condition. A^i, B^i and C_{ik} are known real constant matrices with proper dimensions, and ΔA^i is a piecewise continuous function representing parameter uncertainty with bounded norm.

M. S. Alwan and X. Liu, *Theory of Hybrid Systems: Deterministic and Stochastic*, Nonlinear Physical Science, https://doi.org/10.1007/978-981-10-8046-3_8

System (8.1) can be written in the following form

$$\begin{cases} \dot{x} = (A + \Delta A)x + Bu + F(x) + G(x), & t \neq t_k, \\ \Delta x(t) = \mathscr{I}_k(x(t^-)) = C_k x(t^-), & t = t_k, \quad k \in \mathbb{N}, \\ x(t_0) = x_0, & \end{cases} \tag{8.2}$$

where $x^T = (w^{1^T} \; w^{2^T} \; \cdots \; w^{l^T})$, $((A + \Delta A)x)^T = \Big(((A^1 + \Delta A^1)w^1)^T \; ((A^2 + \Delta A^2)w^2)^T \; \cdots \; ((A^l + \Delta A^l)w^l)^T\Big)$, $(Bu)^T = \Big((B^1u^1)^T \; (B^2u^2)^T \; \cdots \; (B^lu^l)^T\Big)$, $(F(x))^T = \Big(f_1(w^1)^T \; f_2(w^2)^T \; \cdots \; f_l(w^l)^T\Big)$, $(G(x))^T = \Big(g^1(x)^T \; g^2(x)^T \; \cdots \; g^l(x)^T\Big)$, $(C_k x)^T = \Big((C_{1k}w^1)^T \; (C_{2k}w^2)^T \; \cdots \; (C_{lk}w^l)^T\Big)$.

From (8.1), the corresponding isolated subsystems are

$$\begin{cases} \dot{w}^i = (A^i + \Delta A^i)w^i + B^i u^i + f_i(w^i), & t \neq t_k, \\ \Delta w^i(t) = C_{ik} w^i(t^-), & t = t_k, \quad k \in \mathbb{N}, \\ w^i(t_0) = w_0^i, & \end{cases} \tag{8.3}$$

where $i = 1, 2, \ldots, l$ and the corresponding closed-loop system is

$$\begin{cases} \dot{w}^i = (A^i + \Delta A^i + B^i K^i)w^i + f_i(w^i), & t \neq t_k, \\ \Delta w^i(t) = C_{ik} w^i(t^-), & t = t_k, \quad k \in \mathbb{N}, \\ w^i(t_0) = w_0^i. & \end{cases} \tag{8.4}$$

As shown in the last chapter, if we consider the decomposition becomes $B^i = B^i_\sigma + B^i_{\bar{\sigma}}$, then closed-loop systems for the faulty case becomes

$$\begin{cases} \dot{w}^i = (A^i + \Delta A^i + B^i_{\bar{\sigma}} K^i)w^i + f_i(w^i), & t \neq t_k, \\ \Delta w^i(t) = C_{ik} w^i(t^-), & t = t_k, \quad k \in \mathbb{N}, \\ w^i(t_0) = w_0^i. & \end{cases} \tag{8.5}$$

Definition 8.1 The trivial solution, $x \equiv 0$, of system (8.2) is said to be robustly globally exponentially stable if there exist positive constants λ and $\bar{\lambda}$ such that

$$||x|| \leq \bar{\lambda}||x_0||e^{-\lambda(t-t_0)}, \quad \forall t \geq t_0$$

for any solution $x(t) = x(t; t_0, x_0)$ of (8.2) with $t_0 \in \mathbb{R}^+$ and $x_0 \in \mathbb{R}^n$.

Throughout this chapter, the system uncertainty is assumed to satisfy the following assumption.

Assumption A For $i = 1, 2, \ldots, l$, the admissible parameter uncertainties are defined by

$$\Delta A^i(t) = D^i \mathscr{U}^i(t) H^i, \quad \forall t \in \mathbb{R}_+$$

with D^i and H^i being known real matrices with appropriate dimensions that give the structure of the uncertainty and $\mathscr{U}^i(t)$ being unknown real time-varying matrix representing the uncertain parameter and satisfying $||\mathscr{U}^i(t)|| \leq 1$.

Lemma 8.1 *For any arbitrary positive constants ϵ_1, ξ_1 and a positive-definite matrix P, we have*

(i) $2x^T P(\Delta A)x \leq x^T(\epsilon_1 PDD^T P + \frac{1}{\epsilon_1}H^T H)x$*; and*
(ii) $2x^T Pf(x) \leq x^T(\xi_1 P^2 + \frac{\delta I}{\xi_1})x$ *such that* $||f(x)||^2 \leq \delta||x||^2$ *with* $\delta > 0$.

8.2 Reliable Control

In this section, we address the problems of stability and stabilization by robust controllers for isolated impulsive subsystems in case of operational actuators (Theorem 8.1) and faulty actuators (Theorem 8.2.) Then, these results will be applied to the large-scale interconnected system (8.2) as presented by Theorem 8.3 (operational actuators) and Theorem 8.4 (faulty actuators).

Theorem 8.1 *Let the control gain K^i be given and assume that Assumption A holds. Then, the trivial solution, $w^i \equiv 0$, of system* (8.4) *is robustly globally exponentially stable if the following inequality holds*

$$\ln \alpha_{ik} - \nu_i(t_k - t_{k-1}) \leq 0, \quad k \in \mathbb{N}, \tag{8.6}$$

where $\alpha_{ik} = \frac{\lambda_{\max}[(I+C_{ik})^T P^i (I+C_{ik})]}{\lambda_{\min}(P^i)}$, *with P^i being a positive-definite matrix satisfying the Riccati-like equation*

$$(A^i + B^i K^i)^T P^i + P^i(A^i + B^i K^i) + \epsilon_{1i} P^i D^i {D^i}^T P^i + \frac{1}{\epsilon_{1i}} {H^i}^T H^i + \xi_{1i} {P^i}^2$$
$$+ \frac{\delta_i I}{\xi_{1i}} - \sigma_i P^i = 0, \tag{8.7}$$

where ϵ_{1i} and ξ_{1i} are any positive constants, $0 < \nu_i < -\sigma_i$, $\sigma_i < 0$ and δ_i is a positive constant such that

$$||f_i(w^i)||^2 \leq \delta_i ||w^i||^2. \tag{8.8}$$

Proof For all $t \geq t_0$ with $t_0 \in \mathbb{R}_+$, let $w^i(t) = w^i(t; t_0, w^i_0)$ be the solution of the ith isolated system in (8.4). For $i = 1, 2, \ldots, l$, define $V^i(w^i) = {w^i}^T P^i w^i$ as a Lyapunov function candidate for the ith subsystem. Then

$$\begin{aligned}\dot{V}^i(w^i) &= {w^i}^T\Big[(A^i + B^iK^i)^TP^i + P^i(A^i + B^iK^i)\Big]w^i + 2{w^i}^TP^i(\Delta A^i)w^i \\ &\quad + 2{w^i}^TP^if_i(w^i) \\ &\le {w^i}^T\Big[(A^i + B^iK^i)^TP^i + P^i(A^i + B^iK^i) + \epsilon_{1i}P^iD^i{D^i}^TP^i \\ &\quad + \frac{1}{\epsilon_{1i}}{H^i}^TH^i + \xi_{1i}{P^i}^2 + \frac{\delta_i I}{\xi_{1i}}\Big]w^i \\ &= \sigma_i V^i(w^i),\end{aligned}$$

where we used (8.8) and Lemma 8.1 in the second bottom line and condition (8.7) in the last line. Then, for all $t \in (t_{k-1}, t_k]$ with $k \in \mathbb{N}$, one may have

$$V^i(w^i(t)) \le V^i(w^i(t_{k-1}^+))e^{\sigma_i(t-t_{k-1})}. \tag{8.9}$$

At $t = t_k^+$, we have

$$\begin{aligned}V^i(w^i(t_k^+)) &\le \lambda_{\max}(L_{ik}){w^i}^T(t_k)w^i(t_k) &(8.10)\\ &\le \alpha_{ik}V^i(w^i(t_k^-)), &(8.11)\end{aligned}$$

where $\alpha_{ik} = \frac{\lambda_{\max}(L_{ik})}{\lambda_{\min}(P^i)}$ and $L_{ik} = [I + C_{ik}]^TP^i[I + C_{ik}]$. From (8.9) and (8.10), we have for $t \in [t_0, t_1]$,

$$V^i(w^i(t)) \le V^i(w_0^i)e^{\sigma_i(t-t_0)},$$

and for $t \in (t_1, t_2]$

$$V^i(w^i(t_1^+)) \le \alpha_{i1}V^i(w_0^i)e^{\sigma_i(t_1-t_0)},$$

$$V^i(w^i(t)) \le V^i(w^i(t_1^+))e^{\sigma_i(t-t_1)},$$

which leads to

$$\begin{aligned}V^i(w^i(t)) &\le \alpha_{i1}V^i(w_0^i)e^{\sigma_i(t_1-t_0)}e^{\sigma_i(t-t_1)} \\ &= \alpha_{i1}V^i(w_0^i)e^{\sigma_i(t-t_0)}, \quad \text{for } t \in [t_0, t_2].\end{aligned}$$

Generally, for $t \in (t_{k-1}, t_k]$, we have

$$\begin{aligned}V^i(w^i(t)) &\le V^i(w_0^i)\,\alpha_{i1}\,\alpha_{i2}\,\cdots\,\alpha_{ik}\,e^{\sigma_i(t-t_0)} \\ &= V^i(w_0^i)\,\alpha_{i1}e^{-\nu_i(t_1-t_0)}\,\cdots\,\alpha_{ik}e^{-\nu_i(t_k-t_{k-1})}e^{(\sigma_i+\nu_i)(t-t_0)} \\ &\le V^i(w_0^i)e^{(\sigma_i+\nu_i)(t-t_0)}, \quad t \ge t_0,\end{aligned}$$

where $0 < \nu_i < -\sigma_i$ and we used condition (8.6) to get the last inequality. The foregoing inequality implies that

$$||w^i|| \le \gamma_i||w_0^i||e^{(\sigma_i+\nu_i)(t-t_0)/2}, \quad t \ge t_0,$$

where $\gamma_i = \sqrt{\frac{\lambda_{\max}(P^i)}{\lambda_{\min}(P^i)}}$. This completes the proof of global exponential stability of trivial solution $w^i \equiv 0$.

Remark 8.1 Theorem 8.1 provides sufficient conditions to ensure robust global exponential stability for each isolated impulsive subsystem (8.4). The time between impulses has to be bounded. This condition is summarized in (8.6). The nonlinearity is assumed to be bounded by some linear growth bound. Condition (8.7) guarantees that the Lyapunov function be decreasing along the trajectory of system (8.4); that is, the continuous system is stabilized by the feedback controller.

The following theorem gives sufficient conditions to guarantee robust global exponential stability for each isolated impulsive subsystems when some control actuators experience failure.

Theorem 8.2 *The trivial solution, $w^i \equiv 0$, of system (8.5) is robustly globally exponentially stable if Assumption A and condition (8.6) hold with P^i being a positive-definite matrix satisfying the Riccati-like equation*

$$A^{i^T}P^i + P^iA^i + P^i(\epsilon_{1i}D^iD^{i^T} - \epsilon_{2i}B^i_{\bar{\Sigma}}B^{i^T}_{\bar{\Sigma}} + \xi_{1i}I)P^i + \frac{1}{\epsilon_{1i}}H^{i^T}H^i + \frac{\delta_i I}{\xi_{1i}} - \sigma_i P^i = 0 \tag{8.12}$$

where ϵ_{1i}, ϵ_{2i}, and ξ_{1i} are positive constants, $0 < \nu_i < -\sigma_i$ with $\sigma_i < 0$, δ_i is a positive constant such that condition (8.8) holds and the control gain is given by $K^i = -\frac{1}{2}\epsilon_{2i}B^{i^T}_{\bar{\sigma}}P^i$.

Proof For all $t \geq t_0$ with $t_0 \in \mathbb{R}_+$, let $w^i(t) = w^i(t; t_0, w^i_0)$ be the solution of system (8.5). Define $V^i(w^i) = w^{i^T}P^iw^i$ as the Lyapunov candidate. Then

$$\begin{aligned}
\dot{V}^i(w^i) &\leq w^{i^T}\Big[A^{i^T}P^i + P^iA^i + P^i(\epsilon_{1i}D^iD^{i^T} - \epsilon_{2i}B^i_{\bar{\sigma}}B^{i^T}_{\bar{\sigma}} + \xi_{1i}I)P^i \\
&\quad + \frac{1}{\epsilon_{1i}}H^{i^T}H^i + \frac{\delta_i I}{\xi_{1i}}\Big]w^i \\
&\leq w^{i^T}\Big[A^{i^T}P^i + P^iA^i + P^i(\epsilon_{1i}D^iD^{i^T} - \epsilon_{2i}B^i_{\bar{\Sigma}}B^{i^T}_{\bar{\Sigma}} + \xi_{1i}I)P^i \\
&\quad + \frac{1}{\epsilon_{1i}}H^{i^T}H^i + \frac{\delta_i I}{\xi_{1i}}\Big]w^i \\
&= \sigma_i V^i(w^i(t)),
\end{aligned}$$

where we used the fact that [1] $B^i_{\bar{\Sigma}}B^{i^T}_{\bar{\Sigma}} \leq B^i_{\bar{\sigma}}(B^i_{\bar{\sigma}})^T$ in the second last line and condition (8.12) in the last line. As done in the last theorem, we can show that the trivial solution, $w^i \equiv 0$, of the closed-loop impulsive system (8.5) is robustly globally exponentially stable.

Having proved the stabilizability of the isolated subsystem in Theorems 8.1 and 8.2, we consider the same properties for the interconnected systems.

Definition 8.2 System (8.4) (or (8.5)) is said to possess property A (or B) if it satisfies the conditions in Theorem 8.1 (or Theorem 8.2).

Remark 8.2 Property A implies that all the impulsive isolated subsystems are robustly globally exponentially stable in the normal actuators case, while Property B implies the same result is held in the faulty case.

Theorem 8.3 *Assume that system* (8.4) *possesses property A. Suppose further that, for any $i, j = 1, 2, \ldots, l$, there exist positive constants b_{ij} such that*

$$2{w^i}^T P^i g^i(w^1, w^2, \cdots, w^i, \cdots, w^l) \leq ||w^i|| \Sigma_{j=1}^l b_{ij} ||w^j|| \tag{8.13}$$

and the test matrix $\mathscr{S} = [s_{ij}]_{l\times l}$ is negative definite where

$$s_{ij} = \begin{cases} \beta_i(\sigma_i^* + b_{ii}), & i = j \\ \frac{1}{2}(\beta_i b_{ij} + \beta_j b_{ji}), & i \neq j \end{cases} \tag{8.14}$$

for some constant $\sigma_i^* = \sigma_i \lambda_{\max}(P^i) < 0$ and positive constant β_i. Then, the trivial solution of system (8.2) *is robustly globally exponentially stable if the following inequality holds*

$$\ln \alpha_k - \phi(t_k - t_{k-1}) \leq 0, \quad k \in \mathbb{N}, \tag{8.15}$$

for $0 < \phi < \theta$ where $\theta = -\frac{\lambda_{\max}(\mathscr{S})}{\bar{\lambda}\beta^*}$ with $\bar{\lambda} = \min\{\lambda_{\max}(P^i) \mid i = 1, 2, \ldots, l\}$ and $\beta^* = \min\{\beta_i \mid i = 1, 2, \ldots, l\}$, $\alpha_k = \big[\max\{\lambda_{\max}[(I + C_{ik})^T P^i (I + C_{ik})] \mid i = 1, 2, \ldots, l\}\big]/\lambda^*$ with $\lambda^* = \min\{\lambda_{\min}(P^i) \mid i = 1, 2, \ldots, l\}$ and P^i being a positive-definite matrix defined in Property A.

Proof For all $t \geq t_0$ with $t_0 \in \mathbb{R}$, let $x(t) = x(t; t_0, x_0)$ be the solution of system (8.2). Define the composite Lyapunov function $V(x(t)) = \Sigma_{i=1}^l \beta^i V^i(w^i)$ as a Lyapunov function candidate for interconnected system (8.2) where β^i is a positive constant and $V^i(w^i)$ is a Lyapunov function for the ith isolated subsystem. Then, along the trajectory of (8.2), we have

$$\begin{aligned} \dot{V}(x) = \Sigma_{i=1}^l \beta^i \dot{V}^i(w^i) &\leq \Sigma_{i=1}^l \beta^i \{\sigma_i ||w^i||^2 + 2{w^i}^T P^i g^i(w^1, w^2, \cdots, w^i, \cdots, w^l)\} \\ &\leq \Sigma_{i=1}^l \beta^i \{\sigma_i ||w^i||^2 + ||w^i|| \Sigma_{j=1}^l b_{ij} ||w^j||\} = z^T \mathscr{S} z, \end{aligned}$$

where $z^T = \big(||w^1||\ ||w^2|| \ \cdots\ ||w^i|| \ \cdots\ ||w^l||\big)$ and $\mathscr{S}$ is a negative-definite matrix with the maximum eigenvalue $\lambda_{\max}(\mathscr{S})$. Then, one can write

$$\dot{V}(x) \leq -\theta V(x),$$

where $\theta = -\frac{\lambda_{\max}(\mathscr{L})}{\bar{\lambda}\beta^*}$ with $\bar{\lambda} = \min\{\lambda_{\max}(P^i) \mid i = 1, 2, \ldots, l\}$ and $\beta^* = \min\{\beta_i \mid i = 1, 2, \ldots, l\}$. The last inequality implies that, for all $t \in (t_{k-1}, t_k]$,

$$V(x(t)) \leq V(x(t_{k-1}^+))e^{-\theta(t-t_{k-1})} \tag{8.16}$$

and, at $t = t_k^+$,

$$\begin{aligned} V(x(t_k^+)) &= \Sigma_{i=1}^l \beta_i w^{i^T}(t_k)[(I + C_{ik})^T P^i (I + C_{ik})]w^i(t_k) \\ &\leq \frac{L^{**}}{\lambda^*} \Sigma_{i=1}^l \beta^i V^i(w^i) \\ &= \alpha_k V(x(t)), \end{aligned} \tag{8.17}$$

where $\alpha_k = \frac{L^{**}}{\lambda^*}$, $L^{**} = \max\{\lambda_{\max}(L^i) \mid i = 1, \cdots, l\}$ and $\lambda^* = \min\{\lambda_{\min}(P^i) \mid i = 1, \cdots, l\}$. From (8.16) and (8.17), we have for $t \in [t_0, t_1]$,

$$V(x(t)) \leq V(x_0)e^{-\theta(t-t_0)}$$

and for $t \in (t_1, t_2]$, we have

$$V(x(t_1^+)) \leq \alpha_1 V(x(t_1)) \leq \alpha_1 V(x_0)e^{-\theta(t_1-t_0)}$$

and

$$V(x(t)) \leq V(x(t_1^+))e^{-\theta(t-t_1)} \leq \alpha_1 V(x_0)e^{-\theta(t_1-t_0)}e^{-\theta(t-t_1)};$$

that is

$$V(x(t)) \leq \alpha_1 V(x_0)e^{-\theta(t-t_0)}, \quad t \in [t_0, t_2].$$

Therefore, for all $t \in (t_{k-1}, t_k]$,

$$\begin{aligned} V(x(t)) \leq& V(x_0)\, \alpha_1\, \alpha_2 \cdots \alpha_k\, e^{-\theta(t-t_0)} \\ &\leq V(x_0)\, \alpha_1 e^{-\phi(t_1-t_0)}\, \alpha_2 e^{-\phi(t_2-t_1)} \cdots \alpha_k\, e^{-\phi(t_k-t_{k-1})} e^{-(\theta-\phi)(t-t_0)} \\ &\leq V(x_0)\, e^{-(\theta-\phi)(t-t_0)}, \quad t \geq t_0, \end{aligned} \tag{8.18}$$

where $0 < \phi < \theta$. The forgoing inequality together with

$$C^*||x||^2 \leq V(x) \leq C^{**}||x||^2,$$

where $C^* = \lambda^*\beta^*$ and $C^{**} = \lambda^{**}\beta^{**}$ with $\lambda^{**} = \max\{\lambda_{\max}(P^i) \mid i = 1, \cdots, l\}$ and $\beta^{**} = \max\{\beta_i \mid i = 1, \cdots, l\}$ implies that

$$||x(t)|| \leq E||x_0||\, e^{-(\theta-\phi)(t-t_0)/2}, \quad \forall t \geq t_0,$$

where $E = \sqrt{\frac{C^{**}}{C^{*}}}$. That is, the trivial solution, $x \equiv 0$, of the composite system (8.2) is robustly globally exponentially stable.

Remark 8.3 Theorem 8.3 shows that the interconnected system can be robustly exponentially stabilized by the sub-controllers of the isolated subsystems in the case where all the actuators are operational. Condition (8.13) estimates the interconnection, which is viewed as a perturbation, by an upper bound. The test matrix $\mathscr{S}$ is needed to guarantee that the degree of stability be greater than the interconnection.

The following theorem shows that the proposed reliable controllers are robust even in the presence of the interconnection effect. The proof is similar to that of Theorem 8.3; thus, it is left here as an exercise.

Theorem 8.4 *Assume that system* (8.5) *possesses property B and suppose that for any $i, j = 1, 2, \ldots, l$, there exist positive constants b_{ij} such that the condition in (8.13) holds, the test matrix $\mathscr{S} = [s_{ij}]_{l \times l}$ defined in Theorem 8.3 is negative definite and ϵ_{2i} is a positive constant such that $K^i = -\frac{1}{2}\epsilon_{2i} B_{\bar{\sigma}}^{i\;T} P^i$. Then, the trivial solution, $x \equiv 0$, of system* (8.2) *is robustly globally exponentially stable if* (8.15) *holds with P^i being a positive-definite matrix defined in Property B.*

Example 8.1 Consider the composite system with $l = 2$ and the following information for the subsystems

$$A^1 = \begin{bmatrix} 0 & 1 \\ -11 & 0 \end{bmatrix}, A^2 = \begin{bmatrix} 0 & 1 \\ -10 & 0 \end{bmatrix}, B^1 = \begin{bmatrix} -5 & 3 \\ -1 & 2 \end{bmatrix}, B^2 = \begin{bmatrix} 1 & -3 \\ 0.1 & -4 \end{bmatrix},$$

$$D^1 = \begin{bmatrix} 1 \\ 0 \end{bmatrix}, D^2 = \begin{bmatrix} 0 \\ 1 \end{bmatrix}, H^1 = \begin{bmatrix} 0 & 1 \end{bmatrix}, H^2 = \begin{bmatrix} 1 & 0 \end{bmatrix}, \mathscr{U}_1 = \mathscr{U}_2 = \sin(t),$$

$$f_1 = 0.5\begin{bmatrix} 0 \\ \sin(w_2) \end{bmatrix}, f_2 = 1.5\begin{bmatrix} 0 \\ \sin(w_4) \end{bmatrix}, C_{1k} = \begin{bmatrix} 2 & 0 \\ 0 & 2 \end{bmatrix}, C_{2k} = \begin{bmatrix} 3 & 0 \\ 0 & 3 \end{bmatrix},$$

for all $k = 1, 2, \ldots$, $\sigma_1 = -2$, $\sigma_2 = -2.5$, $\epsilon_{11} = 2$, $\epsilon_{12} = 0.5$, $\xi_{11} = 1$, $\xi_{12} = 1$, $\epsilon_{21} = 1$, $\epsilon_{22} = 0.7$, $\beta_1 = 1$, $\beta_2 = 2$, $b_{11} = 0.3$, $b_{22} = 1.5$, $b_{12} = 0.5$, $b_{21} = 0.3$ and $t_0 = 0$. From (8.8), one may get $\delta_1 = 0.25$ and $\delta_2 = 2.25$.

Cases 1 When all the control actuators are operational, we have from Riccati-like equation,

$$P^1 = \begin{bmatrix} 0.5427 & -0.2419 \\ -0.2419 & 0.1955 \end{bmatrix} \text{and } P^2 = \begin{bmatrix} 2.9461 & -1.2229 \\ -1.2229 & 0.7834 \end{bmatrix}$$

with $\lambda_{\min}(P^1) = 0.0713$, $\lambda_{\max}(P^1) = 0.6669$, $\lambda_{\min}(P^2) = 0.2323$ and $\lambda_{\max}(P^2) = 3.4971$, so that $\lambda^* = 0.0713$, $\lambda^{**} = 3.4971$ and the control gain matrices are

$$K^1 = \begin{bmatrix} 1.2358 & -0.5071 \\ -0.5722 & 0.1674 \end{bmatrix} \text{ and } K^2 = \begin{bmatrix} -0.9883 & 0.4006 \\ 1.3814 & -0.1873 \end{bmatrix}.$$

Thus, $A^i + B^i K^i$ (for $i = 1, 2$) are Hurwitz, and the time intervals $t_k - t_{k-1} \geq 2.3328$ for the first subsystem and $t_k - t_{k-1} \geq 2.7421$ for the second subsystem. The test matrix here is given by

$$\mathscr{S} = \begin{bmatrix} -1.0338 & 0.55 \\ 0.55 & -14.4855 \end{bmatrix},$$

which is negative-definite and $t_k - t_{k-1} \geq 4.4142$ for the interconnected system.

Cases 2 When there is a failure in the second actuator in the first subsystem and first actuator in the second subsystem, i.e., $\Sigma^1 = \{2\}$ and $B^1_{\bar{\Sigma}} = \begin{bmatrix} -5 & 0 \\ -1 & 0 \end{bmatrix}$ and $\Sigma^2 = \{1\}$ and $B^2_{\bar{\Sigma}} = \begin{bmatrix} 0 & -3 \\ 0 & -4 \end{bmatrix}$, we have from Riccati-like equation,

$$P^1 = \begin{bmatrix} 0.5806 & -0.2330 \\ -0.2330 & 0.2008 \end{bmatrix} \text{ and } P^2 = \begin{bmatrix} 3.0616 & -1.2448 \\ -1.2448 & 0.7834 \end{bmatrix},$$

with $\lambda_{\min}(P^1) = 0.0901$, $\lambda_{\max}(P^1) = 0.6913$, $\lambda_{\min}(P^2) = 0.2351$ and $\lambda_{\max}(P^2) = 3.6099$, so $\lambda^* = 0.0901$, $\lambda^{**} = 3.6099$, and the control gain matrices are

$$K^1 = \begin{bmatrix} 1.3351 & -0.4820 \\ 0 & 0 \end{bmatrix} \text{ and } K^2 = \begin{bmatrix} 0 & 0 \\ 1.4719 & -0.2103 \end{bmatrix}.$$

Thus, $A^i + B^i_{\bar{\sigma}} K^i$ (for $i = 1, 2$) are Hurwitz and the time intervals $t_k - t_{k-1} \geq 2.2286$ for the first subsystem and $t_k - t_{k-1} \geq 2.7519$ for the second subsystem.

The interconnected system is shown in Figs. 8.1 and 8.2 for the operational and faulty cases, respectively.

If we consider $f_1^T = 0.5[w_1 \ (w_2)^2]$ and $f_2^T = 1.5[w_3 \ (w_4)^2]$, one can show that condition (8.8) is satisfied only inside the region $\mathscr{D} = \{(w_1 \ w_2 \ w_3 \ w_4)^T \in \mathbb{R}^4 \mid w_1 \in \mathbb{R}, \ -2 \leq w_2 \leq 2, \ w_3 \in \mathbb{R}, \ -1.5 \leq w_4 \leq 1.5\}$. Thus, $x \equiv 0$ is locally exponentially stable. The local stability and instability of the trivial solution are shown in Figs. 8.3 and 8.4, respectively.

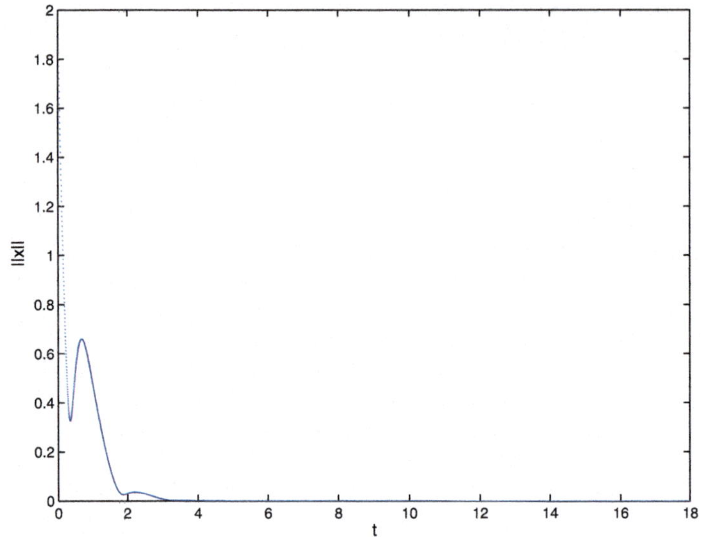

Fig. 8.1 Interconnected system: Operational actuators

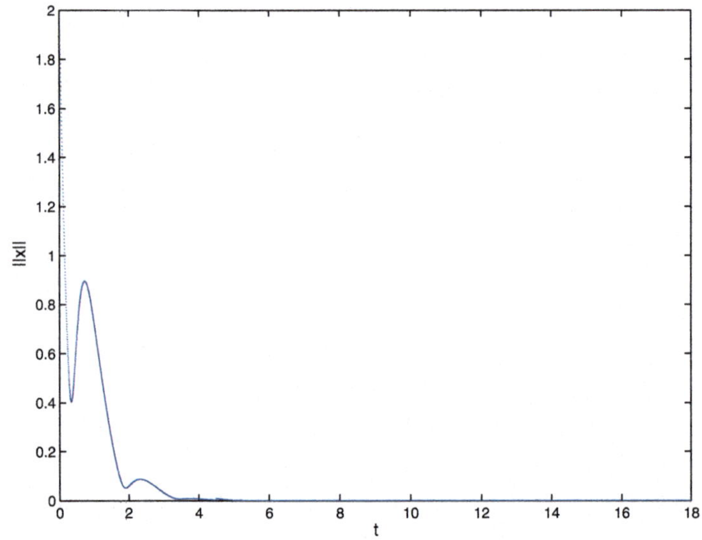

Fig. 8.2 Interconnected system: Faulty actuators

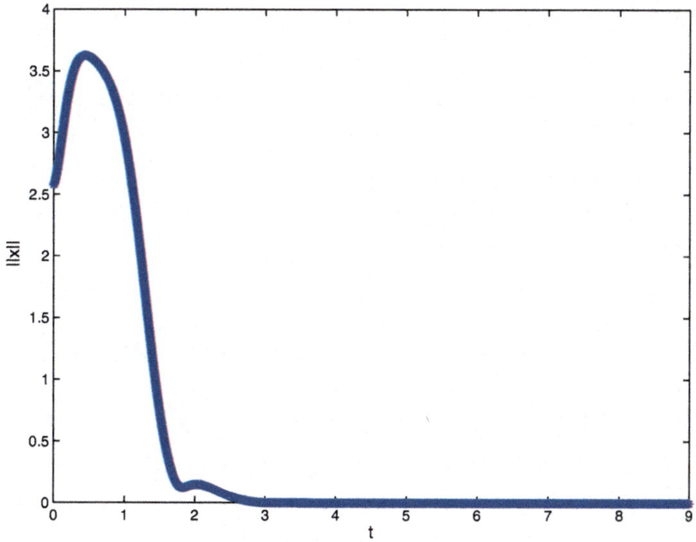

Fig. 8.3 Local stability of the trivial solution

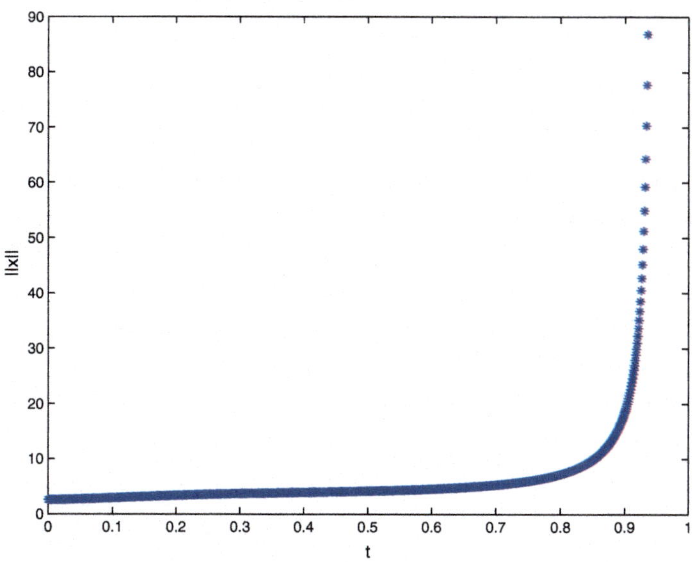

Fig. 8.4 Instability of the trivial solution

8.3 State Estimation

In this section, we address the problem of state estimation for the ILSS. Consider the isolated input–output impulsive subsystem

$$\begin{cases} \dot{w}^i = (A^i + \Delta A^i)w^i + B^i u^i + f_i(w^i), \; t \neq t_k, & k \in \mathbb{N}, \\ \Delta w^i(t) = C_{ik} w^i(t^-), & t = t_k, \\ y^i(t) = C^i w^i(t), \\ w^i(t_0) = w_0^i, \end{cases} \tag{8.19}$$

where $y^i \in \mathbb{R}^{n_i}$ is the measured output vector. Define the Luenberger observer by

$$\begin{cases} \dot{\hat{w}}^i = (A^i + \Delta A^i)\hat{w}^i + B^i u^i + f_i(\hat{w}^i) + \mathscr{L}^i(y^i - C^i \hat{w}^i), \; t \neq t_k, & k \in \mathbb{N}, \\ \Delta \hat{w}^i(t) = C_{ik} \hat{w}^i(t^-), & t = t_k, \\ \hat{w}^i(t_0) = \hat{w}_0^i, \end{cases} \tag{8.20}$$

where $\mathscr{L}^i \in \mathbb{R}^{n_i \times n_i}$ is the observer gain matrix. Define the state estimation error vector by $e^i = w^i - \hat{w}^i$. Then, the closed-loop error system becomes

$$\begin{cases} \dot{e}^i = (A^i + \Delta A^i - \mathscr{L}^i C^i)e^i + f_i(w^i) - f_i(\hat{w}^i), \; t \neq t_k, & k \in \mathbb{N}, \\ \Delta e^i(t) = C_{ik} e^i(t^-), & t = t_k, \\ e^i(t_0) = w_0^i - \hat{w}_0^i =: e_0^i \end{cases} \tag{8.21}$$

Definition 8.3 The pair (A, B) is said to be detectable if there exists a matrix F such that $A - FB$ is Hurwitz.

We will adopt the same stability/stabilization analysis followed in the last section to establish the observability problem of system (8.19).

Theorem 8.5 *Let the observer gain matrix, $\mathscr{L}^i$, be given, Assumption A hold and the matrix pair (A^i, C^i) be detectable. Then, the trivial solution of error system* (8.21) *is robustly globally exponentially stable if the following inequality holds*

$$\ln \alpha_{ik} - \nu_i (t_k - t_{k-1}) \leq 0, \quad k \in \mathbb{N}, \tag{8.22}$$

where $0 < \nu_i < -\sigma_i, \sigma_i < 0$ and $\alpha_{ik} = \frac{\lambda_{\max}[(I+C_{ik})^T P^i (I+C_{ik})]}{\lambda_{\min}(P^i)}$ with P^i being a positive-definite matrix satisfying the Riccati-like equation

$$(A^i - \mathscr{L}^i C^i)^T P^i + P^i (A^i - \mathscr{L}^i C^i) + \epsilon_{1i} P^i D^i D^{i^T} P^i + \frac{1}{\epsilon_{1i}} H^{i^T} H^i + a^i I - \sigma_i P^i = 0, \tag{8.23}$$

where ϵ_{1i} is any positive constants and $a^i > 0$ such that

$$2e^{i^T} P^i[f_i(w^i) - f_i(\hat{w}^i)] \leq a^i ||e^i||^2. \tag{8.24}$$

Proof For all $t \geq t_0$ with $t_0 \in \mathbb{R}_+$, let $e^i(t) = e^i(t; t_0, e^i_0)$ be the solution of the error system (8.21). For $i = 1, 2, \ldots, l$, define $V^i(e^i) = e^{i^T} P^i e^i$ as the Lyapunov function candidate for the ith subsystem. Then,

$$\begin{aligned}\dot{V}^i(e^i) &= e^{i^T}[(A^i - \mathscr{L}^i C^i)^T P^i + P^i(A^i - \mathscr{L}^i C^i)]e^i + 2e^{i^T} P^i \Delta A^i e^i + 2e^{i^T} P^i f_i e^i \\ &\leq e^{i^T}[(A^i - \mathscr{L}^i C^i)^T P^i + P^i(A^i - \mathscr{L}^i C^i) + \epsilon_{1i} P^i D^i D^{i^T} P^i + \frac{1}{\epsilon_{1i}} H^{i^T} H^i \\ &\quad + a^i I]e^i = \sigma_i V^i(e^i),\end{aligned}$$

where we used (8.24) and Lemma 8.1 in the second bottom line and (8.23) in the last line. The last inequality implies that, for all $t \in (t_{k-1}, t_k]$ with $k \in \mathbb{N}$,

$$V^i(e^i(t)) \leq V^i(e^i(t^+_{k-1}))e^{\sigma_i(t-t_{k-1})} \tag{8.25}$$

and, at $t = t_k$, we have

$$V^i(e^i(t^+_k)) \leq \alpha_{ik} V^i(e^i(t^-_k)), \tag{8.26}$$

where $\alpha_{ik} = \lambda_{\max}(L_{ik})/\lambda_{\min}(P^i)$ with $L_{ik} = [I + C_{ik}]^T P^i [I + C_{ik}]$. From (8.22), (8.25) and (8.26), we have for $t \geq t_0$,

$$V^i(e^i(t)) \leq V^i(e^i_0)e^{(\sigma_i+\nu_i)(t-t_0)},$$

where $0 < \nu_i < -\sigma_i$. The last inequality implies that

$$||e^i|| \leq \gamma_i ||e^i_0|| e^{-(\xi_i-\nu_i)(t-t_0)/2}, \quad t \geq t_0,$$

where $\gamma_i = \sqrt{\frac{\lambda_{\max}(P^i)}{\lambda_{\min}(P^i)}}$. Then, the trivial solution is globally exponentially stable. This completes the proof.

As done in the analysis of reliable stabilization, for $i = 1, 2, \ldots, l$, consider the decomposition of the observer matrix $C^i = C^i_\Omega + C^i_{\bar{\Omega}}$ where C^i_Ω and $C^i_{\bar{\Omega}}$ are the observer matrices associated with Ω and $\bar{\Omega}$, respectively, and C^i_Ω and $C^i_{\bar{\Omega}}$ are generated by zeroing out the columns corresponding to $\bar{\Omega}$ and Ω, respectively. For any fixed $i \in \{1, 2, \ldots, l\}$, let $\omega \subseteq \Omega$ correspond to some of the sensors that experience failure and assume that the output of faulty sensors is zero. Then, the decomposition becomes $C^i = C^i_\omega + C^i_{\bar{\omega}}$ where C^i_ω and $C^i_{\bar{\omega}}$ have the same definition of C^i_Ω and $C^i_{\bar{\Omega}}$, respectively. Then, the closed-loop impulsive error system for the faulty case becomes

$$\begin{cases} \dot{e}^i = (A^i + \Delta A^i - \mathscr{L}^i C^i_{\bar{\omega}})e^i + f_i(w^i) - f_i(\hat{w}^i), & t \neq t_k,\ k \in \mathbb{N} \\ \Delta e^i(t) = C_{ik}e^i(t^-), & t = t_k, \\ e^i(t_0) = w^i_0 - \hat{w}^i_0 =: e^i_0 \end{cases} \tag{8.27}$$

In the following theorem, we state and prove the robust global exponential stability for all isolated impulsive subsystems when some control components experience failure.

Theorem 8.6 *The trivial solution of system* (8.27) *is robustly globally exponentially stable if Assumption A holds, the matrix pair* $(A^i, C^i_{\bar{\omega}})$ *is detectable and condition* (8.22) *holds with* P^i *being a positive-definite matrix satisfying the Riccati-like equation*

$$A^{i^T} P^i + P^i A^i + P^i \left[\epsilon_{1i} D^i D^{i^T} - \epsilon_{2i} C^i_{\bar{\Omega}} C^{i^T}_{\bar{\Omega}}\right] P^i + \frac{1}{\epsilon_{1i}} H^{i^T} H^i + a^i I - \sigma_i P^i = 0, \tag{8.28}$$

where ϵ_{1i} *and* ϵ_{2i} *are positive constants such that the observer gain matrix* $\mathscr{L}^i = \frac{1}{2}\epsilon_{2i} C^{i^T}_{\bar{\omega}} P^i$, $0 < \nu_i < -\sigma_i$ *with* $\sigma_i < 0$, *the matrices* P^i *and* $C^i_{\bar{\omega}}$ *are commutative and* $a^i > 0$ *such that* (8.24) *holds.*

Proof For all $t \geq t_0$ with $t_0 \in \mathbb{R}_+$ and $i = 1, 2, \ldots, l$, let $e^i(t) = e^i(t; t_0, e^i_0)$ be the solution of system (8.27). Define $V^i(e^i) = (e^i)^T P^i e^i$ as the Lyapunov function candidate for the ith subsystem. Then, one may have

$$\begin{aligned} \dot{V}^i(e^i) &\leq e^{i^T}\Big[(A^i - \mathscr{L}^i C^i)^T P^i + P^i(A^i - \mathscr{L}^i C^i) + \epsilon_{1i} P^i D^i D^{i^T} P^i \\ &\quad + \frac{1}{\epsilon_{1i}} H^{i^T} H^i + a^i I\Big] e^i \\ &\leq e^{i^T}\Big[A^{i^T} P^i + P^i A^i + P^i(\epsilon_{1i} D^i D^{i^T} + \epsilon_{2i} C^i_{\bar{\omega}} C^{i^T}_{\bar{\omega}})P^i + \frac{1}{\epsilon_{1i}} H^{i^T} H^i \\ &\quad + a^i I\Big] e^i \\ &\leq e^{i^T}\Big[A^{i^T} P^i + P^i A^i + P^i(\epsilon_{1i} D^i D^{i^T} - \epsilon_{2i} C^i_{\bar{\Omega}} C^{i^T}_{\bar{\Omega}})P^i + \frac{1}{\epsilon_{1i}} H^{i^T} H^i \\ &\quad + a^i I\Big] e^i = \sigma_i V^i(e^i(t)), \end{aligned}$$

where we used the fact that [1] $C^i_{\bar{\Omega}}(C^i_{\bar{\Omega}})^T \leq C^i_{\bar{\omega}}(C^i_{\bar{\omega}})^T$ in the second last line and condition (8.28) in the last line. Finally, following the analysis used in the previous theorem shows that the trivial solution of the closed-loop impulsive error system in (8.27) is robustly globally exponentially stable.

Definition 8.4 System (8.21)(or (8.27)) is said to possess property C(or D) if it satisfies the conditions in Theorem 8.5 (or Theorem 8.6), respectively.

Remark 8.4 Property C implies that all the impulsive error isolated subsystems are robustly globally exponentially stable in the normal actuators case, while Property D implies the same result is held in the faulty case.

Considering the interconnection g^i in system (8.19) results in the composite system

$$\begin{cases} \dot{w}^i = (A^i + \Delta A^i)w^i + B^i u^i + f_i(w^i) \\ \qquad +g^i(w^1, w^2, \cdots, w^l), & t \neq t_k, \\ \Delta w^i(t) = C_{ik} w^i(t^-), & t = t_k,\ k \in \mathbb{N}, \\ y^i(t) = C^i w^i(t), \\ w^i(t_0) = w_0^i. \end{cases} \tag{8.29}$$

Similarly, we define the response system as follows

$$\begin{cases} \dot{\hat{w}}^i = (A^i + \Delta A^i)\hat{w}^i + B^i u^i + f_i(\hat{w}^i) \\ \qquad +g^i(\hat{w}^1, \hat{w}^2, \cdots, \hat{w}^l) + \mathscr{L}^i(y^i - C^i\hat{w}^i), & t \neq t_k, \\ \Delta \hat{w}^i(t) = C_{ik}\hat{w}^i(t^-), & t = t_k,\ k \in \mathbb{N}, \\ \hat{w}^i(t_0) = \hat{w}_0^i. \end{cases} \tag{8.30}$$

Then, the closed-loop error system becomes

$$\begin{cases} \dot{e}^i = (A^i + \Delta A^i - \mathscr{L}^i C^i)e^i + f_i(w^i) - f_i(\hat{w}^i) \\ \qquad +g^i(w^1, w^2, \cdots, w^l) - g^i(\hat{w}^1, \hat{w}^2, \cdots, \hat{w}^l), & t \neq t_k, \\ \Delta e^i(t) = C_{ik}e^i(t^-), & t = t_k,\ k \in \mathbb{N}, \\ e^i(t_0) = w_0^i - \hat{w}_0^i =: e_0^i. \end{cases} \tag{8.31}$$

This system can be re-written in the following form

$$\begin{cases} \dot{e}_c = (A + \Delta A - \mathscr{L}C)e_c + F(x) - F(\hat{x}) \\ \qquad +G(x) - G(\hat{x}), & t \neq t_k,\ k \in \mathbb{N}, \\ \Delta e_c(t) = I_k(e_c(t^-)) = C_k e_c(t^-), & t = t_k, \\ e_c(t_0) = e_{c0}, \end{cases} \tag{8.32}$$

where $x^T = ({w^1}^T\ {w^2}^T\ \cdots\ {w^l}^T)$, $\hat{x}^T = \Big((\hat{w}^1)^T\ (\hat{w}^2)^T\ \cdots\ (\hat{w}^l)^T\Big)$,
$e_c^T = ({e^1}^T\ {e^2}^T\ \cdots\ {e^l}^T)$,
$((A + \Delta A - \mathscr{L}C)e_c)^T = \Big[\big[(A^1 + \Delta A^1 - \mathscr{L}^1C^1)e^1\big]^T \quad \big[(A^2 + \Delta A^2 - \mathscr{L}^2C^2) e^2\big]^T \cdots \big[(A^l + \Delta A^l - \mathscr{L}^lC^l)e^l\big]^T\Big]$,
$(F(x))^T = \big({f_1}^T(w^1)\ {f_2}^T(w^2)\ \cdots\ {f_l}^T(w^l)\big)$,
$(F(\hat{x}))^T = \big({f_1}^T(\hat{w}^1)\ {f_2}^T(\hat{w}^2)\ \cdots\ {f_l}^T(\hat{w}^l)\big)$
$(G(x))^T = \Big({g^1}^T(x)\ {g^2}^T(x)\ \cdots\ {g^l}^T(x)\Big)$, $(G(\hat{x}))^T = \Big({g^1}^T(\hat{x})\ {g^2}^T(\hat{x})\ \cdots\ {g^l}^T(\hat{x})\Big)$,
$(C_k e_c)^T = \big((C_{1k}e^1)^T\ (C_{2k}e^2)^T\ \cdots\ (C_{lk}e^l)^T\big)$.

Theorem 8.7 *Assume that system* (8.21) *possesses property C and the observer gain matrix* $\mathscr{L}$ *is given. Suppose further that for any* $i, j = 1, 2, \ldots, l$*, there exists a positive constant* b_{ij} *such that*

$$2e^{i^T}P^i[g^i(w^1, w^2, \cdots, w^l) - g^i(\hat{w}^1, \hat{w}^2, \cdots, \hat{w}^l)] \leq ||e^i||\Sigma_{j=1}^{l} b_{ij}||e^j|| \quad (8.33)$$

and the test matrix $S = [s_{ij}]_{l\times l}$ *is negative definite where*

$$s_{ij} = \begin{cases} \beta_i(\sigma_i^* + b_{ii}), & i = j \\ \frac{1}{2}(\beta_i b_{ij} + \beta_j b_{ji}), & i \neq j \end{cases}, \quad (8.34)$$

for some constant $\sigma_i^* = \sigma_i \lambda_{\max}(P^i) < 0$ *and a positive constant* β_i*. Then, the trivial solution of system* (8.32) *is robustly globally exponentially stable if the following inequality holds*

$$\ln \alpha_k - \phi(t_k - t_{k-1}) \leq 0, \; k \in \mathbb{N}, \quad (8.35)$$

for $0 < \phi < \theta$ *where* $\theta = -\frac{\lambda_{\max}(S)}{\bar{\lambda}\beta^*}$ *with* $\bar{\lambda} = \min\{\lambda_{\max}(P^i) \mid i = 1, 2, \ldots, l\}$ *and* $\beta^* = \min\{\beta_i \mid i = 1, 2, \ldots, l\}$, $\alpha_k = \big[\max\{\lambda_{\max}[(I + C_{ik})^T P^i (I + C_{ik})] \mid i = 1, 2, \ldots, l\}\big]/\lambda^*$ *with* $\lambda^* = \min\{\lambda_{\min}(P^i) \mid i = 1, 2, \ldots, l\}$ *and* P^i *being a positive-definite matrix defined in Property C.*

Proof For all $t \geq t_0$ with $t_0 \in \mathbb{R}_+$, let $e_c(t) = e_c(t; t_0, e_{c0})$ be the solution of system (8.32). Define the composite Lyapunov function $V(e_c(t)) = \Sigma_{i=1}^{l}\beta^i V^i(e^i)$ with $V^i(e^i)$ being the Lyapunov function related to the ith isolated subsystem and $\beta^i > 0$. Then, one may get after using property C and (8.33)

$$\begin{aligned}\dot{V}(e_c) &\leq \Sigma_{i=1}^{l}\beta^i\left\{\sigma_i||e^i||^2 + 2e^{i^T}P^i[g^i(w^1, \cdots, w^l) - g^i(\hat{w}^1, \cdots, \hat{w}^l)]\right\} \\ &\leq \Sigma_{i=1}^{l}\beta^i\left\{\sigma_i||e^i||^2 + ||e^i||\Sigma_{j=1}^{l} b_{ij}||e^j||\right\} \\ &= z^T \mathscr{S} z,\end{aligned}$$

where $z^T = \big(||e^1|| \; ||e^2|| \; \cdots \; ||e^l||\big)$. Then, $\dot{V}(e_c) \leq -\theta V(e_c)$ where $\theta = -\frac{\lambda_{\max}(\mathscr{S})}{\bar{\lambda}\beta^*}$ with $\bar{\lambda} = \min\{\lambda_{\max}(P^i) \mid i = 1, 2, \ldots, l\}$ and $\beta^* = \min\{\beta_i \mid i = 1, 2, \ldots, l\}$. The rest of the proof is similar to that of Theorem 8.3; thus, it is left here as an exercise.

In the following theorem, we state that the proposed reliable sensors are also robust in the presence of the interconnection effect. The proof of this theorem can be achieved in the same way we proved the analogue stability theorem in the last section; thus, it is left here as an exercise.

Theorem 8.8 *Assume that system* (8.27) *possesses property D. Suppose further that, for any* $i, j = 1, 2, \ldots, l$*, there exist positive constants* b_{ij} *such that the condition in* (8.33) *holds, the test matrix* $S = [s_{ij}]_{l\times l}$ *defined in Theorem 8.7 is negative-definite,* *the observer gain matrix* $\mathscr{L}^i = \frac{1}{2}\epsilon_{2i} {C^i_{\bar{\omega}}}^T P^i$ *where* ϵ_{2i} *is a positive constant, and* P^i *and* $C^i_{\bar{\omega}}$ *are commutative. Then, the trivial solution of system* (8.32) *is robustly globally exponentially stable if* (8.35) *holds with* P^i *being a positive-definite matrix defined in Property D.*

Example 8.2 Consider the composite system with $l = 2$, where

$$A^1 = \begin{bmatrix} -4 & 0 \\ 0 & 4 \end{bmatrix}, A^2 = \begin{bmatrix} 5 & 0 \\ 0 & -5 \end{bmatrix}, C^1 = \begin{bmatrix} 1.5 & 0 \\ 0 & 1.5 \end{bmatrix}, C^2 = \begin{bmatrix} 3 & 0 \\ 0 & 3 \end{bmatrix},$$

$$D^1 = \begin{bmatrix} 1 \\ 0 \end{bmatrix}, D^2 = \begin{bmatrix} 0 \\ 1 \end{bmatrix}, H^1 = \begin{bmatrix} 0 & 1 \end{bmatrix}, H^2 = \begin{bmatrix} 1 & 0 \end{bmatrix}, \mathscr{U}_1 = \mathscr{U}_2 = \sin(t),$$

$$f_1 = 0.5 \begin{bmatrix} 0 \\ \sin(w^2) \end{bmatrix}, f_2 = 1.5 \begin{bmatrix} 0 \\ \sin(w^4) \end{bmatrix}, C_{1k} = \begin{bmatrix} 2 & 0 \\ 0 & 2 \end{bmatrix}, C_{2k} = \begin{bmatrix} 3 & 0 \\ 0 & 3 \end{bmatrix},$$

for all $k \in \mathbb{N}$, $\sigma_1 = -2$, $\sigma_2 = -2.5$, $\epsilon_{11} = 2$, $\epsilon_{12} = 0.5$, $\epsilon_{21} = 1$, $\epsilon_{22} = 0.7$, $\beta_1 = 1$, $\beta_2 = 2$, $b_{11} = 1$, $b_{22} = 1.5$, $b_{12} = 0.5$ and $b_{21} = 0.3$. From (8.8), one may get $\delta_1 = 0.25$ and $\delta_2 = 2.25$.

Cases 1 When all the control sensors are operational, we have from Riccati-like equation,

$$P^1 = \begin{bmatrix} 0.0416 & 0 \\ 0 & 4.5182 \end{bmatrix} \text{ and } P^2 = \begin{bmatrix} 2.2800 & 0 \\ 0 & 0.2512 \end{bmatrix}$$

with $\lambda_{\min}(P^1) = 0.0416$, $\lambda_{\max}(P^1) = 4.5182$, $\lambda_{\min}(P^2) = 0.2512$ and $\lambda_{\max}(P^2) = 2.2800$; so that, $\lambda^* = 0.0416$, $\lambda^{**} = 4.5182$ and the observer gain matrices are

$$\mathscr{L}^1 = \begin{bmatrix} 0.0312 & 0 \\ 0 & 3.3887 \end{bmatrix} \text{ and } \mathscr{L}^2 = \begin{bmatrix} 2.3940 & 0 \\ 0 & 0.2638 \end{bmatrix}.$$

Thus, $A^i - \mathscr{L}^i C^i$ (for $i = 1, 2$) are Hurwitz and the time intervals $t_k - t_{k-1} \geq 3.6238$ for the first subsystem and $t_k - t_{k-1} \geq 2.4891$ for the second subsystem. The test matrix in this case is

$$\mathscr{S} = \begin{bmatrix} -8.364 & 0.55 \\ 0.55 & -8.4 \end{bmatrix}.$$

Cases 2 When there is a failure in the first sensor in the first subsystem and second sensor in the second subsystem, i.e., $\Omega^1 = \{1\}$ and $C^1_{\bar{\Omega}} = \begin{bmatrix} 0 & 0 \\ 0 & 1.5 \end{bmatrix}$, and $\Omega^2 = \{2\}$ and $C^2_{\bar{\Omega}} = \begin{bmatrix} 3 & 0 \\ 0 & 0 \end{bmatrix}$, we have from Riccati-like equation,

$$P^1 = \begin{bmatrix} 0.0411 & 0 \\ 0 & 4.5182 \end{bmatrix} \text{ and } P^2 = \begin{bmatrix} 2.2800 & 0 \\ 0 & 0.2942 \end{bmatrix}$$

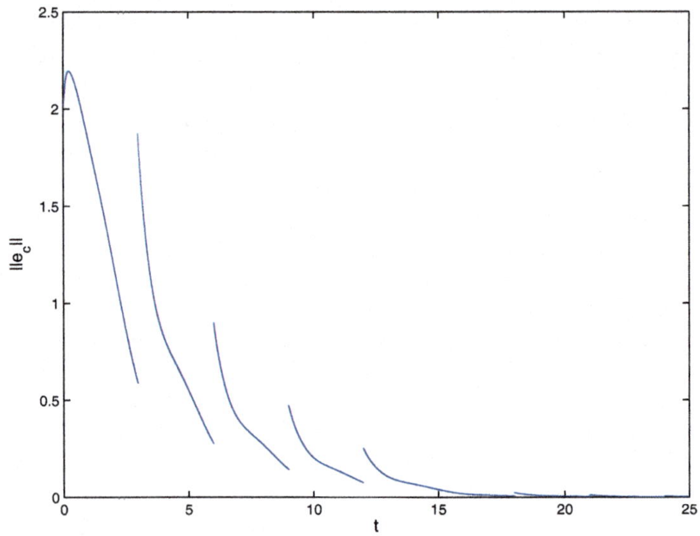

Fig. 8.5 Interconnected system: Operational sensors

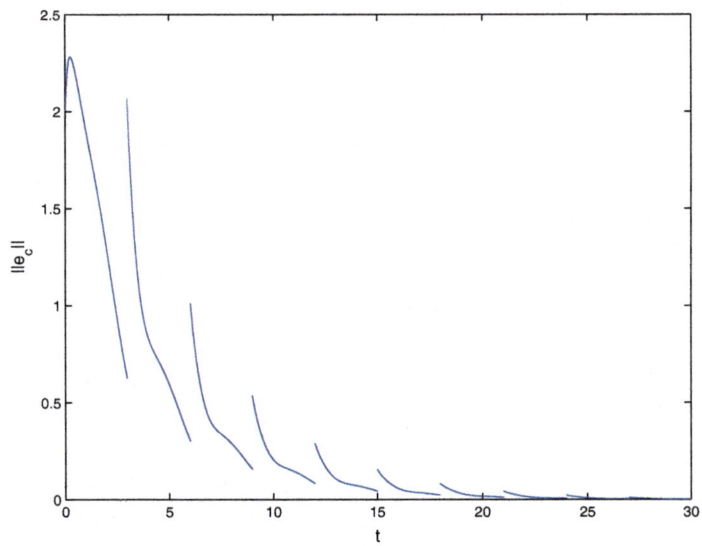

Fig. 8.6 Interconnected system: Faulty sensors

with $\lambda_{\min}(P^1) = 0.0411$, $\lambda_{\max}(P^1) = 4.5182$, $\lambda_{\min}(P^2) = 0.2942$ and $\lambda_{\max}(P^2) = 2.2800$; so that, $\lambda^* = 0.0411$, $\lambda^{**} = 4.5182$ and the observer gain matrices

$$\mathscr{L}^1 = \begin{bmatrix} 0 & 0 \\ 0 & 3.3887 \end{bmatrix} \text{ and } \mathscr{L}^2 = \begin{bmatrix} 2.3940 & 0 \\ 0 & 0 \end{bmatrix}.$$

Thus, $A^i + \mathscr{L}^i C^i_{\bar{\omega}}$ (for $i = 1, 2$) are Hurwitz and the time intervals $t_k - t_{k-1} \geq 3.6300$ and $t_k - t_{k-1} \geq 2.4101$ for the first and second subsystems, respectively. Figures 8.5 and 8.6 show the interconnected error system, $||e_c||$ for the operational and faulty sensors respectively.

8.4 Notes and Comments

This chapter has been devoted to address the problem of designing a robust reliable controller for the uncertain ILSS with fixed impulses. The material of this chapter is taken from [2]. The exponential stability property for such a complex system has been analyzed by decomposing the system into lower order, isolated subsystems and the interconnection was treated as a system perturbation. The isolated subsystems were assumed to be globally exponentially stabilized by the state feedback controllers and the interconnection was estimated by an upper bound which is required to be smaller than the stability degree of the isolated subsystems in order to guarantee the stability of the interconnected system. The scalar Lyapunov functions have been used to achieve our purpose, where this approach has led to solving a Riccati-like equation. In addition, the output of the faulty actuators has been treated as an outage. So that, one may extend these results by considering nonzero outage, where in this case it would be viewed as a perturbation. These results have been carried over to address the problem of state estimation for the input–output ILSS.

References

1. Seo C-J, Kim BK (1996) Robust reliable H_∞ control for linear systems with parameter uncertainty and actuator failure. Automatica 32:465–467
2. Sugati T, Alwan MS, Liu XZ (2015) Robust reliable control for uncertain impulsive large-scale systems. DCDIS Series B Appl Algorithms 22:381–400

Chapter 9
Switched Singularly Perturbed Systems with Time Delay

This chapter deals with exponential stability of a switched system consisting singularly perturbed subsystems with time delay. The multiple Lyapunov functions technique with the dwell-time approach is used to establish stability properties for the switched system.

9.1 Problem Formulation

Consider the following switched singularly perturbed systems with time delay

$$\begin{aligned} \dot{x}(t) &= f_{\sigma(t)}(t, x(t), x_t, z(t), z_t), \\ \varepsilon\dot{z}(t) &= g_{\sigma}(t, x(t), x_t, z(t), z_t), \end{aligned} \tag{9.1}$$

where, for all $t \geq t_0$ with $t_0 \in \mathbb{R}_+$, $x \in \mathbb{R}^m$ and $z \in \mathbb{R}^n$ are, respectively, the slow and fast system states or variables, $0 < \varepsilon \ll 1$ is perturbation parameter and $\sigma : [t_0, \infty) \to \mathscr{S}$ is the switching signal which is a piecewise constant function taking values in $\mathscr{S} = \{1, 2, \ldots, N\}$ with $N \geq 1$. If, for instance, $\sigma = i$ for any $i \in \mathscr{S}$, then system (9.1) becomes

$$\begin{aligned} \dot{x}(t) &= f_i(t, x(t), x_t, z(t), z_t), \quad t \in [t_{k-1}, t_k) \\ \varepsilon\dot{z}(t) &= g_i(t, x(t), x_t, z(t), z_t), \quad t \in [t_{k-1}, t_k). \end{aligned} \tag{9.2}$$

For every $i \in \mathscr{S}$, we assume that f_i and g_i are smooth enough to guarantee the existence of the solution of system (9.2), and $f_i(t, 0, 0, 0, 0) \equiv 0$ and $g_i(t, 0, 0, 0, 0) \equiv 0$ (for all t) to ensure that system (9.2) has the trivial solution $(x\ z)^T \equiv (0^T\ 0^T)$. Let $x_t = x(t + \theta)$ be the delayed state where $\theta \in [-\tau, 0]$ with

M. S. Alwan and X. Liu, *Theory of Hybrid Systems: Deterministic and Stochastic*, Nonlinear Physical Science, https://doi.org/10.1007/978-981-10-8046-3_9

τ being a positive constant representing the time delay, $x(t) = x(t; t_0, x_{t_0}, z_{t_0})$ and $z(t) = z(t; t_0, x_{t_0}, z_{t_0})$ be the solutions of (9.2) with the initial conditions x_{t_0} and z_{t_0}, respectively.

Definition 9.1 The trivial solution of system (9.2) is said to be exponentially stable if there exist positive constants K and λ such that

$$\|x(t)\| + \|z(t)\| \leq K\Big(\|x_{t_0}\|_\tau + \|z_{t_0}\|_\tau\Big)e^{-\lambda(t-t_0)}, \quad t \geq t_0$$

where $x(t)$ and $z(t)$ are any solution of system (9.2).

9.2 Stability Analysis

In this section, we address the exponential stability of switched systems. We consider modes represented by linear time-varying and then a special class of nonlinear system. Let $\mathscr{S}_u = \{1, 2, \ldots, r\}$ and $\mathscr{S}_s = \{r+1, r+2, \ldots, N\}$ be the sets of indices of the unstable and stable modes, respectively.

9.2.1 *Linear Systems*

Consider the following linear time-varying switched singularly perturbed systems with time delay

$$\begin{aligned}\dot{x} &= A_{11_i}(t)x + A_{12_i}(t)x_t + B_{11_i}(t)z + B_{12_i}(t)z_t,\\ \varepsilon\dot{z} &= A_{21_i}(t)x + A_{22_i}(t)x_t + B_{21_i}(t)z, \qquad t \in [t_{k-1}, t_k),\end{aligned} \tag{9.3}$$

where for each $i \in \mathscr{S} = \mathscr{S}_u \cup \mathscr{S}_s$, $A_{rs_i}(t)$, $B_{1s_i}(t)$ and $B_{21_i}(t)(r, s = 1, 2)$ are continuous matrices where $A_{1s_i} \in \mathbb{R}^{m\times m}$, $B_{1s_i} \in \mathbb{R}^{m\times n}$, $A_{2s_i} \in \mathbb{R}^{n\times m}$ and $B_{21_i} \in \mathbb{R}^{n\times n}$. The matrices $A_{22_i}(t)$ and $B_{21_i}(t)$ are continuously differentiable, $B_{21_i}(t)$ is nonsingular and $0 < \varepsilon \ll 1$. Here, we deal with the subsystems of (9.3) as interconnected systems. As said earlier, this system is viewed as a large-scale system when analyzing its stability properties. So that, we decompose them into small isolated subsystems

$$\begin{aligned}\dot{x} &= A_{11_i}(t)x\\ \dot{z} &= B_{21_i}(t)z\end{aligned}$$

and the rest will be considered the interconnection between them. In the following theorem, we state and prove exponential for the switched system in (9.3).

Theorem 9.1 *The trivial solution of system* (9.3) *is globally exponentially stable if the following assumptions hold:*

A1. There exist positive constants α and β such that

(i) *for any $i \in \mathscr{S}_u$, $Re[\lambda(A_{11_i}(t))] > 0$ and $Re[\lambda(B_{21_i}(t))] \le -\alpha$,*
(ii) *for any $i \in \mathscr{S}_s$, $Re[\lambda(A_{11_i}(t))] \le -\alpha$ and $Re[\lambda(B_{21_i}(t))] \le -\alpha$,*
(iii) *for any $i \in \mathscr{S}$ and $t \in [t_{k-1}, t_k)$, $k \in \mathbb{N}$*

$$\max\{\|A_{11_i}(t)\|, \|\dot{A}_{11_i}(t)\|, \|B_{21_i}(t)\|, \\ \|\dot{B}_{21_i}(t)\|, \|B_{21_i}^{-1}(t)A_{21_i}(t)\|, \|B_{21_i}^{-1}(t)A_{22_i}(t)\|\} \le \beta;$$

A2. let $h_i(t) = -B_{21_i}^{-1}(t)[A_{21_i}(t)x + A_{22_i}(t)x_t]$, and $P_{1_i}(t)$ and $P_{2_i}(t)$ be, respectively, the solutions of the Lyapunov matrix equations

$$A_{11_i}^T(t)P_{1_i}(t) + P_{1_i}(t)A_{11_i}(t) = -I_m, \\ B_{21_i}^T(t)P_{2_i}(t) + P_{2_i}(t)B_{21_i}(t) = -I_n,$$

with I_q being the identity matrix of dimension $q \times q$. Assume that there exist bounded functions $a_{rs_i}(t)$ and $b_{rs_i}(t)$ (for $r, s = 1, 2$) satisfying

$$2x^T P_{1_i}(t)[A_{12_i}(t)x_t + B_{11_i}(t)z + B_{12_i}(t)z_t] + x^T \dot{P}_{1_i}(t)x \le \\ a_{11_i}(t)\|x\|^2 + a_{12_i}(t)\|x_t\|_\tau^2 + b_{11_i}(t)\|(z - h_i)\|^2 + b_{12_i}(t)\|(z - h_i)_t\|_\tau^2,$$

$$-2(z - h_i)^T P_{2_i}(t)\dot{h}_i + (z - h_i)^T \dot{P}_{2_i}(t)(z - h_i) \le \\ a_{21_i}(t)\|x\|^2 + a_{22_i}(t)\|x_t\|_\tau^2 + b_{21_i}(t)\|(z - h_i)\|^2 + b_{22_i}(t)\|(z - h_i)_t\|_\tau^2;$$

A3. (i) *for any $i \in \mathscr{S}_u$, let $\alpha(t) = \lambda(\widetilde{A}_i^T(t) + \widetilde{A}_i(t))$ and $\|\widetilde{B}_i(t)\| \le \beta_1$ where*

$$\widetilde{A}_i(t) = \begin{pmatrix} \frac{2\gamma + a_{11_i}(t)}{\lambda_{1m}} & \frac{b_{11_i}(t)}{\lambda_{2m}} \\ \frac{a_{21_i}(t)}{\lambda_{1m}} & -\frac{1-\varepsilon_i b_{21_i}(t)}{\varepsilon_i \lambda_{2m}} \end{pmatrix}, \quad \widetilde{B}_i(t) = \begin{pmatrix} \frac{a_{12_i}(t)}{\lambda_{1m}} & \frac{b_{12_i}(t)}{\lambda_{2m}} \\ \frac{a_{22_i}(t)}{\lambda_{1m}} & \frac{b_{22_i}(t)}{\lambda_{2m}} \end{pmatrix},$$

and γ is a positive constant such that the matrix $A_{11_i} - \gamma I$ has eigenvalues with negative real parts. Assume that $\alpha(t) + \|\widetilde{B}_i(t)\| \le \beta_2$, $(\beta_2 > 0)$;
(ii) *for any $i \in \mathscr{S}_s$ there exist positive constants ε_i^* and η such that $-\widetilde{A}_i(t)$ is an M-matrix and $\lambda(\widetilde{A}_i(t) + \widetilde{A}_i^T(t)) + 2\|\widetilde{B}_i(t)\| \le -\eta < 0$ where*

$$\widetilde{A}_i(t) = \begin{pmatrix} -\frac{1-a_{11_i}(t)}{\lambda_{1M}} & \frac{b_{11_i}(t)}{\lambda_{2m}} \\ \frac{a_{21_i}(t)}{\lambda_{1m}} & -\frac{1-\varepsilon_i^* b_{21_i}(t)}{\varepsilon_i^* \lambda_{2M}} \end{pmatrix}, \quad \widetilde{B}_i(t) = \begin{pmatrix} \frac{a_{12_i}(t)}{\lambda_{1m}} & \frac{b_{12_i}(t)}{\lambda_{2m}} \\ \frac{a_{22_i}(t)}{\lambda_{1m}} & \frac{b_{22_i}(t)}{\lambda_{2m}} \end{pmatrix}$$

where $\lambda_{rm} = min\{\lambda_{min}(P_{r_i}), i \in S\}$ and $\lambda_{rM} = max\{\lambda_{max}(P_{r_i}), i \in S\}$ for $r = 1, 2$; and

A4. let $\lambda^+ = \max\{\xi_i \mid \forall i \in \mathscr{S}_u\}$ *and* $\lambda^- = \min\{\zeta_i \mid \forall i \in \mathscr{S}_s\}$ *with* ξ_i *and* ζ_i *being, respectively, the growth rates of the unstable subsystems and the decay rates of the stable subsystems, and* $T^+(t_0, t)$ *and* $T^-(t_0, t)$ *denote the total activation times of the unstable and stable modes, respectively. Assume that, for any* $t_0 \in \mathbb{R}_+$*, the switching law guarantees that*

$$\inf_{t\geq t_0} \frac{T^-(t_0, t)}{T^+(t_0, t)} \geq \frac{\lambda^+ + \lambda^*}{\lambda^- - \lambda^*} \tag{9.4}$$

where $\lambda^* \in (0, \lambda^-)$*; furthermore, there exists* $0 < \nu < \lambda^*$ *such that*

(i) for $i \in \mathscr{S}_u$

$$\ln \mu - \nu(t_k - t_{k-1}) \leq 0, \qquad k = 1, 2, \ldots, r \tag{9.5}$$

(ii) for $i \in \mathscr{S}_s$

$$\ln \mu + \zeta_i \tau - \nu(t_k - t_{k-1}) \leq 0, \qquad k = r + 1, r + 2, \ldots, N, \tag{9.6}$$

where ζ_i *is a unique positive solution of*

$$\zeta_i = \lambda(\widetilde{A}_i^T + \widetilde{A}_i) + \|\widetilde{B}_i\| + \|\widetilde{B}_i\|e^{\zeta_i\tau}.$$

Proof For all $t \geq t_0$ with $t_0 \in \mathbb{R}_+$, let $x(t) = x(t; t_0, x_{t_0}, z_{t_0})$ and $z(t) = z(t; t_0, x_{t_0}, z_{t_0})$ be the solutions of (9.2). For each $i \in \mathscr{S}$, define $V_i(x(t)) = x^T(t)P_{1_i}(t)x(t)$ and $W_i((z - h_i)(t)) = (z - h_i)^T(t)P_{2_i}(t)(z - h_i)(t)$ as the Lyapunov function candidate for the ith slow and fast subsystems, respectively. Then, for any $i \in \mathscr{S}_u$, the derivative of $V_i(t) = V_i(x(t))$ along the trajectories of the slow state x is given by

$$\begin{aligned}
\dot{V}_i(x(t)) &= \dot{x}^T P_{1_i}(t)x + x^T P_{1_i}(t)\dot{x} + x^T \dot{P}_{1_i}(t)x \\
&\leq (2\gamma + \frac{a_{11_i}(t)}{\lambda_{1m}})V_i(t) + \frac{b_{11_i}(t)}{\lambda_{2m}}W_i(t) + \frac{a_{12_i}(t)}{\lambda_{1m}}\| V_{i_t} \|_\tau \\
&\quad + \frac{b_{12_i}(t)}{\lambda_{2m}}\| W_{i_t} \|_\tau
\end{aligned}$$

where $W_i(t) = W_i((z - h_i)(t))$. Similarly, for any $i \in \mathscr{S}$, the derivative of $W_i(t)$ along the trajectories of the fast state z is given by

$$\begin{aligned}
\dot{W}_i((z - h_i)(t)) &= (\dot{z} - \dot{h}_i)^T P_{2_i}(t)(z - h_i) + (z - h_i)^T P_{2_i}(t)(\dot{z} - \dot{h}_i) \\
&\quad + (z - h_i)^T \dot{P}_{2_i}(t)(z - h_i) \\
&= \left(\frac{1}{\varepsilon}(A_{21_i}(t)x + A_{22_i}(t)x_t + B_{21_i}(t)z) - \dot{h}_i\right)^T P_{2_i}(t)(z - h_i) \\
&\quad + (z - h_i)^T P_{2_i}(t) \times \left(\frac{1}{\varepsilon}(A_{21_i}(t)x + A_{22_i}(t)x_t + B_{21_i}(t)z) - \dot{h}_i\right) \\
&\quad + (z - h_i)^T \dot{P}_{2_i}(z - h_i)
\end{aligned}$$

$$
\begin{aligned}
&\le \frac{a_{21_i}(t)}{\lambda_{1m}} V_i(t) - \frac{1-\varepsilon_i b_{21_i}(t)}{\varepsilon_i \lambda_{2M}} W_i(t) + \frac{a_{22_i}(t)}{\lambda_{1m}} \| V_{i_t} \|_\tau \\
&\quad + \frac{b_{22_i}(t)}{\lambda_{2m}} \| W_{i_t} \|_\tau \\
&\le \frac{a_{21_i}(t)}{\lambda_{1m}} V_i(t) - \frac{1-\varepsilon_i^* b_{21_i}(t)}{\varepsilon_i^* \lambda_{2M}} W_i(t) + \frac{a_{22_i}(t)}{\lambda_{1m}} \| V_{i_t} \|_\tau \\
&\quad + \frac{b_{22_i}(t)}{\lambda_{2m}} \| W_{i_t} \|_\tau
\end{aligned}
$$

where $\varepsilon_i^* \ge \varepsilon_i > 0$.

Combining $\dot{V}(t)$ and $\dot{W}(t)$ in a vector form yields

$$
\begin{aligned}
\begin{pmatrix} \dot{V}(t) \\ \dot{W}(t) \end{pmatrix} \le & \begin{pmatrix} \frac{2\gamma + a_{11_i}(t)}{\lambda_{1m}} & \frac{b_{11_i}(t)}{\lambda_{2m}} \\ \frac{a_{21_i}(t)}{\lambda_{1m}} & -\frac{1-\varepsilon_i b_{21_i}(t)}{\varepsilon_i \lambda_{2m}} \end{pmatrix} \begin{pmatrix} V(t) \\ W(t) \end{pmatrix} \\
& + \begin{pmatrix} \frac{a_{12_i}(t)}{\lambda_{1m}} & \frac{b_{12_i}(t)}{\lambda_{2m}} \\ \frac{a_{22_i}(t)}{\lambda_{1m}} & \frac{b_{22_i}(t)}{\lambda_{2m}} \end{pmatrix} \begin{pmatrix} \| V_{i_t} \|_\tau \\ \| W_{i_t} \|_\tau \end{pmatrix}
\end{aligned}
$$

Then, by $A3$(i) and Lemma 2.4, there exists a positive constant ξ_i such that

$$
\begin{aligned}
V_i(x(t)) &\le (\| V_{i_{t_{k-1}}} \|_\tau + \| W_{i_{t_{k-1}}} \|_\tau) e^{\xi_i (t - t_{k-1})} \\
W_i((z-h_i)(t)) &\le (\| V_{i_{t_{k-1}}} \|_\tau + \| W_{i_{t_{k-1}}} \|_\tau) e^{\xi_i (t - t_{k-1})}.
\end{aligned}
$$

For $i \in \mathscr{S}_s$, we have

$$
\dot{V}_i(x(t)) \le -\frac{1 - a_{11_i}(t)}{\lambda_{1M}} V_i(t) + \frac{b_{11_i}(t)}{\lambda_{2m}} W_i(t) + \frac{a_{12_i}(t)}{\lambda_{1m}} \| V_{i_t} \|_\tau + \frac{b_{12_i}(t)}{\lambda_{2m}} \| W_{i_t} \|_\tau,
$$

and

$$
\begin{aligned}
\begin{pmatrix} \dot{V}(t) \\ \dot{W}(t) \end{pmatrix} \le & \begin{pmatrix} -\frac{1 - a_{11_i}(t)}{\lambda_{1M}} & \frac{b_{11_i}(t)}{\lambda_{2m}} \\ \frac{a_{21_i}(t)}{\lambda_{1m}} & -\frac{1-\varepsilon_i^* b_{21_i}(t)}{\varepsilon_i^* \lambda_{2M}} \end{pmatrix} \begin{pmatrix} V(t) \\ W(t) \end{pmatrix} \\
& + \begin{pmatrix} \frac{a_{12_i}(t)}{\lambda_{1m}} & \frac{b_{12_i}(t)}{\lambda_{2m}} \\ \frac{a_{22_i}(t)}{\lambda_{1m}} & \frac{b_{22_i}(t)}{\lambda_{2m}} \end{pmatrix} \begin{pmatrix} \| V_{i_t} \|_\tau \\ \| W_{i_t} \|_\tau \end{pmatrix}
\end{aligned}
$$

Then, by $A3$(ii) and Lemma 2.3, there exists a positive constant ζ_i such that

$$
\begin{aligned}
V_i(t) &\le (\| V_{i_{t_{k-1}}} \|_\tau + \| W_{i_{t_{k-1}}} \|_\tau) e^{-\zeta_i (t - t_{k-1})} \\
W_i(t) &\le (\| V_{i_{t_{k-1}}} \|_\tau + \| W_{i_{t_{k-1}}} \|_\tau) e^{-\zeta_i (t - t_{k-1})}.
\end{aligned}
$$

Recall that for any $i, j \in \mathscr{S}$ we have

$$V_j(t) \leq \mu V_i(t)$$
$$W_j(t) \leq \mu W_i(t)$$

where $\mu = \max\{\mu_l = \frac{\lambda_{lM}}{\lambda_{lm}} \text{ for } l = 1, 2\}$. Then

$$V_N(x(t)) \leq \prod_{i=1}^{r} 2\mu e^{\xi_i(t_i - t_{i-1})} \times \prod_{j=r+1}^{N-r-1} 2\mu e^{\zeta_j \tau} e^{-\zeta_j(t_j - t_{j-1})}$$
$$\times (\| V_{1_{t_0}} \|_\tau + \| W_{1_{t_0}} \|_\tau) e^{-\zeta_N(t - t_{N-1})}$$

Making use of Assumption $A4$, we get

$$V_N(t) \leq (\| V_{1_{t_0}} \|_\tau + \| W_{1_{t_0}} \|_\tau) e^{-(\lambda^* - \nu)(t - t_0)}, \quad t \in [t_0, \infty)$$

Similarly, we have

$$W_N((z - h_N)(t)) \leq (\| V_{1_{t_0}} \|_\tau + \| W_{1_{t_0}} \|_\tau) e^{-(\lambda^* - \nu)(t - t_0)}, \quad t \in [t_0, \infty).$$

Then, there exists K_1 such that

$$\|x(t)\| \leq K_1(\|x_{t_0}\|_\tau + \|z_{t_0}\|_\tau) e^{-(\lambda^* - \nu)(t - t_0)/2},$$

and, by the fact that,

$$\|z\| - \|h_i\| \leq \|z - h_i\| \leq \frac{1}{\sqrt{\lambda_{2m}}} W_i^{1/2},$$

there exists K_2 such that

$$\|z(t)\| \leq K_2(\|x_{t_0}\|_\tau + \|z_{t_0}\|_\tau) e^{-(\lambda^* - \nu)(t - t_0)/2}.$$

Hence,

$$\|x(t)\| + \|z(t)\| \leq K(\|x_{t_0}\|_\tau + \|z_{t_0}\|_\tau) e^{-(\lambda^* - \nu)(t - t_0)/2}.$$

where $K = K_1 + K_2$. This shows that the trivial solution of (9.3) is exponentially stable. This completes the proof.

Remark 9.1 Assumptions $A1$[(i),(iii)], $A2$ and $A3$(i) are to ensure that, for $i \in \mathscr{S}_u$, the ith mode is unstable, while Assumptions $A1$[(ii),(iii)] and $A3$(ii) are to guarantee that, for $i \in \mathscr{S}_s$, the ith mode is exponentially stable. As well known in switched systems, the exponential stability of each subsystem is insufficient to guarantee exponential stability of the entire switched systems; so that, an additional condition is required; namely, the total activation time of stable subsystem is larger than that of unstable ones as represented in Assumption $A4$.

The following example shows these results.

Example 9.1 Consider the following switched system with the unstable and stable modes

$$\begin{aligned} \dot{x} &= 0.1x + 0.04z(t-1) \\ \varepsilon\dot{z} &= 0.45x(t-1) - 0.1z \end{aligned}$$

$$\begin{aligned} \dot{x} &= -1.5x - 0.1z(t-1) \\ \varepsilon\dot{z} &= -x(t-1) - 2z \end{aligned}$$

with the initial condition $(x_{t_0}, z_{t_0})^T = (t+0.3, t+0.3)$ for $t \in [-1, 0]$. For the unstable mode, taking $\gamma = 0.2$, $\varepsilon = 0.3$, $Q_{1_1} = 0.1$ and $Q_{2_1} = 0.07$ gives $P_{1_1} = 0.5$, $P_{2_1} = 0.35$,

$\widetilde{A}_u = \begin{pmatrix} 0.44 & 0 \\ 0 & -0.0948 \end{pmatrix}$ and $\widetilde{B}_u = \begin{pmatrix} 0 & 0.0571 \\ 0.315 & 0 \end{pmatrix}$. Clearly, Assumption $A3$(i) holds which affirms the instability of mode 1; so that, by Lemma 2.4, the growth rates are $\xi = \{0.2207, 0.755\}$, while for the stable mode, taking $Q_{1_2} = 1.5$ and $Q_{2_2} = 5$ gives $P_{1_2} = 0.5$, $P_{2_2} = 1.25$ and $\mu = 3.5714$; from $A3$(ii), we get $\varepsilon^* = 1.5238$. Thus

$\widetilde{A}_s = \begin{pmatrix} -2.9 & 0 \\ 0 & -2.9297 \end{pmatrix}$ and $\widetilde{B}_s = \begin{pmatrix} 0 & 0.1429 \\ 1.875 & 0 \end{pmatrix}$; hence, by Lemma 2.3, the decay rates are $\zeta = \{0.5927, 0.5791\}$. Taking $\lambda^+ = 0.755$, $\lambda^- = 0.5791$ and $\lambda^* = 0.25$ gives $T^- \geq 3.05T^+$. Taking $\nu = 0.2$, by $A4$[(i),(ii)] we get, respectively, $T^+ = 9.9$ and $T^- = 29.5$. Figure 9.1 shows these results where unstable and stable modes are activated alternatively. In this example, the unstable mode is activated on subintervals [0,9.9) and [39.4,49.3). This result illustrates the necessity for running stable modes longer than the unstable ones.

9.2.2 Nonlinear Systems

Consider the following nonlinear switched DSPSs

$$\begin{aligned} \dot{x} &= f_i(x, x_t, z, z_t) \\ \varepsilon\dot{z} &= B_{21_i} z + B_i(x, x_t), \qquad t \in [t_{k-1}, t_k), \end{aligned} \tag{9.7}$$

where $f_i(x, x_t, z, z_t) = A_{11_i} x + g_i(x, x_t, z, z_t)$ for any $i \in \mathscr{S}$. We assume that system (9.7) has a unique equilibrium point at the origin and the matrix B_{21_i} is nonsingular.

In the following theorem, we state the sufficient conditions to guarantee exponential stability of system (9.7).

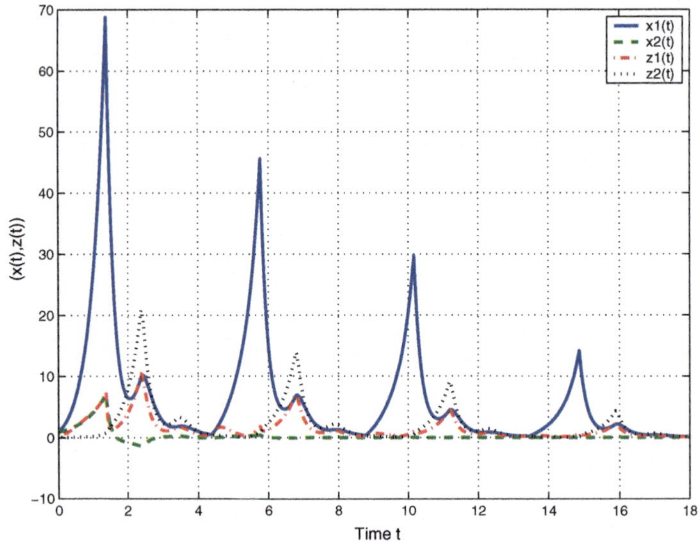

Fig. 9.1 Linear system with unstable and stable modes

Theorem 9.2 *The trivial solution of system* (9.7) *with* $\mathscr{S} = \mathscr{S}_u \cup \mathscr{S}_s$ *is globally exponentially stable if the following assumptions hold:*

A1. *(i) for any* $i \in \mathscr{S}_u$, $Re[\lambda(A_{11_i})] > 0$ *and* B_{21_i} *is Hurwitz;*
(ii) for any $i \in \mathscr{S}_s$, A_{11_i} *and* B_{21_i} *are Hurwitz;*

A2. *for any* $i \in \mathscr{S}$*, there exist positive constants* a_{rs_i} *and* b_{rs_i} *(for* $r, s = 1, 2$*) such that*

$$2x^T P_{1_i} g_i(x, x_t, z, z_t) \le a_{11_i} \|x\|^2 + a_{12_i} \|x_t\|_\tau^2 + b_{11_i} \|z - h_i\|^2 + b_{12_i} \|(z - h_i)_t\|_\tau^2,$$
$$-2(z - h_i)^T P_{2_i} \dot{h}_i \le a_{11_i} \|x\|^2 + a_{12_i} \|x_t\|_\tau^2 + b_{11_i} \|z - h_i\|^2 + b_{12_i} \|(z - h_i)_t\|_\tau^2,$$

where $h_i = -B_{21_i}^{-1} B_i(x, x_t)$*, and* P_{1_i} *and* P_{2_i} *are positive-definite matrices satisfying Lyapunov matrix equations*

$$A_{11_i}^T P_{1_i} + P_{1_i} A_{11_i} = -Q_{1_i}$$
$$B_{21_i}^T P_{2_i} + P_{2_i} B_{21_i} = -Q_{2_i}$$

with Q_{1_i} *and* Q_{2_i} *being given positive-definite matrices;*

A3. *(i) for any* $i \in \mathscr{S}_u$*, let* $\gamma > 0$ *be a positive constant such that the matrix* $A_{11_i} - \gamma I$ *has eigenvalues with negative real parts and assume that* $\beta_{2_i} = \alpha_i + \beta_{1_i} > 0$ *where* $\beta_{1_i} = \|\widetilde{B}_i\|$, $\alpha_i = \lambda(\widetilde{A}_i^T + \widetilde{A}_i)$,

$$\widetilde{A}_i = \begin{pmatrix} 2\gamma + \frac{a_{11_i}}{\lambda_{1m}} & \frac{b_{11_i}}{\lambda_{2m}} \\ \frac{a_{21_i}}{\lambda_{1m}} & -\frac{\lambda_{min}(Q_{2_i})-\varepsilon_i b_{21_i}}{\varepsilon_i \lambda_{2M}} \end{pmatrix} \text{ and } \widetilde{B}_i = \begin{pmatrix} \frac{a_{12_i}}{\lambda_{1m}} & \frac{b_{12_i}}{\lambda_{2m}} \\ \frac{a_{22_i}}{\lambda_{1m}} & \frac{b_{22_i}}{\lambda_{2m}} \end{pmatrix}.$$

(ii) for any $i \in \mathscr{S}_s$, there exist positive constants ε_i^* and η such that $-\widetilde{A}_i$ is an M−matrix and $\lambda(\widetilde{A}_i + \widetilde{A}_i^T) + 2\|\widetilde{B}_i\| \le -\eta < 0$ where

$$\widetilde{A}_i = \begin{pmatrix} -\frac{\lambda_{min}(Q_{1_i})-a_{11_i}}{\lambda_{1M}} & \frac{b_{11_i}}{\lambda_{2m}} \\ \frac{a_{21_i}}{\lambda_{1m}} & -\frac{\lambda_{min}(Q_{2_i})-\varepsilon_i^* b_{21_i}}{\varepsilon_i^* \lambda_{2M}} \end{pmatrix}, \widetilde{B}_i = \begin{pmatrix} \frac{a_{12_i}}{\lambda_{1m}} & \frac{b_{12_i}}{\lambda_{2m}} \\ \frac{a_{22_i}}{\lambda_{1m}} & \frac{b_{22_i}}{\lambda_{2m}} \end{pmatrix}; \text{ and}$$

A4. Assumption A4 of Theorem 9.1 holds.

Proof For all $t \ge t_0$ with $t_0 \in \mathbb{R}_+$, let $x(t) = x(t; t_0, x_{t_0}, z_{t_0})$ and $z(t) = z(t; t_0, x_{t_0}, z_{t_0})$ be the solutions of (9.2). For each $i \in \mathscr{S}$, define $V_i(x) = x^T P_{1_i} x$ and $W_i(z - h_i) = (z - h_i)^T P_{2_i}(z - h_i)$. Then, for any $i \in \mathscr{S}_u$, the time derivative of V_i along the trajectories of the state x is given by

$$\begin{aligned} \dot{V}_i(t) &= \dot{x}^T P_{1_i} x + x^T P_{1_i} \dot{x} \\ &\le (2\gamma + \frac{a_{11_i}}{\lambda_{1m}}) V_i(t) + \frac{b_{11_i}}{\lambda_{2m}} W_i(t) + \frac{a_{12_i}}{\lambda_{1m}} \| V_{i_t} \|_\tau + \frac{b_{12_i}}{\lambda_{2m}} \| W_{i_t} \|_\tau \end{aligned}$$

and for any $i \in \mathscr{S}_s$, we have

$$\dot{V}_i(t) \le -\frac{\lambda_{min}(Q_{1_i}) - a_{11_i}}{\lambda_{1M}} V_i(t) + \frac{b_{11_i}}{\lambda_{2m}} W_i(t) + \frac{a_{12_i}}{\lambda_{1m}} \| V_{i_t} \|_\tau + \frac{b_{12_i}}{\lambda_{2m}} \| W_{i_t} \|_\tau.$$

Similarly, the time derivative of W_i along the trajectories of the state z is given by

$$\begin{aligned} \dot{W}_i(t) &= (\dot{z} - \dot{h}_i)^T P_{2_i}(z - h_i) + (z - h_i)^T P_{2_i}(\dot{z} - \dot{h}_i) \\ &= \left[\frac{1}{\varepsilon}(B_{21_i} z + B_i(x, x_t)) - \dot{h}_i\right]^T P_{2_i}(z - h_i) \\ &\quad + (z - h_i)^T P_{2_i} \left[\frac{1}{\varepsilon}(B_{21_i} z + B_i(x, x_t)) - \dot{h}_i\right] \\ &\le \frac{a_{21_i}}{\lambda_{1m}} V_i(t) - \frac{\lambda_{min}(Q_{2_i}) - \varepsilon_i^* b_{21_i}}{\varepsilon_i^* \lambda_{2M}} W_i(t) + \frac{a_{22_i}}{\lambda_{1m}} \| V_{i_t} \|_\tau + \frac{b_{22_i}}{\lambda_{2m}} \| W_{i_t} \|_\tau \end{aligned}$$

where $\varepsilon^* \ge \varepsilon > 0$. Then, there exists a positive constant ξ_i such that

$$\begin{aligned} V_i(x(t)) &\le (\| V_{i_{t_{k-1}}} \|_\tau + \| W_{i_{t_{k-1}}} \|_\tau) e^{\xi_i (t - t_{k-1})} \\ W_i((z - h_i)(t)) &\le (\| V_{i_{t_{k-1}}} \|_\tau + \| W_{i_{t_{k-1}}} \|_\tau) e^{\xi_i (t - t_{k-1})}. \end{aligned}$$

While for stable modes, there exists a positive constant ζ_i such that

$$V_i(x(t)) \leq (\| V_{i_{t_{k-1}}} \|_\tau + \| W_{i_{t_{k-1}}} \|_\tau) e^{-\zeta_i(t-t_{k-1})}$$
$$W_i((z-h_i)(t)) \leq (\| V_{i_{t_{k-1}}} \|_\tau + \| W_{i_{t_{k-1}}} \|_\tau) e^{-\zeta_i(t-t_{k-1})}.$$

The rest of the proof is similar to that of Theorem 9.1.

Example 9.2 Consider the switched system with following unstable and stable modes

$$\dot{x} = 0.1x + \sin z(t-1)$$
$$\varepsilon\dot{z} = 0.1x - z$$

and

$$\dot{x} = -10x + \ln(1 + x^2(t-1)) + z$$
$$\varepsilon\dot{z} = x - 2z.$$

For the unstable mode, we take $V_u = 0.5x^2$ and $W_u = 0.5(z-h)^2$ where $h = 0.1x$. With little effort, one may find $\dot{V}_u \leq 1.4V_u + \| W_{u_t} \|_\tau$ and $\dot{W}_u \leq (-\frac{2}{\varepsilon} + 0.12)W_u + 0.01V_u + 0.01\| V_{u_t} \|_\tau + 0.1\| W_{u_t} \|_\tau$, so that $\widetilde{A}_u = \begin{pmatrix} 1.4 & 0 \\ 0.01 & -\frac{2}{\varepsilon} + 0.12 \end{pmatrix}$,

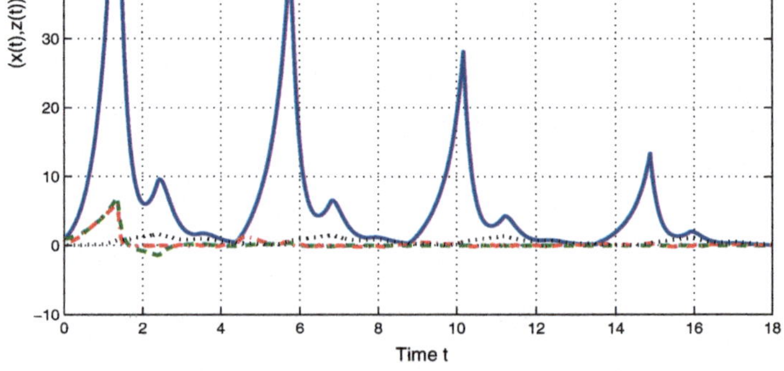

Fig. 9.2 Nonlinear system with unstable and stable modes

$\widetilde{B}_u = \begin{pmatrix} 0 & 1 \\ 0.01 & 0.1 \end{pmatrix}$. When $\varepsilon = 0.1$, the growth rates are $\xi = \{1.8499, 4.81\}$. For the stable mode, taking $V_s = 0.5x^2$ and $W_s = 0.5(z-h)^2$ where $h = 0.5x$ gives $\dot{V}_s \le -14V_s + W_s + 4\|V_{s_t}\|_\tau$ and $\dot{W}_s \le (-\frac{4}{\varepsilon^*} + 11.5)W_s + 5.5V_s + 2\|V_{s_t}\|_\tau$; thus, $\widetilde{A}_s = \begin{pmatrix} -14 & 1 \\ 5.5 & -\frac{4}{\varepsilon^*} + 11.5 \end{pmatrix}$, $\widetilde{B}_s = \begin{pmatrix} 4 & 0 \\ 2 & 0 \end{pmatrix}$; by $A3(\text{ii})$, we get $\varepsilon^* = 0.2341$. When $\varepsilon = 0.1 \in (0, 0.2341]$, the decay rates are $\zeta = \{1.5279, 2.4432\}$. Take $\lambda^+ = 4.81$ and $\lambda^- = 1.5279$. If we choose $\lambda^* = 0.52$ and $\nu = 0.5$, then we get $T^- \ge 5.3T^+$. In this example, we have $\mu = 1$, and from $A4$, $T^+ = 1.38$ and $T^- = 8$. The unstable mode is run on $[0, 1.38)$ and $[9.38, 10.76)$. These results are illustrated in Fig. 9.2.

9.3 Notes and Comments

Stability of switched systems incorporating unstable, stable singularly perturbed systems with time delay has been established. The material of this chapter is taken from [1]. Particularly, linear time-varying and a special class of nonlinear systems are considered. Multiple Lyapunov functions technique along with a dwell-time switching signal is used to analyze the stability of these systems. We have shown that, when stable subsystems are run longer than unstable modes, exponential stability of the entire switched systems is guaranteed. For further reading about singularly perturbed system, one may refer to [2–11]. Particularly, [3] is concerned with reviewing several control problems including optimal controls of various systems with singular perturbations.

References

1. Alwan MS, Liu XZ, Ingalls B (2008) Exponential stability of singularly switched systems with time delay. Nonlinear Anal Hybrid Syst 2:913–921
2. Chow JH (ed) (1982) Time-scale modelling of dynamic networks with applications to power systems. In: Number 46 in Lecture notes in control and information sciences. Springer, New York
3. Dmitriev MG, Kurina GA (2006) Singular perturbations in control problems. Autom Remote Control 67(1):1–43
4. Hoppensteadt F (1968) Asymptotic stability in singular perturbation problems. J Differ Equ 4:350–358
5. Hoppensteadt F (1974) Asymptotic stability in singular perturbation problems. II: problems having matched asymptotic expansion solutions. J Differ Equ 15:510–521
6. Kokotovic PV, Khalil HK (eds) (1986) Singular perturbations in systems and control. IEEE Press Inc., New York
7. Kokotovic PV, Khalil HK, O'Reilly J (1986) Singular perturbation methods in control: analysis and design. Academic Press Inc., London
8. Kokotovic P, Bensoussan A, Blankenship G (eds) (1987) Singular perturbations and asymptotic analysis in control systems. Lecture notes in control and information sciences. Springer, New York

9. Khalil HK (2002) Nonlinear systems. Prentice-Hall Inc., NJ
10. Naidu DS (1988) Singular perturbation methodology in control systems. Peter Peregrinus Ltd
11. Saberi A, Khalil H (1984) Quadratic-type Lyapunov functions for singularly perturbed systems. IEEE Trans Autom Control AC 29(6):542–550

Chapter 10
Singularly Perturbed Impulsive-Switched Systems with Time Delay

This chapter deals with some stability notions for impulsive-switched systems with time-delayed singularly perturbed subsystems. The technique of multiple Lyapunov functions and dwell-time switching signal are used to analyze the stability properties. As will be seen, the impulses can contribute to obtain stability properties even when the system consists of only unstable subsystems.

10.1 Problem Formulation

Consider the following impulsive-switched delay systems

$$\begin{aligned} \dot{x} &= f_{\sigma(t)}(t, x, x_t, z, z_t), \quad t \neq t_k \\ \varepsilon \dot{z} &= g_{\sigma(t)}(t, x, x_t, z, z_t), \quad t \neq t_k \\ \Delta x &= B_k x(t^-), \quad t = t_k \\ \Delta z &= C_k z(t^-), \quad t = t_k \end{aligned} \tag{10.1}$$

where $x \in \mathbb{R}^m$ and $z \in \mathbb{R}^n$ are, respectively, the slow and fast states of the system, and ε is a small positive perturbation parameter. For $t_0 \in \mathbb{R}_+$ and $\mathscr{S} = \{1, 2, \ldots, N\}$ with $N > 1$ being the number of subsystems, $\sigma : [t_0, \infty) \to \mathscr{S}$, which is represented by $\{i_k\}$ according to $[t_{k-1}, t_k) \to i_k \in \mathscr{S}$, is a piecewise constant function switching signal. Here, i_k (or i for simplicity of notation) means the ith subsystem is activated on the subinterval $[t_{k-1}, t_k)$. The discontinuities of σ form a strictly increasing sequence of impulsive-switching times $\{t_k\}_{k=1}^{\infty}$ satisfying $t_{k-1} < t_k$ with $\lim_{k\to\infty} t_k = \infty$; that is, the impulses are here a consequences of the switchings. For any switching signal σ, we denote by $T^+(t_0, t)$ and $T^-(t_0, t)$ the total activation time of unstable and stable subsystems, respectively, over the time interval $[t_0, t)$.

M. S. Alwan and X. Liu, *Theory of Hybrid Systems: Deterministic and Stochastic*, Nonlinear Physical Science, https://doi.org/10.1007/978-981-10-8046-3_10

We assume that $x(t_k^+) = x(t_k)$ and $z(t_k^+) = z(t_k)$, meaning that the solution of (10.1) is right continuous. In the difference equation, $\Delta y = y(t) - y(t^-)$, where $y(t^-) = \lim_{s \to t^-} y(s)$, represents the state just before and after the impulse action. The vector field functions, f_i and g_i, are assumed to be smooth enough to guarantee that system (10.1) has a unique solution, and $f_i(t, 0, 0, 0, 0) \equiv 0$ and $g_i(t, 0, 0, 0, 0) \equiv 0$; that is, system (10.1) admits a trivial solution, $(x\ z)^T \equiv (0^T\ 0^T)$.

10.2 Stability Analysis

In this section, we write some Lyapunov-type sufficient conditions to guarantee some stability properties of linear and a special class of nonlinear systems. Throughout this chapter, we denote by $\mathscr{S}_u = \{1, 2, \ldots, q\}$ and $\mathscr{S}_s = \{q+1, q+2, \ldots, N\}$ the sets of indices of the unstable and stable subsystems, respectively.

10.2.1 Linear Systems

Consider the following linear impulsive-switched system

$$\begin{aligned} \dot{x} &= A_{11_i} x + A_{12_i} x_t + B_{11_i} z + B_{12_i} z_t, & t \neq t_k \\ \varepsilon_i \dot{z} &= A_{21_i} x + A_{22_i} x_t + B_{21_i} z, & t \neq t_k \\ \Delta x &= B_k x(t^-), \qquad t = t_k \\ \Delta z &= C_k z(t^-), \qquad t = t_k \end{aligned} \tag{10.2}$$

where, for any $i \in \mathscr{S} = \mathscr{S}_u \cup \mathscr{S}_s$, $A_{11_i}, A_{12_i} \in \mathbb{R}^{m \times m}$, $B_{11_i}, B_{12_i} \in \mathbb{R}^{m \times n}$, $A_{2s_i}, A_{22_i} \in \mathbb{R}^{n \times m}$, $B_{21_i} \in \mathbb{R}^{n \times n}$ and B_{21_i} is nonsingular and Hurwitz. Let the uncoupled slow and fast subsystems of (11.2) be, respectively, given by

$$\dot{x} = A_{11_i} x \quad \text{and} \quad \varepsilon_i \dot{z} = B_{21_i} z.$$

Theorem 10.1 *The trivial solution of* (10.2) *is exponentially stable if the following assumptions hold:*

A1. *for any $i \in \mathscr{S}_u$, A_{11_i} has eigenvalues with positive real parts and, for any $i \in \mathscr{S}_s$, A_{11_i} is Hurwitz;*

A2. *for any $i \in \mathscr{S}$ and all $t \in [t_{k-1}, t_k)$, there exist positive constants a_{11_i}, a_{12_i}, a_{21_i}, a_{22_i}, b_{11_i}, b_{12_i}, b_{21_i} and b_{22_i} such that*

$$\begin{aligned} 2x^T P_{1_i} [A_{12_i} x_t + B_{11_i} z + B_{12_i} z_t] &\le a_{11_i} \|x\|^2 + a_{12_i} \|x_t\|_\tau^2 + b_{11_i} \|(z - h_i)\|^2 \\ &\quad + b_{12_i} \|(z - h_i)_t\|_\tau^2, \\ -2(z - h_i)^T P_{2_i} \dot{h}_i &\le a_{21_i} \|x\|^2 + a_{22_i} \|x_t\|_\tau^2 + b_{21_i} \|(z - h_i)\|^2 + b_{22_i} \|(z - h_i)_t\|_\tau^2, \end{aligned}$$

where $h_i(t) = -B_{21_i}^{-1}[A_{21_i}x + A_{22_i}x_t]$ *and* P_{1_i} *and* P_{2_i} *are positive-definite matrices satisfying the Lyapunov matrix equations*

$$\begin{aligned} A_{11_i}^T P_{1_i} + P_{1_i} A_{11_i} &= -Q_{1_i}, \\ B_{21_i}^T P_{2_i} + P_{2_i} B_{21_i} &= -Q_{2_i}, \end{aligned}$$

for any given positive-definite matrices Q_{1_i} *and* Q_{2_i}*;*

A3. *(i)* *for any* $i \in \mathscr{S}_u$, *assume that* $\lambda_{\min}(\widetilde{A}_i^T + \widetilde{A}_i) + \|\widetilde{B}_i\| > 0$, *where*

$$\widetilde{A}_i = \begin{pmatrix} 2\gamma^* + \frac{a_{11_i}}{\lambda_{1m}} & \frac{b_{11_i}}{\lambda_{2m}} \\ \frac{a_{21_i}}{\lambda_{1m}} & -\frac{\lambda_{\min}(Q_{2_i}) - \varepsilon_i b_{21_i}}{\varepsilon_i \lambda_{2m}} \end{pmatrix}, \ \widetilde{B}_i = \begin{pmatrix} \frac{a_{12_i}}{\lambda_{1m}} & \frac{b_{12_i}}{\lambda_{2m}} \\ \frac{a_{22_i}}{\lambda_{1m}} & \frac{b_{22_i}}{\lambda_{2m}} \end{pmatrix}$$

and γ^* *is a positive constant such that the matrix* $A_{11_i} - \gamma^* I$ *is Hurwitz;*

(ii) *for any* $i \in \mathscr{S}_s$, *there exist positive constants* ε_i^* *such that* $-\widetilde{A}_i$ *is an* M*-matrix and* $\lambda_{\max}(\widetilde{A}_i + \widetilde{A}_i^T) + 2\|\widetilde{B}_i\| < 0$ *where*

$$\widetilde{A}_i = \begin{pmatrix} -\frac{\lambda_{\min}(Q_{1_i}) - a_{11_i}}{\lambda_{1M}} & \frac{b_{11_i}}{\lambda_{2m}} \\ \frac{a_{21_i}}{\lambda_{1m}} & -\frac{\lambda_{\min}(Q_{2_i}) - \varepsilon_i^* b_{21_i}}{\varepsilon_i^* \lambda_{2M}} \end{pmatrix}, \ \widetilde{B}_i = \begin{pmatrix} \frac{a_{12_i}}{\lambda_{1m}} & \frac{b_{12_i}}{\lambda_{2m}} \\ \frac{a_{22_i}}{\lambda_{1m}} & \frac{b_{22_i}}{\lambda_{2m}} \end{pmatrix},$$

and Q_{1_i} *and* Q_{2_i} *are defined in assumption A2;*

A4. *let* $\lambda^+ = \max\{\xi_i \mid i \in \mathscr{S}_u\}$, $\lambda^- = \min\{\zeta_i \mid i \in \mathscr{S}_s\}$ *with* ξ_i *and* ζ_i *being the growth and decay rates of unstable and stable subsystems, respectively, and, for any* t_0, *assume that the switching signal guarantees that*

$$\inf_{t \ge t_0} \frac{T^-(t_0, t)}{T^+(t_0, t)} \ge \frac{\lambda^+ + \lambda^*}{\lambda^- - \lambda^*}, \tag{10.3}$$

where $T^+(t_0, t)$ *and* $T^-(t_0, t)$ *are defined in the previous section and* $\lambda^* \in (0, \lambda^-)$. *Furthermore, there exists* $0 < \nu < \zeta_i$ *such that*

(i) *for* $i \in \mathscr{S}_u$ *and* $k = 1, 2, \ldots, l$

$$\ln \mu(\alpha_k + \beta_k + \gamma_k + \psi_k) - \nu(t_k - t_{k-1}) \le 0; \tag{10.4}$$

(ii) *for* $i \in \{l+1, l+2, \ldots, N-1\}$ *and* $k = l+1, l+2, \ldots, N-1$

$$\ln \mu(\alpha_k + \beta_k + \gamma_k + \psi_k e^{\zeta_i \tau}) + \zeta_i \tau - \nu(t_k - t_{k-1}) \le 0, \tag{10.5}$$

where ζ_i *is the unique positive solution of*

$$\zeta_i + \lambda_{\max}(\widetilde{A}_i^T + \widetilde{A}_i) + \|\widetilde{B}_i\| + \|\widetilde{B}_i\| e^{\zeta_i \tau} = 0,$$

$\alpha_k = \mu\lambda_{\max}^2([I + B_k])$, $\beta_k = \frac{\lambda_{2M}}{\lambda_{1m}}(\|U_k\| + r_k + s_k)r_k$,
$\gamma_k = \mu(\|U_k\| + r_k + s_k)\|U_k\|$, $\psi_k = \frac{\lambda_{2M}}{\lambda_{1m}}(\|U_k\| + r_k + s_k)s_k$,
$U_k = I + C_k$,
$r_k = \max\{\|R_{ik}\| : R_{ik} = [I + C_k]B_{21_i}^{-1}A_{21_i} - B_{21_i}^{-1}A_{21_i}[I + B_k] \ \ \forall i \in \mathscr{S}\}$,
and
$s_k = \max\{\|S_{ik}\| : S_{ik} = [I + C_k]B_{21_i}^{-1}A_{22_i} - B_{21_i}^{-1}A_{22_i}[I + B_k] \ \ \forall i \in \mathscr{S}\}$.

Proof For all $t \geq t_0$ with $t_0 \in \mathbb{R}_+$, let $x(t)$ and $z(t)$ be the solution of (10.2). Define $V_i(t) = V_i(x(t)) = x^T(t)P_{1_i}x(t)$ and $W_i(t) = W_i((z - h_i)(t)) = (z - h_i)^T(t)P_{2_i}(z - h_i)(t)$. Then, the time derivative of V_i and W_i along the trajectories of x and z are:

(i) for any $i \in \mathscr{S}_u$

$$
\begin{aligned}
\dot{V}_i(t) &\leq (2\gamma^* + \frac{a_{11_i}}{\lambda_{1m}})V_i(t) + \frac{b_{11_i}}{\lambda_{2m}}W_i((t)) + \frac{a_{12_i}}{\lambda_{1m}}\|V_{i_t}\|_\tau + \frac{b_{12_i}}{\lambda_{2m}}\|W_{i_t}\|_\tau, \\
\dot{W}_i(t) &= (\dot{z} - \dot{h}_i)^T P_{2_i}(z - h_i) + (z - h_i)^T P_{2_i}(\dot{z} - \dot{h}_i) \\
&= \left(\frac{1}{\varepsilon}(A_{21_i}x + A_{22_i}x_t + B_{21_i}z) - \dot{h}_i\right)^T P_{2_i}(z - h_i) \\
&\quad + (z - h_i)^T P_{2_i}\left(\frac{1}{\varepsilon}(A_{21_i}x + A_{22_i}x_t + B_{21_i}z) - \dot{h}_i\right) \\
&\leq \frac{a_{21_i}}{\lambda_{1m}}V_i(t) - \frac{\lambda_{\min}(Q_{2_i}) - \varepsilon_i b_{21_i}}{\varepsilon_i\lambda_{2M}}W_i(t) + \frac{a_{22_i}}{\lambda_{1m}}\|V_{i_t}\|_\tau + \frac{b_{22_i}}{\lambda_{2m}}\|W_{i_t}\|_\tau,
\end{aligned}
$$

(ii) for any $i \in \mathscr{S}_s$

$$
\begin{aligned}
\dot{V}_i(t) &\leq -\frac{\lambda_{\min}(Q_{1_i}) - a_{11_i}}{\lambda_{1M}}V_i(t) + \frac{b_{11_i}}{\lambda_{2m}}W_i(t) + \frac{a_{12_i}}{\lambda_{1m}}\|V_{i_t}\|_\tau + \frac{b_{12_i}}{\lambda_{2m}}\|W_{i_t}\|_\tau, \\
\dot{W}_i(t) &\leq \frac{a_{21_i}}{\lambda_{1m}}V_i(t) - \frac{\lambda_{\min}(Q_{2_i}) - \varepsilon_i^* b_{21_i}}{\varepsilon_i^*\lambda_{2M}}W_i(t) + \frac{a_{22_i}}{\lambda_{1m}}\|V_{i_t}\|_\tau + \frac{b_{22_i}}{\lambda_{2m}}\|W_{i_t}\|_\tau,
\end{aligned}
$$

where $\varepsilon_i^* \geq \epsilon_i > 0$. Combining $\dot{V}$ and $\dot{W}$ in a vector form yields, for $i \in \mathscr{S}_u$,

$$
\begin{pmatrix} \dot{V}(t) \\ \dot{W}(t) \end{pmatrix} \leq \begin{pmatrix} 2\gamma^* + \frac{a_{11_i}}{\lambda_{1m}} & \frac{b_{11_i}}{\lambda_{2m}} \\ \frac{a_{21_i}}{\lambda_{1m}} & -\frac{\lambda_{\min}(Q_{2_i}) - \varepsilon_i b_{21_i}}{\varepsilon_i\lambda_{2M}} \end{pmatrix}\begin{pmatrix} V(t) \\ W(t) \end{pmatrix} + \begin{pmatrix} \frac{a_{12_i}}{\lambda_{1m}} & \frac{b_{12_i}}{\lambda_{2m}} \\ \frac{a_{22_i}}{\lambda_{1m}} & \frac{b_{22_i}}{\lambda_{2m}} \end{pmatrix}\begin{pmatrix} \|V_{i_t}\|_\tau \\ \|W_{i_t}\|_\tau \end{pmatrix}
$$

and, by $A3(i)$ and Lemma 2.4, there exists $\xi_i > 0$ such that

$$
\begin{aligned}
V_i(t) &\leq (\|V_{i_{t_{k-1}}}\|_\tau + \|W_{i_{t_{k-1}}}\|_\tau)e^{\xi_i(t - t_{k-1})} \\
W_i(t) &\leq (\|V_{i_{t_{k-1}}}\|_\tau + \|W_{i_{t_{k-1}}}\|_\tau)e^{\xi_i(t - t_{k-1})}.
\end{aligned}
$$

Similarly, for any $i \in \mathscr{S}_s$, we get

$$\begin{pmatrix} \dot{V}(t) \\ \dot{W}(t) \end{pmatrix} \le \begin{pmatrix} -\frac{\lambda_{\min}(Q_{1_i})-a_{11_i}}{\lambda_{1M}} & \frac{b_{11_i}}{\lambda_{2m}} \\ \frac{a_{21_i}}{\lambda_{1m}} & -\frac{\lambda_{\min}(Q_{2_i})-\varepsilon_i^* b_{21_i}}{\varepsilon_i^*\lambda_{2M}} \end{pmatrix} \begin{pmatrix} V(t) \\ W(t) \end{pmatrix} + \begin{pmatrix} \frac{a_{12_i}}{\lambda_{1m}} & \frac{b_{12_i}}{\lambda_{2m}} \\ \frac{a_{22_i}}{\lambda_{1m}} & \frac{b_{22_i}}{\lambda_{2m}} \end{pmatrix} \begin{pmatrix} \|V_{i_t}\|_\tau \\ \|W_{i_t}\|_\tau \end{pmatrix}$$

and, by $A3$(ii) and Lemma 2.3, there exists $\zeta_i > 0$ such that

$$\begin{aligned} V_i(t) &\le (\|V_{i_{t_{k-1}}}\|_\tau + \|W_{i_{t_{k-1}}}\|_\tau)e^{-\zeta_i(t-t_{k-1})} \\ W_i(t) &\le (\|V_{i_{t_{k-1}}}\|_\tau + \|W_{i_{t_{k-1}}}\|_\tau)e^{-\zeta_i(t-t_{k-1})}. \end{aligned}$$

At the impulsive-switching moment, $t = t_k$, we have

$$\begin{aligned} V_i(t_k) &= x(t_k)^T P_{1_i} x(t_k) \\ &\le \lambda_{\max}\Big([I + B_k]^T P_{1_i} [I + B_k]\Big) x^T(t_k^-) x(t_k^-) \\ &= \alpha_k V_i(t_k^-), \end{aligned}$$

where $\alpha_k = \mu\lambda_{\max}^2(I + B_k)$. We also have at $t = t_k$

$$\begin{aligned} W_i(t_k) &= \Big(z(t_k) - h_i(t_k)\Big)^T P_{2_i} \Big(z(t_k) - h_i(t_k)\Big) \\ &= \Big\{z(t_k) + B_{21_i}^{-1}[A_{21_i}x(t_k) + A_{22_i}x_{t_k}]\Big\}^T P_{2_i} \Big\{z(t_k) + B_{21_i}^{-1}[A_{21_i}x(t_k) + A_{22_i}x_{t_k}]\Big\} \\ &= \Big\{[I + C_k]z(t_k^-) + B_{21_i}^{-1}[A_{21_i}[I + B_k]x(t_k^-) + A_{22_i}[I + B_k]x_{t_k^-}]\Big\}^T P_{2_i} \\ &\quad \times \Big\{[I + C_k]z(t_k^-) + B_{21_i}^{-1}[A_{21_i}[I + B_k]x(t_k^-) + A_{22_i}[I + B_k]x_{t_k^-}]\Big\} \\ &= \Big(z(t_k^-) - h_i(t_k^-)\Big)^T U_k^T P_{2_i} U_k \Big(z(t_k^-) - h_i(t_k^-)\Big) + x^T(t_k^-) R_{ik}^T P_{2_i} R_{ik} x(t_k^-) \\ &\quad + x_{t_k^-}^T S_{ik}^T P_{2_i} S_{ik} x_{t_k^-} - 2\Big(z(t_k^-) - h_i(t_k^-)\Big)^T U_k^T P_{2_i} R_{ik} x(t_k^-) \\ &\quad - 2\Big(z(t_k^-) - h_i(t_k^-)\Big)^T U_k^T P_{2_i} S_{ik} x_{t_k^-} - 2x^T(t_k^-) R_{ik}^T P_{2_i} S_{ik} x_{t_k^-} \\ &\le \|U_k\|^2 \|P_{2_i}\| \|z(t_k^-) - h_i(t_k^-)\|^2 + \|R_{ik}\|^2 \|P_{2_i}\| \|x(t_k^-)\|^2 \\ &\quad + \|S_{ik}\|^2 \|P_{2_i}\| \|x_{t_k^-}\|_\tau^2 + \|U_k\| \|P_{2_i}\| \|R_{ik}\| \Big(\|z(t_k^-) - h_i(t_k^-)\|^2 + \|x(t_k^-)\|^2\Big) \\ &\quad + \|U_k\| \|P_{2_i}\| \|S_{ik}\| \Big(\|z(t_k^-) - h_i(t_k^-)\|^2 + \|x_{t_k^-}\|_\tau^2\Big) \\ &\quad + \|R_{ik}\| \|P_{2_i}\| \|S_{ik}\| \Big(\|x(t_k^-)\|^2 + \|x_{t_k^-}\|_\tau^2\Big) \end{aligned}$$

$$\leq \lambda_{\max}(P_{2_i})\Big(\|U_k\| + \|R_{ik}\| + \|S_{ik}\|\Big)\Big\{\frac{\|U_k\|}{\lambda_{\min}(P_{2_i})}W_i(t_k^-) + \frac{\|R_{ik}\|}{\lambda_{\min}(P_{1_i})}V_i(t_k^-)$$
$$+ \frac{\|S_{ik}\|}{\lambda_{\min}(P_{1_i})}\|V_{i_{t_k^-}}\|_\tau\Big\}$$
$$\leq \lambda_{2M}\Big(\|U_k\| + r_k + s_k\Big)\Big\{\frac{\|U_k\|}{\lambda_{2m}}W_i(t_k^-) + \frac{r_k}{\lambda_{1m}}V_i(t_k^-) + \frac{s_k}{\lambda_{1m}}\|V_{i_{t_k^-}}\|_\tau\Big\}$$
$$= \gamma_k W_i(t_k^-) + \beta_k V_i(t_k^-) + \psi_k\|V_{i_{t_k^-}}\|_\tau,$$

where $\beta_k = \lambda_{2M}\Big(\|U_k\| + r_k + s_k\Big)\frac{r_k}{\lambda_{1m}}$, $\gamma_k = \lambda_{2M}\Big(\|U_k\| + r_k + s_k\Big)\frac{\|U_k\|}{\lambda_{2m}}$, $\psi_k = \lambda_{2M}\Big(\|U_k\| + r_k + s_k\Big)\frac{s_k}{\lambda_{1m}}$, $r_k = \max\{\|R_{ik}\| \mid \forall i \in \mathscr{S}\}$ and $s_k = \max\{\|S_{ik}\| \mid \forall i \in \mathscr{S}\}$.

For instance, if we run an unstable subsystem on the first interval and a stable one on the second interval, we get, respectively,

$$V_1(t) \leq \Big(\|V_{1_{t_0}}\|_\tau + \|W_{1_{t_0}}\|_\tau\Big)e^{\xi_1(t-t_0)},$$
$$V_2(t) \leq \Big(\|V_{2_{t_1}}\|_\tau + \|W_{2_{t_1}}\|_\tau\Big)e^{-\zeta_2(t-t_1)},$$

where

$$\|V_{2_{t_1}}\|_\tau \leq \alpha_1\mu\Big(\|V_{1_{t_0}}\|_\tau + \|W_{1_{t_0}}\|_\tau\Big)e^{\xi_1(t_1-t_0)},$$
$$\|W_{2_{t_1}}\|_\tau \leq \mu(\beta_1 + \gamma_1 + \psi_1)\Big(\|V_{1_{t_0}}\|_\tau + \|W_{1_{t_0}}\|_\tau\Big)e^{\xi_1(t_1-t_0)}.$$

Generally, one may have

$$V_N(t) \leq \prod_{i=1}^{l}\mu(\alpha_i + \beta_i + \gamma_i + \psi_i)e^{\xi_i(t_i-t_{i-1})} \times \prod_{j=l+1}^{N-l-1}\mu\big(\alpha_j + \beta_j + \gamma_j$$
$$+ \psi_j e^{\zeta_j\tau}\big)e^{\zeta_j\tau}e^{-\zeta_j(t_j-t_{j-1})} \times \Big(\|V_{1_{t_0}}\|_\tau + \|W_{1_{t_0}}\|_\tau\Big)e^{-\zeta_N(t-t_{N-1})}.$$

Making use of $A4$, we have

$$V_N(t) \leq \Big(\|V_{1_{t_0}}\|_\tau + \|W_{1_{t_0}}\|_\tau\Big)e^{-(\lambda^*-\nu)(t-t_0)}.$$

Similarly, we have

$$W_N(t) \leq \Big(\|V_{1_{t_0}}\|_\tau + \|W_{1_{t_0}}\|_\tau\Big)e^{-(\lambda^*-\nu)(t-t_0)}.$$

Then, there exists K_1 such that

$$\|x(t)\| \leq K_1(\|x_{t_0}\|_\tau + \|z_{t_0}\|_\tau)e^{-(\lambda^*-\nu)(t-t_0)/2}$$

and by the fact that

$$\|z\| - \|h_i\| \le \|z - h_i\| \le \frac{1}{\sqrt{\lambda_{2m}}} W_i^{1/2},$$

there exists K_2 such that

$$\|z(t)\| \le K_2(\|x_{t_0}\|_\tau + \|z_{t_0}\|_\tau)e^{-(\lambda^* - \nu)(t-t_0)/2}.$$

Hence,

$$\|x(t)\| + \|z(t)\| \le K(\|x_{t_0}\|_\tau + \|z_{t_0}\|_\tau)e^{-(\lambda^* - \nu)(t-t_0)/2},$$

where $K = K_1 + K_2$. This shows that the trivial solution of (10.2) is exponential stable.

Example 10.1 Consider the impulsive-switched system (10.2) with the following unstable and stable subsystems

$$\begin{aligned} \dot{x} &= x + 4z(t-1), \\ \varepsilon\dot{z} &= x(t-1) - z, \end{aligned}$$

and

$$\begin{aligned} \dot{x} &= -5x + z(t-1) \\ \varepsilon\dot{z} &= x(t-1) - z \end{aligned}$$

and, in the difference impulsive equations for any $k \in \mathbb{N}$, $B_k = -1/2$ and $C_k = -1/2$. In this example, the switching signal σ takes values in the set $\{1, 2\}$ alternatively.

For the unstable subsystems, when $\gamma = 3$, $\varepsilon = 0.4$, $Q_{1_1} = 13$ and $Q_{2_1} = 1$, then, from the Lyapunov matrix equations, we get $P_{1_1} = 3.25$ and $P_{2_1} = 0.5$, $\widetilde{A}_u = \begin{pmatrix} 10 & 0 \\ 0 & -6 \end{pmatrix}$ and $\widetilde{B}_u = \begin{pmatrix} 0 & 26 \\ 0.31 & 0 \end{pmatrix}$. In this case, the condition in $A3$(i) is satisfied. While for the stable subsystem, when $Q_{1_2} = 44$ and $Q_{2_2} = 8$, then, from the Lyapunov matrix equations, we get $P_{1_2} = 4.4$ and $P_{2_2} = 4$, and, from condition in $A3$(ii), we get $\varepsilon^* = 0.15$, $\widetilde{A}_s = \begin{pmatrix} -9 & 0 \\ 0 & -140 \end{pmatrix}$ and $\widetilde{B}_s = \begin{pmatrix} 0 & 8.8 \\ 6.15 & 0 \end{pmatrix}$. The dwell times for the unstable subsystem is 1.5 and for the stable one is 4. Figure 10.1 illustrates these results where unstable and stable subsystems are run alternatively. The set of switching or impulsive times is $\{t_k\}_{k=1}^{k=8} = \{1.5, 5.5, 7, 11, 12.5, 16.5, 18, 22\}$. For instance, $\sigma(t) = 1$ (or 2) for $t \in [0, 1.5)$(or $[1.5, 5.5)$), respectively.

In the following theorem, we show how impulses can play as a stabilizer in some linear impulsive systems with unstable subsystems.

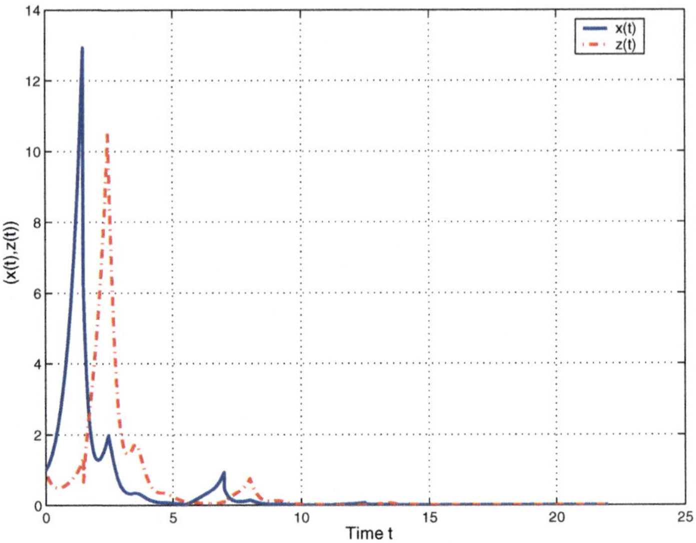

Fig. 10.1 Impulsive-switched system with unstable and stable subsystems

Theorem 10.2 *Consider system* (10.2) *with* $\mathscr{S} = \{1, 2, \ldots, N\}$. *Assume that the following assumptions hold:*

A1. *for any* $i \in \mathscr{S}$, A_{11_i} *has eigenvalues with positive real parts;*
A2. *assumptions A2 and A3(i) of Theorem 10.1 hold;*
A3. *there exists a constant* $\vartheta \geq 1$ *such that*

$$\ln\Big(\vartheta\mu(\alpha_i + \beta_i + \gamma_i + \psi_i)\Big) + \xi_i(t_{k+1} - t_k) \leq 0,$$

where μ, α_i, β_i, γ_i, ψ_i *and* ξ_i *are defined in Theorem 10.1.*

Then, the trivial solution of (10.2) *is stable if* $\vartheta = 1$ *and asymptotically stable if* $\vartheta > 1$.

Proof For all $t \geq t_0$ with $t_0 \in \mathbb{R}_+$, let $x(t)$ and $z(t)$ be the solution of (10.2). Define $V_i(t) = x^T P_{1_i} x$ and $W_i(t) = (z - h_i)^T P_{2_i}(z - h_i)$. Then, the time derivative of V_i and W_i along the trajectories of system (11.2) are

$$\dot{V}_i(t) \leq (2\gamma + \frac{a_{11_i}}{\lambda_{1m}})V_i(t) + \frac{b_{11_i}}{\lambda_{2m}} W_i(t) + \frac{a_{12_i}}{\lambda_{1m}} \|V_{i_t}\|_\tau + \frac{b_{12_i}}{\lambda_{2m}} \|W_{i_t}\|_\tau$$
$$\dot{W}_i(t) \leq \frac{a_{21_i}}{\lambda_{1m}} V_i(t) - \frac{\lambda_{\min}(Q_{2_i}) - \varepsilon_i b_{21_i}}{\varepsilon_i \lambda_{2M}} W_i(t) + \frac{a_{22_i}}{\lambda_{1m}} \|V_{i_t}\|_\tau + \frac{b_{22_i}}{\lambda_{2m}} \|W_{i_t}\|_\tau.$$

Then, there exists a positive constant ξ_i such that

$$V_i(t) \leq (\|V_{i_{t_{k-1}}}\|_\tau + \|W_{i_{t_{k-1}}}\|_\tau)e^{\xi_i(t-t_{k-1})}$$
$$W_i(t) \leq (\|V_{i_{t_{k-1}}}\|_\tau + \|W_{i_{t_{k-1}}}\|_\tau)e^{\xi_i(t-t_{k-1})}.$$

From Theorem 10.1, we have at $t = t_k$

$$V_i(t_k) \leq \alpha_k V_i(t_k^-)$$
$$W_i(t_k) \leq \beta_k V_i(t_k^-) + \gamma_k W_i(t_k^-) + \psi_k \|V_{i_{t_k^-}}\|_\tau.$$

We also have, for $t \in [t_k, t_{k+1})$,

$$\begin{aligned}
V_i(t) &\leq (\|V_{1_{t_0}}\|_\tau + \|W_{1_{t_0}}\|_\tau)e^{\xi_1(t_1-t_0)}\mu(\alpha_1+\beta_1+\gamma_1+\psi_1)e^{\xi_2(t_2-t_1)} \\
&\quad \times \mu(\alpha_2+\beta_2+\gamma_2+\psi_2)e^{\xi_2(t_3-t_2)} \cdots \mu(\alpha_k+\beta_k+\gamma_k+\psi_k)e^{\xi_i(t_{k+1}-t_k)} \\
&= (\|V_{1_{t_0}}\|_\tau + \|W_{1_{t_0}}\|_\tau)\frac{1}{\vartheta k}e^{\xi_1(t_1-t_0)}\vartheta\mu(\alpha_1+\beta_1+\gamma_1+\psi_1)e^{\xi_2(t_2-t_1)} \\
&\quad \times \vartheta\mu(\alpha_2+\beta_2+\gamma_2+\psi_2)e^{\xi_2(t_3-t_2)} \cdots \vartheta\mu(\alpha_k+\beta_k+\gamma_k+\psi_k)e^{\xi_i(t_{k+1}-t_k)} \\
&\leq (\|V_{1_{t_0}}\|_\tau + \|W_{1_{t_0}}\|_\tau)\frac{1}{\vartheta k}e^{\xi_1(t_1-t_0)}.
\end{aligned}$$

Similarly,

$$W_i(t) \leq (\|V_{1_{t_0}}\|_\tau + \|W_{1_{t_0}}\|_\tau)\frac{1}{\vartheta k}e^{\xi_1(t_1-t_0)}.$$

From Theorem 10.1, there exists a positive constant K such that

$$\|x(t)\| + \|z(t)\| \leq \frac{K}{\sqrt{\vartheta k}}(\|x_{t_0}\|_\tau + \|z_{t_0}\|_\tau)e^{\xi_1(t_1-t_0)/2}.$$

Clearly, if $\vartheta = 1$, then the trivial solution of system (10.2) is stable and, if $\vartheta > 1$ and $k \to \infty$, then the trivial solution of the system is asymptotically stable. This completes the proof.

Example 10.2 Consider the impulsive-switched system (10.2) with the following unstable subsystems

$$\dot{x} = x + 3z(t-1),$$
$$\varepsilon\dot{z} = 2x(t-1) - 2z, \quad \varepsilon = 0.7$$

and

$$\dot{x} = x + 2z(t-1),$$
$$\varepsilon\dot{z} = 4x(t-1) - 2z, \quad \varepsilon = 0.7$$

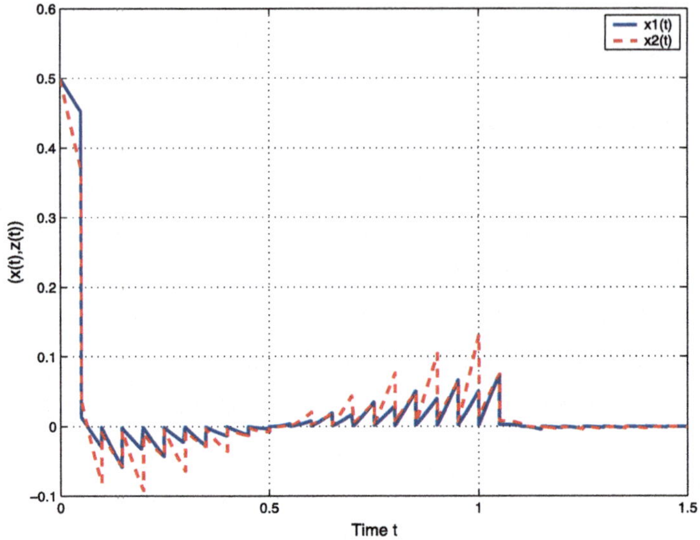

Fig. 10.2 Impulsive-switched system with all unstable subsystems

and difference equations are $\Delta x = -0.97x(t)$ and $\Delta z = -0.9z(t)$. The switching signal σ takes values in $\{1, 2\}$ alternatively.

When $\gamma = 2$, $Q_{1_1} = 2$ and $Q_{2_1} = 2$, then we get $P_{1_1} = 1$ and $P_{2_1} = 0.5$, while, if $Q_{1_2} = 3$ and $Q_{2_2} = 1$, then we get $P_{1_2} = 1.5$ and $P_{2_2} = 0.25$, so that $\mu = 2$. We also get, for the first subsystem, $\lambda(\widetilde{A}_1^T + \widetilde{A}_1) = \{-2.3571, 14\}$ and $\|\widetilde{B}_1\| = 12$, so that the growth rates are $\xi_1 = \{19, 10.8214\}$. For the second subsystem, $\lambda(\widetilde{A}_2^T + \widetilde{A}_2) = \{-0.9286, 14\}$ and $\|\widetilde{B}_2\| = 12$, so that the growth rates are $\xi_2 = \{19, 11.5357\}$. The impulse parameters are, for any $k \in \mathbb{N}$, $\alpha_k = 0.0018$, $\beta_k = 0$, $\gamma_k = 0.048$ and $\psi_k = 0.0672$. A simple check shows that $A3$ holds if $\vartheta \in [1, 4.2735)$. Taking $\vartheta = 2$ and $\xi_1 = 10.8214$ gives $t_{k+1} - t_k \leq 0.0702$ and $\xi_2 = 11.5357$ gives $t_{k+1} - t_k \leq 0.0658$. Thus, if we choose $T_D = 0.0658$, then the switching or impulsive times are $\{t_k\}_{k=1}^{k=30} = kT_D$. Figure 10.2 shows the simulation results.

10.2.2 Nonlinear Systems

Consider the following nonlinear impulsive-switched system

$$
\begin{aligned}
\dot{x} &= A_{11_i} + g_i(x, x_t, z, z_t), \qquad t \neq t_k \\
\varepsilon \dot{z} &= B_{21_i} z + B_i(x, x_t), \qquad t \neq t_k \\
\Delta x &= B_k x(t), \qquad t = t_k \\
\Delta z &= C_k z(t), \qquad t = t_k
\end{aligned}
\tag{10.6}
$$

where $i \in \mathscr{S} = \mathscr{S}_u \cup \mathscr{S}_s$, and the $n \times n$ matrix B_{21_i} is nonsingular and Hurwitz.

In the following theorem, we state and prove exponential stability of the trivial solution of system (10.6).

Theorem 10.3 *The trivial solution of* (10.6) *is exponentially stable if the following assumptions hold:*

A1. *assumption A*1 *of Theorem 10.1 holds;*

A2. *(i) there exist positive constants* a_{11_i}, a_{12_i}, a_{21_i}, a_{22_i}, b_{11_i}, b_{12_i}, b_{21_i} *and* b_{22_i} *such that*

$$2x^T P_{1_i} g_i(x, x_t, z, z_t) \le a_{11_i}\|x\|^2 + a_{12_i}\|x_t\|_\tau^2 + b_{11_i}\|z - h_i\|^2 + b_{12_i}\|(z - h_i)_t\|_\tau^2,$$
$$-2(z - h_i)^T P_{2_i}\dot{h}_i \le a_{21_i}\|x\|^2 + a_{22_i}\|x_t\|_\tau^2 + b_{21_i}\|z - h_i\|^2 + b_{22_i}\|(z - h_i)_t\|_\tau^2,$$

where $h_i(t) = -B_{21_i}^{-1}B_i(x(t), x_t)$*, and* P_{1_i} *and* P_{2_i} *are positive-definite matrices satisfying Lyapunov matrix equations*

$$A_{11_i}^T P_{1_i} + P_{1_i}A_{11_i} = -Q_{1_i},$$
$$B_{21_i}^T P_{2_i} + P_{2_i}B_{21_i} = -Q_{2_i},$$

for any given positive-definite matrices Q_{1_i} *and* Q_{2_i}*;*

(ii) there exist positive constants a, b and c such that

$$2\Big(z(t_k^-) - h_i(t_k^-)\Big)^T [I + C_k]^T P_{2_i}\Big\{[I + C_k]h_i(t_k^-) - h_i(t_k)\Big\}$$
$$+ \Big\{[I + C_k]h_i(t_k^-) - h_i(t_k)\Big\}^T P_{2_i}\Big\{[I + C_k]h_i(t_k^-) - h_i(t_k)\Big\}$$
$$\le a\|z(t_k^-) - h_i(t_k^-)\|^2 + b\|x(t_k^-)\|^2 + c\|x_{t_k^-}\|_\tau^2,$$

where $h_i(t_k) = -B_{21_i}^{-1}B_i(x(t_k), x_{t_k})$*;*

A3. *assumption A*3 *of Theorem 10.1 holds; and*

A4. *assumption A*4 *of Theorem 10.1 holds where* $\alpha_k = \mu_1\lambda_{\max}^2([I + B_k])$, $\beta_k = b/\lambda_{1m}$, $\gamma_k = \mu_2\lambda_{\max}^2([I + C_k]) + a$ *and* $\psi_k = c/\lambda_{1m}$.

Proof For all $t \ge t_0$ with $t_0 \in \mathbb{R}_+$, let $x(t)$ and $z(t)$ be the solution of (10.6). Define $V_i(t) = x^T(t)P_{1_i}x(t)$ and $W_i(t) = (z - h_i)^T(t)P_{2_i}(z - h_i)(t)$. Then, the time derivative of V_i and W_i along the trajectories of x and z are:
(i) for any $i \in \mathscr{S}_u$,

$$\dot{V}_i(t) \le (2\gamma + \frac{a_{11_i}}{\lambda_{1m}})V_i(t) + \frac{b_{11_i}}{\lambda_{2m}}W_i(t) + \frac{a_{12_i}}{\lambda_{1m}}\|V_{i_t}\|_\tau + \frac{b_{12_i}}{\lambda_{2m}}\|W_{i_t}\|_\tau$$
$$\dot{W}_i(t) \le \frac{a_{21_i}}{\lambda_{1m}}V_i(t) - \frac{\lambda_{\min}(Q_{2_i}) - \varepsilon_i b_{21_i}}{\varepsilon_i\lambda_{2M}}W_i(t) + \frac{a_{22_i}}{\lambda_{1m}}\|V_{i_t}\|_\tau + \frac{b_{22_i}}{\lambda_{2m}}\|W_{i_t}\|_\tau,$$

(ii) for any $i \in \mathscr{S}_s$,

$$\dot{V}_i(t) \le -\frac{\lambda_{\min}(Q_{1_i}) - a_{11_i}}{\lambda_{1M}} V_i(t) + \frac{b_{11_i}}{\lambda_{2m}} W_i(t) + \frac{a_{12_i}}{\lambda_{1m}} \|V_{i_t}\|_\tau + \frac{b_{12_i}}{\lambda_{2m}} \|W_{i_t}\|_\tau$$
$$\dot{W}_i(t) \le \frac{a_{21_i}}{\lambda_{1m}} V_i(t) - \frac{\lambda_{\min}(Q_{2_i}) - \varepsilon_i^* b_{21_i}}{\varepsilon_i^* \lambda_{2M}} W_i(t) + \frac{a_{22_i}}{\lambda_{1m}} \|V_{i_t}\|_\tau + \frac{b_{22_i}}{\lambda_{2m}} \|W_{i_t}\|_\tau.$$

Then, as done before there exists positive constants ξ_i (for any $i \in \mathscr{S}_u$) such that

$$V_i(t) \le (\|V_{i_{t_{k-1}^+}}\|_\tau + \|W_{i_{t_{k-1}^+}}\|_\tau) e^{\xi_i (t - t_{k-1})}$$
$$W_i(t) \le (\|V_{i_{t_{k-1}^+}}\|_\tau + \|W_{i_{t_{k-1}^+}}\|_\tau) e^{\xi_i (t - t_{k-1})}$$

and ζ_i (for $i \in \mathscr{S}_s$) such that

$$V_i(t) \le (\|V_{i_{t_{k-1}}}\|_\tau + \|W_{i_{t_{k-1}}}\|_\tau) e^{-\zeta_i (t - t_{k-1})}$$
$$W_i(t) \le (\|V_{i_{t_{k-1}}}\|_\tau + \|W_{i_{t_{k-1}}}\|_\tau) e^{-\zeta_i (t - t_{k-1})}.$$

At the impulsive-switching moments $t = t_k$, we have

$$V_i(t_k) \le \alpha_k V_i(t_k^-),$$

where $\alpha_k = \mu \lambda_{\max}^2 (I + B_k)$, and

$$\begin{aligned}
W_i(t_k) &= \left(z(t_k) - h_i(t_k)\right)^T P_{2_i} \left(z(t_k) - h_i(t_k)\right) \\
&= \left(z(t_k^-) - h_i(t_k^-)\right)^T [I + C_k]^T P_{2_i} [I + C_k] \left(z(t_k^-) - h_i(t_k^-)\right) \\
&\quad + 2\left(z(t_k^-) - h_i(t_k^-)\right)^T [I + C_k]^T P_{2_i} \left\{[I + C_k] h_i(t_k^-) - h_i(t_k)\right\} \\
&\quad + \left\{[I + C_k] h_i(t_k^-) - h_i(t_k)\right\}^T P_{2_i} \left\{[I + C_k] h_i(t_k^-) - h_i(t_k)\right\} \\
&\le \lambda_{\max}\left([I + C_k]^T P_{2_i} [I + C_k]\right) \|z(t_k^-) - h_i(t_k^-)\|^2 + a\|z(t_k^-) - h_i(t_k^-)\|^2 \\
&\quad + b\|x(t_k^-)\|^2 + c\|x_{t_k^-}\|_\tau^2 \\
&= \beta_k V_i(t_k^-) + \gamma_k W_i(t_k^-) + \psi_k \|V_{i_{t_k^-}}\|_\tau,
\end{aligned}$$

where $\beta_k = b/\lambda_{1m}$, $\gamma_k = (\lambda_{2M} \lambda_{\max}^2 [I + C_k] + a)/\lambda_{2m}$ and $\psi_k = c/\lambda_{1m}$. The rest of the proof is similar to that of Theorem 10.1; thus, it is left here as an exercise.

Example 10.3 Consider impulsive-switched system (10.6) with the following unstable and stable subsystems

$$\begin{aligned}\dot{x} &= 0.1x + \sin z(t-1)\\ \varepsilon\dot{z} &= 0.1x - z,\end{aligned}$$

and

$$\begin{aligned}\dot{x} &= -10x + \ln(1 + x^2(t-1)) + z\\ \varepsilon\dot{z} &= x - 2z\end{aligned}$$

and, in the difference equation, $B_k = -1/2$ and $C_k = -1/2$.

For the unstable subsystem, define $V_u(x) = 0.5x^2$ and $W_u(z-h) = 0.5(z-h)^2$, where $h = 0.1x$. Then, one can find $\dot{V}_u(x) \leq 1.4V_u + \|W_{u_t}\|_\tau$ and $\dot{W}_u \leq (-2/\varepsilon + 0.12)W_u + 0.01V_u + 0.01\|V_u\|_\tau + 0.1\|W_{u_t}\|_\tau$, $\widetilde{A}_u = \begin{pmatrix} 1.4 & 0 \\ 0.01 & -2/\varepsilon + 0.12 \end{pmatrix}$ and $\widetilde{B}_u = \begin{pmatrix} 0 & 1 \\ 0.01 & 0.1 \end{pmatrix}$. Taking $\varepsilon = 0.1$, the growth rate are $\xi = \{1.85, 4.8\}$. While for the stable subsystem, defining $V_s(x) = 0.5x^2$, $W_s(z-h) = 0.5(z-h)^2$ where $h = 0.5x$ give $\dot{V}_s \leq -14V_s + W_s + 4\|V_{s_t}\|_\tau$ and $\dot{W}_s \leq (-4/\varepsilon^* + 11.5)W_s + 5.5V_s + 2\|V_{s_t}\|_\tau$; thus, $\widetilde{A}_s = \begin{pmatrix} -14 & 1 \\ 5.5 & -4/\varepsilon^* + 11.5 \end{pmatrix}$ and $\widetilde{B}_s = \begin{pmatrix} 4 & 0 \\ 2 & 0 \end{pmatrix}$. By $A3(ii)$, we get $\varepsilon^* = 0.2341$; if we take $\varepsilon = 0.1 \in (0, 0.2341]$, the decay rates are $\zeta = \{1.5279, 2.4432\}$. The dwell times for the unstable subsystems is 1.1 and the stable one is 5. Figure 10.3 shows these results after running unstable and stable subsystems alternatively.

In the following theorem, we state sufficient conditions to guarantee stability and asymptotic stability of systems (10.6) with all unstable subsystems.

Theorem 10.4 *Consider the impulsive-switched nonlinear system in* (10.6) *with* $\mathscr{S} = \{1, 2, \ldots, N\}$. *Assume that the following assumptions are satisfied:*

A1. *for any* $i \in \mathscr{S}$, A_{11_i} *has eigenvalues with positive real parts;*
A2. *assumption A2 of Theorem 10.3 and A3(i) of Theorem 10.1 hold; and*
A3. *there exists a constant* $\vartheta \geq 1$ *such that*

$$\ln\Big(\vartheta\mu(\alpha_i + \beta_i + \gamma_i + \psi_i)\Big) + \xi_i(t_{k+1} - t_k) \leq 0,$$

where μ *and* ξ_i *are defined in Theorem 10.1 and* α_i, β_i, γ_i *and* ψ_i *are defined in Theorem 10.3.*

Then, the trivial solution of system (10.6) *is stable if* $\vartheta = 1$ *and it is asymptotically stable if* $\vartheta > 1$.

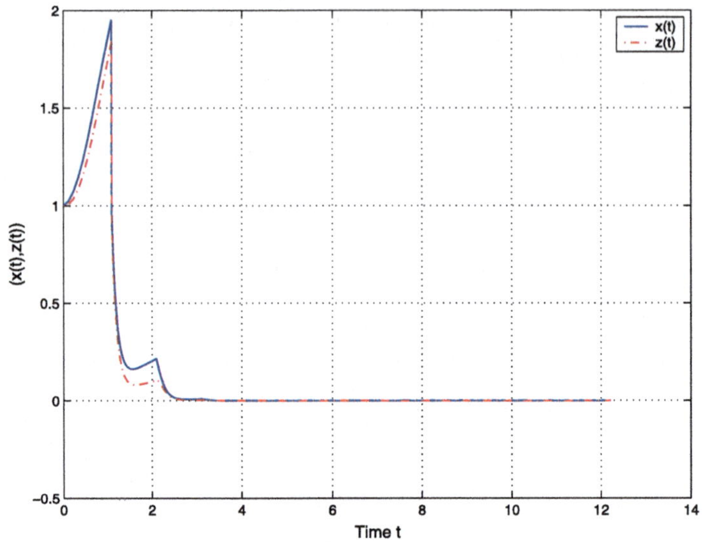

Fig. 10.3 Impulsive-switched system with unstable and stable linear subsystems

The proof of this theorem is a consequence of the previous theorems; thus, it is left here as an exercise.

10.3 Notes and Comments

Throughout this chapter, an impulsive-switched singularly perturbed system with time delay has been addressed. Particularly, some stability properties of the system have been presented. The material of this chapter is taken from [1]. We have shown that exponential stability of the system that consists of unstable and stable subsystems is guaranteed if the total activation time of stable subsystems be larger than that of the unstable ones. We have also explored that impulses do contribute to achieve stability properties of systems consisting of unstable subsystems. Multiple Lyapunov functions technique and dwell-time approach are used to analyze the qualitative properties of the system.

Reference

1. Alwan MS, Liu XZ (2009) Stability of singularly perturbed switched systems with time delay and impulsive effects. Nonlinear Anal Theory Method Appl 71(9):4297–4308

Chapter 11
Stabilization and State Estimation via Sliding Mode Control

This chapter deals with designing a nonlinear sliding mode control (SMC) and nonlinear sliding mode observer (SMO) for a class of linear time-invariant (LTI) singularly perturbed systems (SPS) subject to impulsive effects. As treaded in the last two chapters, the continuous states of the system are viewed as a large-scale interconnected system with two-timescale (slow and fast) subsystems. The impulses are considered as a perturbation to the system. To analyze the stabilization and state estimation problems, Lyapunov function technique is used. As will be seen, the goal is to design a SMC law through the slow *reduced* order subsystems to achieve closed-loop stability of the *full*-order system. This approach in turn results in lessening some unnecessary sufficient conditions on the fast subsystem. Later, assuming that partial output measurement of the slow subsystem is available, a similar control design is adopted to estimate the states of full-order SPS, where a sliding mode modification of a Luenberger observer is used.

11.1 Problem Formulation

Consider the following impulsive singularly perturbed system with control feedback

$$\dot{x} = A_{11}x + A_{12}z + Bu, \qquad t \neq \tau_k, \tag{11.1a}$$
$$\varepsilon\dot{z} = A_{21}x + A_{22}z, \qquad t \neq \tau_k, \tag{11.1b}$$
$$x(t) = (I + E_k)x(t^-), \qquad t = \tau_k, \tag{11.1c}$$
$$z(t) = (I + F_k)z(t^-), \qquad t = \tau_k, \tag{11.1d}$$
$$x(0) = x_0, \qquad z(0) = z_0, \tag{11.1e}$$

where $x \in \mathbb{R}^n$ and $z \in \mathbb{R}^m$ are, respectively, the slow and fast state vectors of the system, $u \in \mathbb{R}^r$ is the system $r-$dimensional feedback control or input of the form

M. S. Alwan and X. Liu, *Theory of Hybrid Systems: Deterministic and Stochastic*, Nonlinear Physical Science, https://doi.org/10.1007/978-981-10-8046-3_11

Kx with $K \in \mathbb{R}^{r\times n}$ being the control gain matrix, ε is a small positive parameter, $A_{11} \in \mathbb{R}^{n\times n}$, $A_{12} \in \mathbb{R}^{n\times m}$, $A_{21} \in \mathbb{R}^{m\times n}$, $A_{22} \in \mathbb{R}^{m\times m}$ are real constant matrices, $B \in \mathbb{R}^{n\times r}$ is the control matrix, and $E_k \in \mathbb{R}^{n\times n}$ and $F_k \in \mathbb{R}^{m\times m}$ are the impulsive gain matrices. For all $k \in \mathbb{N}$, τ_k form a strictly increasing sequence of impulsive moments $\{\tau_k\}_{k\in\mathbb{N}}$ with $\tau_1 > 0$ and $\lim_{k\to\infty} \tau_k = \infty$. Throughout this chapter, we assume that the solution is right-continuous (i.e. $x(\tau_k^+) = x(\tau_k)$ and $z(\tau_k^+) = z(\tau_k)$ for all $k \in \mathbb{N}$) and the matrix A_{22} is nonsingular and Hurwitz.

In the following, we state the definitions of exponential stability of impulsive SPS, where we assume that there is no impulsive action at the initial time.

Definition 11.1 The trivial solution of system (11.1) is said to be globally exponentially stabilized by the feedback control law u if there exist two positive constants K and λ such that

$$\|x(t)\| + \|z(t)\| \le K\big(\|x_0\| + \|z_0\|\big)e^{-\lambda t}, \qquad \forall t \ge t_0 \text{ and } t_0 \in \mathbb{R}_+$$

where $(x(t)\ z(t))^T$ is any solution vector of (11.1). Particularly, if this relation holds with $u \equiv 0$, then the trivial solution of (11.1) is said to be globally exponentially stable.

11.2 Slow Sliding Mode Control Design

In this section, we start with designing the SMC through the slow reduced, nonimpulsive subsystem, then it is carried over to stabilize the full-order SPS under the impulsive effects. Moreover, it is reasonable to assume that the system is impulsive-free during the reachability stage, because the system states reach the sliding surface in a finite, short period of time.

Toward our goal, for any $t \ne \tau_k$, setting $\varepsilon = 0$ in (11.1b) yields

$$\dot{x} = A_{11}x + A_{12}z + Bu, \tag{11.2a}$$

$$0 = A_{21}x + A_{22}z. \tag{11.2b}$$

From (11.2b), we get

$$z = h(x) = -A_{22}^{-1}A_{21}x \tag{11.3}$$

and, by substituting z into (11.2a), we obtain the *reduced* subsystem

$$\dot{x}_s = A_0x_s + B_0u_s, \tag{11.4}$$

where $A_0 = A_{11} - A_{12}A_{22}^{-1}A_{21}$ and $B_0 = B$.

11.2.1 Sliding Mode Control with Multiple Inputs

Consider the $r-$dimensional sliding mode hyper-surface defined by the vector-valued function

$$S_s(x_s) = \begin{bmatrix} s_1(x_s) \\ s_2(x_s) \\ \vdots \\ s_r(x_s) \end{bmatrix}_{r\times 1} = \begin{bmatrix} c_1\, x_s \\ c_2\, x_s \\ \vdots \\ c_r\, x_s \end{bmatrix}_{r\times 1} = \begin{bmatrix} c_1 \\ c_2 \\ \vdots \\ c_r \end{bmatrix}_{r\times n} x_s = C_s\, x_s, \tag{11.5}$$

where $s_i(x_s)$ is a scalar-valued function which represents the ith sliding mode hyper-surface and is defined by $s_i(x_s) = c_i x_s$ with $c_i \in \mathbb{R}^{1\times n}$ $(i = 1, 2, \ldots r)$ and $C_s \in \mathbb{R}^{r\times n}$. Then, the time derivative of each $s_i(x_s)$ is given by

$$\dot{s}_i(x_s(t)) = (\nabla s_i(x_s))^T \dot{x}_s = \frac{\partial s_i(x_s)}{\partial x_s}\frac{dx_s}{dt} = c_i \dot{x}_s, \qquad i = 1, 2, \ldots r \tag{11.6}$$

where ∇s_i stands for the gradient of s_i, i.e. $\nabla s_i(x_s) = \left[\frac{\partial s_{i1}(x_s)}{\partial x_{s1}} \;\; \frac{\partial s_{i2}(x_s)}{\partial x_{s2}} \;\; \cdots \;\; \frac{\partial s_{ir}(x_s)}{\partial x_{sn}}\right]^T$ with $x_s = [x_{s1}\; x_{s2}\; \cdots\; x_{sn}]^T$. In a matrix form, (11.6) becomes

$$\dot{S}_s(x_s(t)) = \begin{bmatrix} (\nabla s_1(x_s))^T \\ (\nabla s_2(x_s))^T \\ \vdots \\ (\nabla s_r(x_s))^T \end{bmatrix}_{r\times n} \dot{x}_s = C_s \dot{x}_s(t). \tag{11.7}$$

Thus, along the trajectories of (11.4), we have

$$\dot{S}_s(x_s) = C_s A_0 x_s + C_s B_0 u_s = 0 \tag{11.8}$$

which leads to the $r-$ *dimensional equivalent control*

$$u_s^{eq} = -(C_s B_0)^{-1} C_s A_0 x_s, \tag{11.9}$$

where $B_0 \in \mathbb{R}^{n\times r}$ and $(C_s B_0)^{-1}$ is the inverse matrix of $C_s B_0 \in \mathbb{R}^{r\times r}$. Substituting u_s^{eq} into (11.4) leads to the corresponding closed-loop *equivalent reduced system*. The $n \times n$ matrix A_{eq} is stable as it has $n - r$ eigenvalues in the left half of the complex plane, and r zero eigenvalues to endure the system's motion on the sliding surface.

11.2.2 Reachability Analysis

To analyze the motion of the reduced slow system with multiple inputs outside the sliding surface (i.e. $S_s(x_s) \neq 0$), we define

$$V\big(S_s(x_s)\big) = \frac{1}{2} S_s^T(x_s) S_s(x_s) \tag{11.10}$$

and require that the time derivative

$$\dot{V}(S_s(x_s)) = \frac{\partial V}{\partial S_s(x_s)} \frac{dS_s(x_s)}{dt} = S_s^T(x_s)\dot{S}_s(x_s) = S_s^T(x_s)\big(C_s A_0 x_s + C_s B_0 u_s\big) < 0$$

which is guaranteed if the r−dimensional control is given by

$$u_s(t) = u_s^{eq}(t) - (C_s B_0)^{-1}\text{diag}(\eta)\text{Sgn}(S_s(x_s)), \tag{11.11}$$

where $\text{diag}(\eta)$ is an $r \times r$ diagonal matrix with diagonal elements being equal to positive constant numbers η_i (for $i = 1, 2, \ldots, r$) and Sgn refers to the r−dimensional signum vector function defined as follows:

$$\text{Sgn}(S_s(x_s)) = \begin{bmatrix} \text{sgn}(s_1(x_s)) \\ \text{sgn}(s_2(x_s)) \\ \vdots \\ \text{sgn}(s_r(x_s)) \end{bmatrix} \text{ where } \text{sgn}(s_i(x_s)) = \begin{cases} 1, & \text{if } s_i(x_s) > 0 \\ 0, & \text{if } s_i(x_s) = 0 \\ -1, & \text{if } s_i(x_s) < 0. \end{cases} \tag{11.12}$$

Therefore, the continuous closed-loop full system outside the sliding surface is given by

$$\dot{x}_s = A_{11}x_s + A_{12}z + Bu_s(t), \qquad t \neq \tau_k, \tag{11.13a}$$

$$\varepsilon\dot{z} = A_{21}x_s + A_{22}z, \qquad t \neq \tau_k, \tag{11.13b}$$

$$x_s(0) = x_{s0}, \qquad z(0) = z_0. \tag{11.13c}$$

Clearly, on the sliding surface $S(x_s) = 0$, we have $\text{Sgn}(S_s(x_s)) = 0$ which leads to $u_s(t) = u_s^{eq}(t)$ and, hence, the continuous closed-loop system during the sliding motion is given by

$$\dot{x}_s = A_{11}x_s + A_{12}z + Bu_s^{eq}(t), \qquad t \neq \tau_k, \tag{11.14a}$$

$$\varepsilon\dot{z} = A_{21}x_s + A_{22}z, \qquad t \neq \tau_k, \tag{11.14b}$$

$$x_s(0) = x_{s0}, \qquad z(0) = z_0. \tag{11.14c}$$

We now aim to apply the designed control law with multiple inputs of the reduced slow system in (11.11) to establish the stability property of the closed-loop system of the full-order impulsive SPS.

Theorem 11.1 *Assume that the following assumptions hold:*

(i) the reduced slow subsystem (or the matrix pair (A_0, B_0)) is stabilizable and A_{22} is Hurwitz;

(ii) for all $t \neq \tau_k$, there exist positive constants a_{21} and a_{22} such that

$$-2\big(z - h(x)\big)^T P\dot{h}(x) \leq a_{21}x^T x + a_{22}\big(z - h(x)\big)^T \big(z - h(x)\big),$$

where P is an $m \times m$ positive-definite matrix satisfying the Lyapunov matrix equation $A_{22}^T P + PA_{22} = -I$ and $\dot{h}(x) = \dfrac{\partial h(x)}{\partial x}\dot{x}$ with $\dfrac{\partial h(x)}{\partial x}$ being the $m \times n$ Jacobian matrix;

(iii) there exists a positive constant ε^* such that $-\tilde{A}$ is an M-matrix, where

$$\tilde{A} = \begin{bmatrix} \frac{1}{\alpha_2}\max\{Re[\lambda(\tilde{A}_{11})]\} & \frac{1}{2\gamma_1\beta_1} \\ \frac{a_{21}}{\alpha_1} & -\left(\frac{1}{\beta_2\varepsilon^*} - \frac{a_{22}}{\beta_1}\right) \end{bmatrix},$$

where $\alpha_1, \alpha_2, \beta_1$ and β_2 are positive constants that will be defined later, $\tilde{A}_{11} = A_{11} + \frac{\gamma_1}{2}A_{12}A_{12}^T - A_{12}A_{22}^{-1}A_{21} - B_1(CB_0)^{-1}CA_0$ with γ_1 being a positive constant and $\max\{Re[\lambda(\tilde{A}_{11})]\}$ is the maximum of real parts of the eigenvalues of $\tilde{A}_{11}$; and

(iv) for any $i = 1, 2, \ldots, k$, the time between impulses satisfy

$$t_i - t_{i-1} > \frac{1}{\vartheta}\ln(\alpha_{1i} + \alpha_{2i} + \beta_{2i}),$$

with ϑ, α_{1i}, α_{2i} and β_{2i} being positive constants such that $\alpha_{1i} + \alpha_{2i} + \beta_{2i} > 1$.

Then, the SMC law (11.11) guarantees that the closed-loop of the full-order impulsive system be globally exponentially stable for $\varepsilon \in (0, \varepsilon^*]$.

Proof For all $t \geq t_0$ with $t_0 \in \mathbb{R}_+$, let $x(t) = x(t; t_0, x_0, z_0)$ and $z(t) = z(t; t_0, x_0, z_0)$ be the solution of (11.1). Define $V(x) = \frac{1}{2}x^T x$ and $W(z - h(x)) = (z - h(x))^T P(z - h(x))$ as Lyapunov function candidates for the slow and fast subsystems, respectively. Then, there exist positive constants $\alpha_1 \leq \frac{1}{2}$, $\alpha_2 \geq \frac{1}{2}$, $\beta_1 \leq \lambda_{\min}(P)$ and $\beta_2 \geq \lambda_{\max}(P)$ such that

$$\alpha_1\|x\|^2 \leq V(x) \leq \alpha_2\|x\|^2, \tag{11.15}$$

$$\beta_1\|(z - h(x))\|^2 \leq W(z - h(x)) \leq \beta_2\|(z - h(x))\|^2. \tag{11.16}$$

The time derivative of V along the continuous trajectories of x during the sliding motion ($S_s = 0$) is given by

$$\begin{aligned}
\dot{V}(x) &= x^T\Big(A_{11}x + A_{12}(z - h(x)) + A_{12}h(x) - B_1(CB_0)^{-1}CA_0x\Big) \\
&\leq x^T\Big(A_{11} + \frac{\gamma_1}{2}A_{12}A_{12}^T - A_{12}A_{22}^{-1}A_{21} - B_1(CB_0)^{-1}CA_0\Big)x \\
&\quad + \frac{1}{2\gamma_1}(z - h(x))^T(z - h(x)) \\
&\leq \frac{1}{\alpha_2}\max\{\mathrm{Re}[\lambda(\tilde{A}_{11})]\}V(x) + \frac{1}{2\gamma_1\beta_1}W(z - h(x)),
\end{aligned} \tag{11.17}$$

where we have used assumption (ii), the fact that $2x^TA_{12}(z - h(x)) \leq \gamma_1 x^TA_{12}A_{12}^T x + \frac{1}{\gamma_1}(z - h(x))^T(z - h(x))$, right inequality in (11.15), and left inequality in (11.16).

Likewise, the time derivative $\dot{W}$ along the trajectories of z during the sliding motion is given by

$$\begin{aligned}
\dot{W}(z - h(x)) &= \Big(\frac{1}{\varepsilon}(A_{21}x + A_{22}z) - \dot{h}(x)\Big)^T P(z - h(x)) \\
&\quad + (z - h(x))^T P\Big(\frac{1}{\varepsilon}(A_{21}x + A_{22}z) - \dot{h}(x)\Big) \\
&= \frac{1}{\varepsilon}\big(A_{22}z - A_{22}h(x)\big)^T P(z - h(x)) - \dot{h}^T(x)P(z - h(x)) \\
&\quad + \frac{1}{\varepsilon}(z - h(x))^T P\big(A_{22}z - A_{22}h(x)\big) - (z - h(x))^T P\dot{h}(x) \\
&= \frac{1}{\varepsilon}(z - h(x))^T\big(A_{22}^TP + PA_{22}\big)(z - h(x)) - 2(z - h(x))^T P\dot{h}(x) \\
&\leq -\Big(\frac{1}{\beta_2\varepsilon} - \frac{a_{22}}{\beta_1}\Big)W(z - h(x)) + \frac{a_{21}}{\alpha_1}V(x),
\end{aligned} \tag{11.18}$$

where we have used the second inequality in assumption (ii), left inequality in (11.15) and right inequality in (11.16).

Combining the last inequalities in (11.17) and (11.18) yields the matrix inequality

$$\begin{bmatrix} \dot{V}(x) \\ \dot{W}(z - h(x)) \end{bmatrix} \leq \tilde{A}\begin{bmatrix} V(x) \\ W(z - h(x)) \end{bmatrix}$$

where $-\tilde{A}$ is an M-matrix for a positive constant ε^*, as defined in assumption (iii). Then, there exists a positive constant ξ such that, for all $t \in [\tau_k, \tau_{k+1})$,

$$V(x(t)) \leq \big(\|V(\tau_k)\| + \|W(\tau_k)\|\big)e^{-\xi(t-\tau_k)}, \tag{11.19}$$

$$W((z - h(x))(t)) \leq \big(\|V(\tau_k)\| + \|W(\tau_k)\|\big)e^{-\xi(t-\tau_k)}, \tag{11.20}$$

where $V(\tau_k) = V(x(\tau_k))$ and $W(\tau_k) = W\big((z - h(x))(\tau_k)\big)$. At the impulsive moments $t = \tau_k$, we have

$$V(x(\tau_k)) \leq \alpha_{1k} V(x(\tau_k^-)), \tag{11.21}$$

where $\alpha_{1k} = \lambda_{\max}^2(I + E_k)$ and

$$\begin{aligned}
W((z - h(x))(\tau_k)) &= \Big(z(\tau_k) + \overbrace{A_{22}^{-1}\big(A_{21} - B_2(C_s B_0)^{-1} C_s A_0\big)}^{=:\bar{A}} x(\tau_k)\Big)^T P \\
&\quad \times \Big(z(\tau_k) + A_{22}^{-1}\big(A_{21} - B_2(C_s B_0)^{-1} C_s A_0\big) x(\tau_k)\Big) \\
&= \Big((I + F_k) z(\tau_k^-) + \bar{A}(I + E_k) x(\tau_k^-)\Big)^T P \Big((I + F_k) z(\tau_k^-) \\
&\quad + \bar{A}(I + E_k) x(\tau_k^-)\Big) \\
&= \Big(\overbrace{(I + F_k)}^{=:\bar{F}_k}(z - h(x))(\tau_k^-) + \big(\overbrace{\bar{A}(I + E_k) - (I + F_k)\bar{A}}^{=:\bar{A}_k}\big) x(\tau_k^-)\Big)^T \\
&\quad \times P\Big((I + F_k)(z - h(x))(\tau_k^-) + \big(\bar{A}(I + E_k) - (I + F_k)\bar{A}\big) x(\tau_k^-)\Big) \\
&= \Big(\bar{F}_k (z - h(x))(\tau_k^-) + \bar{A}_k x(\tau_k^-)\Big)^T P \Big(\bar{F}_k (z - h(x))(\tau_k^-) + \bar{A}_k x(\tau_k^-)\Big) \\
&= \big(z - h(x)\big)^T (\tau_k^-) \bar{F}_k^T P \bar{F}_k \big(z - h(x)\big)(\tau_k^-) + x(\tau_k^-)^T \bar{A}_k^T P \bar{A}_k x(\tau_k^-) \\
&\quad + 2\big(z - h(x)\big)(\tau_k^-) \bar{F}_k^T P \bar{A}_k \\
&\leq \big(z - h(x)\big)^T (\tau_k^-) \Big(\overbrace{\bar{F}_k^T P \bar{F}_k + \gamma_3 (\bar{F}_k^T P \bar{A}_k)(\bar{F}_k^T P \bar{A}_k)^T}^{=:\Pi_k}\Big) \\
&\quad \times \big(z - h(x)\big)(\tau_k^-) + x(\tau_k^-)^T \big(\overbrace{\bar{A}_k^T P \bar{A}_k + \frac{1}{\gamma_k} I}^{=:\Xi_k}\big) x(\tau_k^-) \\
&\leq \beta_{2k} W((z - h(x))(\tau_k^-)) + \alpha_{2k} V(x(\tau_k^-)),
\end{aligned} \tag{11.22}$$

where we have used $h(x(\tau_k^-)) = -A_{22}^{-1}\big(A_{21} - B_2(C_s B_0)^{-1} C_s A_0\big) x(\tau_k^-) = -\bar{A} x(\tau_k^-)$, $\alpha_{2k} = \|\Xi_k\|/\alpha_1 > 0$ and $\beta_{2k} = \|\Pi_k\|/\beta_1 > 0$.

For instance, for $t \in [\tau_0, \tau_1)$ with $t_0 = \tau_0$, we have

$$\begin{aligned}
V(x(t)) &\leq \big(\|V(t_0)\| + \|W(t_0)\|\big) e^{-\xi(t - t_0)}, \\
W((z - h(x))(t)) &\leq \big(\|V(t_0)\| + \|W(t_0)\|\big) e^{-\xi(t - t_0)},
\end{aligned}$$

and for $t \in [\tau_1, \tau_2)$, we have

$$\begin{aligned}
V(x(t)) &\leq \big(\|V(\tau_1)\| + \|W(\tau_1)\|\big) e^{-\xi(t - \tau_1)}, \\
W((z - h(x))(t)) &\leq \big(\|V(\tau_1)\| + \|W(\tau_1)\|\big) e^{-\xi(t - \tau_1)}.
\end{aligned}$$

Using (11.21) and (11.22) with $k = 1$ yields, for $t \in [t_0, \tau_2)$,

$$V(x(t)) \leq (\alpha_{11} + \alpha_{21} + \beta_{21})\big(\|V(t_0)\| + \|W(t_0)\|\big)e^{-\xi(t-t_0)},$$
$$W((z - h(x))(t)) \leq (\alpha_{11} + \alpha_{21} + \beta_{21})\big(\|V(t_0)\| + \|W(t_0)\|\big)e^{-\xi(t-t_0)}.$$

By the mathematical induction, one may get, for $t \in [t_0, \tau_k)$,

$$V(x(t)) \leq \prod_{i=1}^{k}(\alpha_{1i} + \alpha_{2i} + \beta_{2i})\big(\|V(t_0)\| + \|W(t_0)\|\big)e^{-\xi(t-t_0)},$$
$$W((z - h(x))(t)) \leq \prod_{i=1}^{k}(\alpha_{1i} + \alpha_{2i} + \beta_{2i})\big(\|V(t_0)\| + \|W(t_0)\|\big)e^{-\xi(t-t_0)}.$$

Choose $0 < \upsilon < \xi$ and provoke the impulsive effects (i.e. assumption (iv)) to obtain, for all $t \geq t_0$,

$$V(x(t)) \leq \big(\|V(t_0)\| + \|W(t_0)\|\big)e^{-(\xi-\upsilon)(t-t_0)},$$
$$W((z - h(x))(t)) \leq \big(\|V(t_0)\| + \|W(t_0)\|\big)e^{-(\xi-\upsilon)(t-t_0)}.$$

As proceeded in Chap. 9, there exists $K > 0$ such that

$$\|x(t)\| + \|z(t)\| \leq K\big(\|x(t_0)\| + \|z(t_0)\|\big)e^{-(\xi-\upsilon)(t-t_0)/2}, \quad \forall t \geq t_0.$$

This completes the proof of exponential stability of the full-order, closed-loop impulsive SPS.

Remark 11.1

(i) To guarantee the exponential stability of the continuous composite SPS, it is required that the degree of stability for the uncoupled slow and fast subsystems be larger than the strength of the interconnection which is treaded as a perturbation to the isolated slow and fast subsystems. This requirement is represented by assumption (iii).
(ii) In the Lyapunov function W related to the fast system, we have considered the vector $(z - h(x))(t) = z(t) - h(x(t))$, but not $z(t)$, to shift the equilibrium state z to the origin [1].

11.3 Sliding Mode Luenberger Observer

In this section, we carry over the control design adopted in the last section to design a sliding mode observer (SMO) to estimate the states of the full-order system.

Consider again the impulsive system in (11.1) and measured outputs $y \in \mathbb{R}^l$ of the slow system $y = Dx$ for some matrix $D \in \mathbb{R}^{l\times n}$. As presented in the last section, the interest here is to design an SMO through the reduced slow system to observe the states of the full-order impulsive system in (11.1). To that goal, we define the state estimate impulsive SPS by

$$\begin{aligned}
\dot{\hat{x}} &= A_{11}\hat{x} + A_{12}\hat{z} + Bu + Lv(\hat{y} - y), & t \neq \tau_k, \quad (11.23a)\\
\varepsilon\dot{\hat{z}} &= A_{21}\hat{x} + A_{22}\hat{z}, & t \neq \tau_k, \quad (11.23b)\\
\hat{x}(t) &= (I + E_k)\hat{x}(t^-), & t = \tau_k, \quad (11.23c)\\
\hat{z}(t) &= (I + F_k)\hat{z}(t^-), & t = \tau_k, \quad (11.23d)\\
\hat{x}(0) &= \hat{x}_0, \qquad \hat{z}(0) = \hat{z}_0, & (11.23e)
\end{aligned}$$

where $\hat{x} \in \mathbb{R}^n$ and $\hat{z} \in \mathbb{R}^m$ are, respectively, the slow and fast state vectors of the estimate system, $A_{22} \in \mathbb{R}^{m\times m}$ is a nonsingluar, Huwritz matrix $L \in \mathbb{R}^{n\times r}$ is the observer gain matrix which plays a similar role as in the traditional linear Luenberger observer and v is a nonlinear vector function of the error between estimated state $\hat{y} = D\hat{x}$ and the available measured output y and satisfies $v(0) = 0$. Here, v is considered the $r-$dimensional observer (or control law) to be designed.

Defining the error states $e_x = \hat{x} - x$ and $e_z = \hat{z} - z$ leads to the corresponding impulsive error system

$$\begin{aligned}
\dot{e}_x &= A_{11}e_x + A_{12}e_z + L\bar{v}(e_x), & t \neq \tau_k, \quad (11.24a)\\
\varepsilon\dot{e}_z &= A_{21}e_x + A_{22}e_z, & t \neq \tau_k, \quad (11.24b)\\
e_x(t) &= (I + E_k)e_x(t^-), & t = \tau_k, \quad (11.24c)\\
e_z(t) &= (I + F_k)e_z(t^-), & t = \tau_k, \quad (11.24d)\\
e_x(0) &= e_{x_0}, \qquad e_z(0) = e_{z_0}, & (11.24e)
\end{aligned}$$

where $\bar{v}(e_x) = v(De_x)$. As proceeded earlier, setting $\varepsilon = 0$ results in the reduced error subsystem

$$\dot{e}_{x_s} = A_e e_{x_s} + L_s\bar{v}(e_{x_s}), \tag{11.25}$$

where $e_z = h(e_x) = -A_{22}^{-1}A_{21}e_x$, $A_e = A_{11} - A_{12}A_{22}^{-1}A_{21}$ and L_s has the definition of L. Define the sliding mode error surface by

$$S_e(e_{x_s}) = C_e e_{x_s}, \tag{11.26}$$

for some matrix $C_e \in \mathbb{R}^{r\times n}$. Then, in the sliding mode the equivalent control becomes

$$\bar{v}_{eq}(e_{x_s}) = -(C_e L_s)^{-1} C_e A_e e_{x_s}$$

and the corresponding equivalent reduced system is given by

$$\dot{e}_{x_s} = \big(I - L_s(C_e L_s)^{-1} C_e\big) A_e e_{x_s} =: A_e^{eq} e_{x_s}.$$

As for the reachability condition, define $V(S_e(e_x)) = \frac{1}{2}S_e^T(e_{x_s})S_e(e_{x_s})$. Then, as done in the last section, along the sliding surface $\dot{V}(S_e) = S_e(e_{x_s})\dot{S}_e(e_{x_s}) < 0$ is guaranteed if

$$\bar{\upsilon}(e_{x_s}) = \bar{\upsilon}_{eq}(e_{x_s}) + \bar{\upsilon}^* \tag{11.27}$$

with

$$\bar{\upsilon}^* = -(C_eL_s)^{-1}\text{diag}(\eta)\text{Sgn}\big(S_e(e_{x_s})\big)$$

where the $r \times r$ matrix $\text{diag}(\eta)$ and $r-$dimensional vector $\text{Sgn}\big(S_e(e_{x_s})\big)$ are as defined in the last subsection.

In the following theorem, we prove that the impulsive error system (11.24) is globally exponentially stabilized by the designed SMC in (11.27).

Theorem 11.2 *Assume that the following assumptions hold:*

(i) the reduced slow and fast subsystems are observable;
(ii) there exists a positive constant ε^* such that the matrix $-\tilde{A}$ is an M-matrix where

$$\tilde{A} = \begin{bmatrix} a_{11} & a_{12} \\ a_{21} & -(\frac{1}{\varepsilon^*} - a_{22}) \end{bmatrix},$$

with a_{11}, a_{12}, a_{21} and a_{22} being some constants defined later;
(iii) for any $i = 1, 2, \ldots, k$,

$$t_i - t_{i-1} > \frac{1}{\vartheta}\ln(\alpha_{1i} + \alpha_{2i} + \beta_{2i}),$$

for some positive constants ϑ, α_{1i}, α_{2i} and β_{2i} such that $\alpha_{1i} + \alpha_{2i} + \beta_{2i} > 1$.

Then, the sliding mode control law in (11.27) guarantees that the closed-loop full-order error system (11.24) is globally exponentially stable for $\varepsilon \in (0, \varepsilon^*]$.

Proof For all $t \geq t_0$ with $t_0 \in \mathbb{R}_+$, let $e_x(t) = e_x(t; t_0, e_{x_0}, e_{z_0})$ and $e_z(t) = e_z(t; t_0, e_{x_0}, e_{z_0})$ be the solution of (11.24). To analyze the exponential stabilization of the full order of the impulsive error system in (11.24) during the sliding mode, define $V(e_x) = \frac{1}{2}e_x^T e_x$ and $W(e_z - h(e_x)) = (e_z - h(e_x))^T P(e_z - h(e_x))$ as Lyapunov function candidates, where the subscript s of state x is dropped for simplicity of notation. Then,

$$\begin{aligned}
\dot{V}(e_x) &\leq e_x^T\Big(\overbrace{A_{11} + \gamma_1 A_{12}A_{12}^T + A_{12}A_{22}^{-1}A_{12} - L_s(C_eL_s)^{-1}(C_eA_e)}^{=:G}\Big)e_x \\
&\quad + \frac{1}{\gamma_1}(e_z - h(e_x))^T(e_z - h(e_x)) \\
&\leq a_{11}V(e_x) + a_{12}W(e_z - h(e_x))
\end{aligned} \tag{11.28}$$

where $a_{11} = \frac{1}{\alpha_2}\max\{\mathrm{Re}[\lambda(G)]\}$, $a_{12} = \frac{1}{\gamma_1\beta_1}$ and the fact that $2e_x^T A_{12}(e_z - h(e_x)) \leq \gamma_1 e_x^T A_{12}A_{12}^T e_x + \frac{1}{\gamma_1}(e_z - h(e_x))^T(e_z - h(e_x))$. Similarly,

$$\begin{aligned}\dot{W}(e_z - h(e_x)) &= \frac{1}{\varepsilon}(e_z - h(e_x))^T\big(A_{22}^T P + PA_{22}\big)(e_z - h(e_x))\\ &\quad - 2(e_z - h(e_x))^T P\dot{h}(e_x)\\ &\leq -\frac{1}{\varepsilon}(e_z - h(e_x))^T(e_z - h(e_x)) + \frac{2}{\gamma_2}e_x^T e_x\\ &\quad + a_{22}(e_z - h(e_x))^T(e_z - h(e_x))\\ &\leq -(\frac{1}{\varepsilon} - a_{22})W(e_z - h(e_x)) + a_{21}V(e_x), \end{aligned} \tag{11.29}$$

where $a_{21} = \frac{2}{\alpha_1\gamma_2}$, and $a_{22} = 2\|P\bar{A}_1\bar{A}_1^T P + P\bar{A}_2\|$ with $\bar{A} = A_{11} - A_{12}A_{22}^{-1}A_{21} - L_s(C_eL_s)^{-1}C_eA_e$, $\bar{A}_1 = A_{22}^{-1}A_{21}\bar{A}$ and $\bar{A}_2 = A_{22}^{-1}A_{21}A_{12}$.

Combining the inequalities in (11.28) and (11.29) in a matrix inequality yields

$$\begin{bmatrix}\dot{V}(e_x)\\ \dot{W}(e_z - h(e_x))\end{bmatrix} \leq \tilde{A}\begin{bmatrix}V(e_x)\\ W(e_z - h(e_x))\end{bmatrix},$$

where $\tilde{A}$ is defined in assumption (ii) with $-\tilde{A}$ being assumed to be an $M-$matrix for some positive number ε^*. Then, there exists $\xi > 0$ such that, for all $t \neq \tau_k$,

$$V(e_{x(t)}) \leq \big(\|V(t_k)\| + \|W(t_k)\|\big)e^{-\xi(t-\tau_k)}, \tag{11.30}$$

$$W((e_z - h(e_x))(t)) \leq \big(\|V(t_k)\| + \|W(t_k)\|\big)e^{-\xi(t-\tau_k)}, \tag{11.31}$$

where $V(\tau_k) = V(e_{x(\tau_k)})$ and $W(\tau_k) = W\big((e_z - h(e_x))(\tau_k)\big)$. As achieved in Theorem 11.1, at the impulsive moments $t = \tau_k$, we have

$$V(e_{x(\tau_k)}) \leq \alpha_{1k}V(e_{x(\tau_k^-)}), \tag{11.32}$$

$$W\big((e_z - h(e_x))(\tau_k)\big) \leq \beta_{2k}W((e_z - h(e_x))(\tau_k^-)) + \alpha_{2k}V(e_{x(\tau_k^-)}), \tag{11.33}$$

where $\alpha_{1k} = \lambda_{\max}^2(I + E_k)$, $\alpha_{2k} = \|C_k^* + \frac{1}{\gamma_3}I\|/\alpha_1$, and $\beta_{2k} = \|B_{k1}^* + B_{k2}^*\|/\beta_1$, with $C_k^* = (A_{k1}^* - A_{k2}^*)^T(A_{k1}^* - A_{k2}^*)$, $B_{k1}^* = (I + F_k)^T P(I + F_k)$, $B_{k2}^* = \Big((I + F_k)^T P(A_{k1}^* - A_{k2}^*)\Big)^T\Big((I + F_k)^T P(A_{k1}^* - A_{k2}^*)\Big)$, $A_{k1}^* = A_{22}^{-1}A_{21}(I + E_k)$ and $A_{k2}^* = (I + E_k)A_{22}^{-1}A_{21}$.

Considering the impulsive effects in (11.24c) and (11.24d) results in that the full-order impulsive error system (11.24) is globally exponentially stabilized by the SMC in (11.27). This completes the proof.

11.4 Numerical Examples

We present some numerical examples to illustrate the developed results.

Example 11.1 Consider the impulsive SPS in (11.1) where

$$A_{11} = \begin{bmatrix} -1 & 0 \\ 1 & 1 \end{bmatrix}, \quad A_{12} = \begin{bmatrix} -0.08 \\ 0 \end{bmatrix}, \quad B = \begin{bmatrix} 0 \\ 0.2 \end{bmatrix}, \quad E_k = -\frac{1}{2.5^k} I_{2\times 2},$$

$$A_{21} = \begin{bmatrix} 0.1 & 1 \end{bmatrix}, \quad A_{22} = -0.1, \quad F_k = -\frac{1}{2.5^k} I_{2\times 2},$$

where $k \in \mathbb{N}$. Setting $\varepsilon = 0$ gives $z = h(x) = 0.025x_1 + 0.5x_2$. Then, the reduced slow system becomes

$$\dot{x}_s = \begin{bmatrix} -1.9975 & 0.5 \\ 0.1025 & 0.15 \end{bmatrix} x_s + \begin{bmatrix} 2 \\ 0.5 \end{bmatrix} u,$$

where $x^T = (x_1 \; x_2)$. Choosing $C_s = \begin{bmatrix} -1 & -0.1 \end{bmatrix}$, the equivalent control is

$$u_s^{eq} = K x_s = \begin{bmatrix} 0.758 & -0.08 \end{bmatrix} x_s,$$

the corresponding equivalent system is

$$\dot{x}_s = A_{eq} x_s = \begin{bmatrix} -0.4815 & -0.11 \\ 0.4815 & 0.11 \end{bmatrix} x_s,$$

which is only stable, where $\lambda(A_{eq}) = -0.3715, 0$. Thus, the feedback control law is given by

$$u_s(t) = u_s^{eq}(t) - 4\,\mathrm{diag}(\eta)\,\mathrm{Sgn}(S_s(x_s)),$$

where $\eta = \mathrm{diag}(8, 5, 0.2)$. Define $V(x) = 0.5x^T x$ and $W(z - h(x)) = 0.05(z - h(x))^T (z - h(x))$. Then, one can show that

$$-\tilde{A} = -\begin{bmatrix} -0.212 & 4 \\ 1 & -(\frac{2}{\varepsilon} - 0.062) \end{bmatrix}$$

is an M-matrix if $\varepsilon \in (0, 0.1057]$, i.e. $\varepsilon^* = 0.1057$ where we have taken $\alpha_1 = \alpha_2 = 0.5$, $\beta_1 = 0.125$, $\beta_2 = 0.5$, $\gamma_1 = 1$, $\gamma_2 = 2$ and $\gamma_3 = 5$. We also found, from assumption (iv), $\tau_{k+1} - \tau_k > 0.975$. The simulation results of this system are shown in Fig. 11.1.

Example 11.2 Consider the impulsive SPS in (11.24) where A_{11}, A_{12}, A_{21}, A_{22} and B are given in Example 11.1 and the impulsive state estimate system is as defined

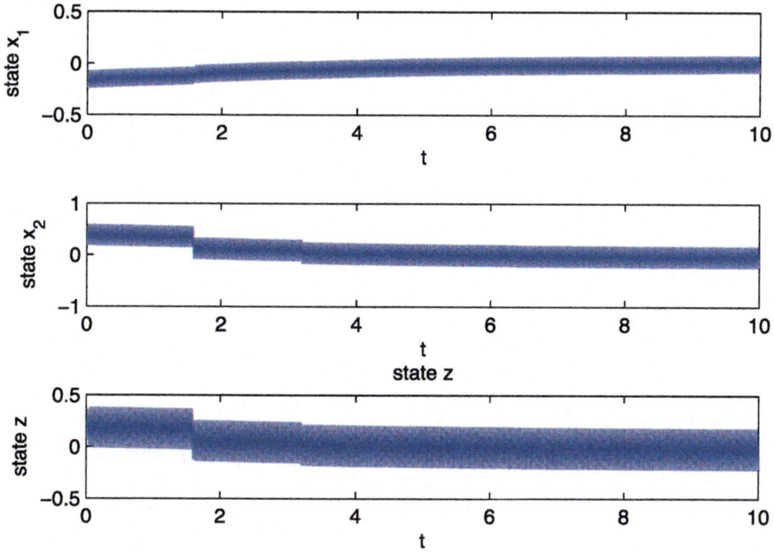

Fig. 11.1 Impulsive system states, x_1, x_2, and z

in (11.23) with $D = I_{2\times 2}$. The corresponding impulsive error system is defined in (11.24).

Let $L^T = \begin{bmatrix} 0.02 & -0.026 \end{bmatrix}$. Then, the equivalent control system is $\bar{v}_s^{eq}(e_{x_s}) = \begin{bmatrix} 315.8333 & -33.3333 \end{bmatrix} e_{x_s}$ and the corresponding equivalent reduced system is

$$\dot{e}_{x_s} = A_e^{eq} e_{x_s} = \begin{bmatrix} -8.3142 & 0.7167 \\ 8.3142 & -0.7167 \end{bmatrix} e_{x_s},$$

which is a stable system with the eigenvalues being -9.0308 and 0. The control law is given by

$$\bar{v}_s(e_{x_s}) = \bar{v}_s^{eq}(e_{x_s}) - 1.6667 \times 10^{-3} \mathrm{diag}(\eta)\, \mathrm{Sgn}(S_e(e_{x_s})).$$

Adopting the same Lyapunov functions, one can show that

$$-\tilde{A} = -\begin{bmatrix} -0.3328 & -2.0 \\ -2.0 & -16.3213 \end{bmatrix},$$

with $\varepsilon^* = 0.0505$, where we have taken $\alpha_1 = \alpha_2 = 0.5$, $\beta_1 = 0.125$, $\beta_2 = 0.5$, $\gamma_1 = 1$, $\gamma_2 = 2$ and $\gamma_3 = 5$. We have also found that, from assumption (iv), the $\tau_{k+1} - \tau_k > 4.5348$. Clearly, the time between impulses is larger than that in Example 11.2 due to the small decay rate of the interconnected full-order system. The simulation results of the error SPS states are shown in Fig. 11.2, and control inputs are shown in Figs. 11.3 and 11.4.

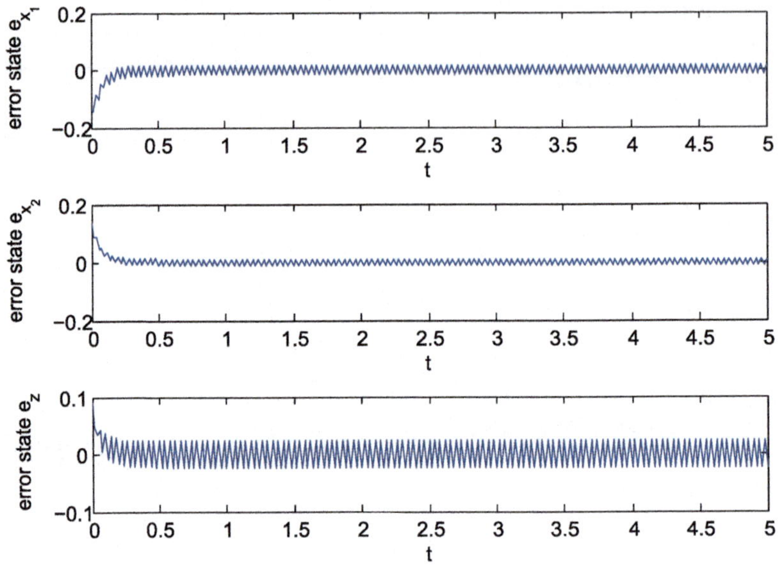

Fig. 11.2 Impulsive error system states, e_{x_1}, e_{x_2}, and e_z

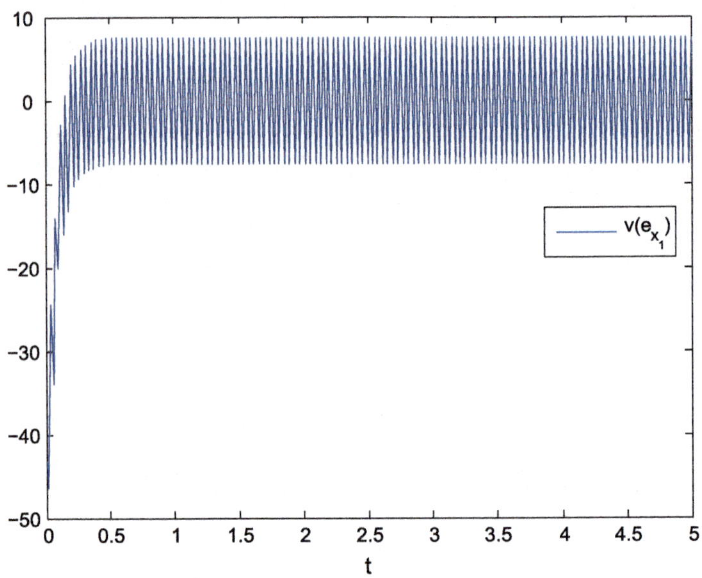

Fig. 11.3 Control input $v(e_{x_1})$

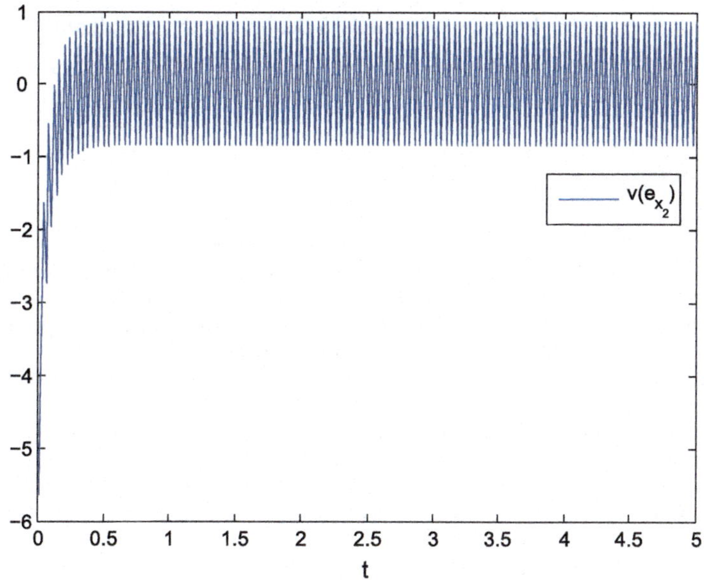

Fig. 11.4 Control input $v(e_{x_2})$

11.5 Notes and Comments

The state feedback law represented by an SMC is an efficient designing tool for stabilizing closed-loop variable structure systems undergoing matched uncertainties and external input disturbances. That is, SMC provides robust stabilization for systems with uncertainties because of its fast response, good transient performance and its tolerance to model uncertainty and perturbations. Throughout this chapter, we have addressed the problems of stabilization and state estimation for impulsive singularly perturbed systems via a sliding mode control. The material of this chapter is adapted from [2]. The continuous SPS has been viewed as a large-scale interconnected system for which state feedback control laws are synthesized. In this chapter, however, the controller has been designed through the dominating reduced order subsystem to stabilize the full-order system. This approach has lessened some unnecessary sufficient conditions imposed on the fast subsystem. Along this line of design, one can also see [3]. The impulsive effects of fixed types were considered as a perturbation to the system. This results in that the time between impulses is bounded below. In analyzing the stabilization and state estimation, the classical Lyapunov function technique has been used. The general theory and design of SMC have been addressed in several works, readers may refer to [4–9] and many references therein. The modified Luenberger observer, which is an efficient estimator to provide output approximation, can be read, for instance, in [10]. Due to the system complexity, the stabilization problem of these systems by such a state feedback control law is actively researched; see

[11–13]. A part of this literature focused on designing decentralized controllers for the slow and fast subsystems [11, 13].

References

1. Khalil HK (2002) Nonlinear systems. Prentice-Hall Inc, NJ
2. Alwan MS (2017) On stabilization and state estimation of impulsive singularly perturbed systems via sliding mode control. J Appl Nonlinear Dyn 6(1):105–119
3. Ahmed A, Schwartz H, Aitken V (2004) Sliding mode control for singularly perturbed system. In: 5th Asian Control Conference, vol 3, pp 1946–1950
4. Edwards C, Spuegeon SK (1998) Sliding mode control: theory and applications. Taylor & Francis, London
5. Ryvkin S, Palomar Lever E (2012) Sliding mode control for synchronous electric drives. CRC Press/Balkema
6. Sinha A (2007) Linear systems: optimal and robust control. Taylor & Francis, CRC Press
7. Tan S-C, Lai Y-M, Tse CK (2012) Sliding mode control of switching power converters: techniques and implementation. CRC Press
8. Utkin VI (1992) Sliding mode in control and optimization. Springer, New York
9. Utkin VI, Guldner J, Shi J (2009) Sliding mode control in electromechanical systems. Taylor & Francis, London
10. Drakunov SV (1992) Sliding-mode observers-based on equivalent control method. In: Proceedings of the 31st IEEE conference on decision and control (CDC), Tucson, Arizona, December 1618, pp 2368–2370
11. Nguyen T, Gajic Z, Su W-C (2010) Sliding mode control for singularly perturbed linear continuous time systems: composite control approach. In: (2010) IEEE international symposium on computer-aided control system design, 2011–2016. Yokohama, Japan, p 2010
12. Su W-C (1998) Sliding surface design for singularly perturbed systems. In: Proceedings of the american control conference, Philadelphia, Pennsylvania, USA, pp 1508–1512
13. Zhang X, Qian X, Lu G (2011) Robust decentralized sliding mode control for uncertain stochastic systems with time delay. Chin Control Decis Conf 2011:116–121

Chapter 12
Comparison Method and Stability of EPCA

This chapter deals with systems of nonlinear differential equations with piecewise constant arguments (EPCAs). We start by developing a comparison principle for this system. Then, this result will be used later to establish some stability properties of the system. As will be seen, the piecewise (constant) arguments can play as a stabilizing role in some cases where the underlying systems are unstable. A class of linear retarded EPCA is also considered in this chapter. Numerical examples and an application to a single-species logistic growth model with density-dependence harvesting are presented to show the effectiveness of the theoretical results.

12.1 Introduction

By EPCA, we mean differential equations with piecewise constant arguments over certain intervals. The arguments can be delay, advanced, or a mix of these two types. The dynamics of these differential equations generally depend on both continuous and discrete arguments. Hence, such equations can form a special class of hybrid systems. From the functional differential equation theory perspective, EPCA is special equations where the state history is given at certain individual points, rather than on intervals. Typically, nonlinear EPCA has the form

$$\dot{x}(t) = f(t, x(t), x(\gamma(t))), \tag{12.1}$$

where the argument γ is a piecewise constant function defined on intervals with a certain length, and it may be defined by $\gamma(t) = [t], [t-n], t-n[t], [t+1]$, for all t and a positive integer n, where $[\cdot]$ is the greatest integer function [1–3].

These differential equations have a similar structure to those seen in some "sequential-continuous" disease models treated by Busenberg and Cooke [4]. Also, the system of differential equations having the form

M. S. Alwan and X. Liu, *Theory of Hybrid Systems: Deterministic and Stochastic*, Nonlinear Physical Science, https://doi.org/10.1007/978-981-10-8046-3_12

$$\dot{x}(t) = f\Big(t, x(t), \lambda_k(x_k))\Big), \quad t \in [t_k, t_{k+1}]$$

was considered in [5] where, for some nonnegative integer number k, $x_k = x(t_k)$ and λ_k are some continuous functions. The system state experiences impulsive effects due to the switching in the arguments λ_k and x_k.

In this chapter, the system of nonlinear EPCA is being viewed as a hybrid, particularly switched, system, which allows us to apply the theory of continuous ordinary differential equations to each individual subsystem. This approach motivates concept of dwell time.

12.2 Problem Formulation

Let $\{t_k\}_{k=0}^{\infty}$ and $\{\xi_k\}_{k=0}^{\infty}$ be sequences of nonnegative real numbers such that $\lim_{k\to\infty} t_k = \infty$. Generally, ξ_k is defined such that $t_{k-1} < \xi_k \leq t_k$ for any $k \in \mathbb{N}$ with $\xi_0 = t_0$.

Consider the EPCA of the form

$$\dot{x}(t) = f\Big(t, x(t), \lambda_{\varrho(t)}(x(\gamma(t)))\Big), \tag{12.2a}$$

where $x \in \mathbb{R}^n$ is the system state, and for all $t \geq t_0$ with $t_0 \in \mathbb{R}_+$, $\varrho(t)$ and $\gamma(t)$ are functions taking values in $\{k\}_{k=0}^{\infty}$ and $\{\xi_k\}_{k=0}^{\infty}$, respectively. More specifically, for $t \in [t_k, t_{k+1}]$, we define $\varrho(t) = k$ and $\gamma(t) = \xi_k$. These piecewise constant functions, ϱ and γ, represent the switching signals whose roles of switching between the vector field function arguments λ_k and the values of its state argument x, respectively. Obviously, if, for any k, λ_k is an identity function, the EPCA (12.2a) reduces to (12.1). Also, when $k = 0$, then we have $\xi_0 = t_0$, $t \in [t_0, t_1]$ and the differential equation in (12.2a) is an ordinary one. Thus, for $k > 0$ and $t \in [t_k, t_{k+1}]$, the system state is allowed to be fed back with some historic data evaluated at individual moments $\xi_k \in (t_{k-1}, t_k]$. In addition, since the solution depends on the past history through an individual point, the initial state, in contrast to the functional differential equation case, is given at a specific time rather than over an interval, i.e.,

$$x(t_0) = x_0, \tag{12.2b}$$

for some $x_0 \in \mathbb{R}^n$.

In the following definition, we state the solution of the initial-value problem (IVP) given in (12.2).

Definition 12.1 A function $x : (\alpha, \beta) \to \mathbb{R}$ is said to be a solution of (12.2) if the following conditions hold:

(i) $x(t)$ is continuous for all $t \in (\alpha, \beta)$;
(ii) the derivative of $x(t)$ exists and is continuous for all $t \in (\alpha, \beta)$ except at $t \neq \xi_k$ ($k \in \mathbb{N}$), where at $t = \xi_k$ the one-sided derivative exists;
(iii) the derivative of $x(t)$, wherever exists, satisfies the EPCA in (12.2a); and
(iv) $x(t)$ satisfies the initial condition in (12.2b) at $t = t_0$.

System (12.2) may be rewritten in the form

$$\dot{x}(t) = f\Big(t, x(t), \lambda_k(x_{\xi_k})\Big), \qquad t \in [t_k, t_{k+1}), \quad k = 0, 1, 2, \ldots \tag{12.3a}$$

$$x(t_0) = x_0, \tag{12.3b}$$

where $x_{\xi_k} = x(\xi_k)$ and $\lambda_k(x_{\xi_k}) = \lambda_k(x(\xi_k))$ being constants. Throughout this chapter, we assume that the function $f(t, x, y)$ is continuous in its variables, i.e., $f \in \mathscr{C}(\mathbb{R}_+ \times \mathbb{R}^n \times \mathbb{R}^m; \mathbb{R}^n)$, and is globally Lipschitz in x and y.

In fact, the dependence of the solution, x, of the IVP (12.2) or (12.3) on the initial state at $t = t_0$ allows us to employ the theory of ordinary differential equations. For instance, for $k = 0$, and $t \in [t_0, t_1)$, the IVP

$$\begin{aligned} \dot{x}(t) &= f(t, x(t), \lambda_0(x_{\xi_0})), \\ x(t_0) &= x_0, \quad \text{with } \xi_0 = t_0 \end{aligned}$$

has a unique solution, say $x_0(t)$, for all $t \in [t_0, t_1)$ and $\lim_{t \to t_1^-} x_0(t) = x_0(t_1^-) \in \mathbb{R}^n$. Similarly, for $k = 1$ and $t \in [t_1, t_2)$, we have the IVP[1]

$$\begin{aligned} \dot{x}(t) &= f(t, x(t), \lambda_1(x_{\xi_1})), \\ x(t_1) &= x_0(t_1^-), \end{aligned}$$

which has a unique solution, say $x_1(t)$, for all $t \in [t_1, t_2)$ and $\lim_{t \to t_2^-} x_1(t) = x_1(t_2^-)$. By induction, for any k and all $t \in [t_k, t_{k+1})$, $x_k(t)$ is a unique solution and $\lim_{t \to t_{k+1}^-} x_k(t)$ exists. Define the solution x by

$$x(t) = \begin{cases} x_0, & t = t_0 \\ x_0(t, t_0, x_0), & t \in [t_0, t_1) \\ x_1(t, t_1, x_1), & t \in [t_1, t_2), \ \text{where } x_1 = x_0(t_1^-, t_0, x_0) \\ \cdots \\ x_k(t, t_k, x_k), & t \in [t_k, t_{k+1}), \ \text{where } x_k = x_{k-1}(t_k^-, t_{k-1}, x_{k-1}) \\ \cdots . \end{cases}$$

Since $\lim_{t \to t_{k+1}^-} x(t)$ exists for any k, the solution x must exist over a right-maximal interval $[t_0, \infty)$. These solution steps represent the proof of the following proposition.

[1] We should remark that, in the unified notation of the solution x, the initial condition $x(t_1) = x_0(t_1^-)$ becomes $x(t_1) = x(t_1^-)$, by our definition of x.

Proposition 12.1 *For all $k = 0, 1, \ldots$, let $\varrho : [t_k, t_{k+1}) \to \{k\}_{k=0}^{\infty}$ and $\gamma : [t_k, t_{k+1}) \to \{\xi_k\}_{k=0}^{\infty}$, where ξ_k is as defined earlier. Assume that $f \in \mathscr{C}(\mathbb{R}_+ \times \mathbb{R}^n \times \mathbb{R}^m; \mathbb{R}^n)$ and $f(t, x, y)$ is globally Lipschitz in x and y for all t. Then, the IVP (12.2) or (12.3) has a unique solution x defined over the right-maximal interval $[t_0, \infty)$.*

The auxiliary scalar initial-value problem can be defined analogously:

$$\dot{u}(t) = g(t, u(t), \sigma_k(u_{\xi_k})), \tag{12.4a}$$

$$u(t_0) = u_0, \tag{12.4b}$$

where $u \in \mathbb{R}_+$, $u_{\xi_k} = u(\xi_k)$, $\sigma_k \in \mathscr{C}(\mathbb{R}_+; \mathbb{R})$ and $g \in \mathscr{C}(\mathbb{R}_+^2 \times \mathbb{R}; \mathbb{R})$.

Moreover, assume that $f(t, 0, \lambda_k(0)) = 0$ and $g(t, 0, \sigma_k(0)) = 0$ for all $t \in \mathbb{R}_+$, and then systems (12.3) and (12.4) admit trivial solutions $x \equiv 0$ and $u \equiv 0$, respectively.

Definition 12.2 Let $x, y \in \mathbb{R}^n$ and $t \in [t_k, t_{k+1})$, for $k = 0, 1, 2, \ldots$. Then, if $V \in \mathscr{C}([t_k, t_{k+1}) \times \mathbb{R}^n; \mathbb{R}_+)$, the upper right-hand (Dini) derivative of V is defined by

$$D^+V(t, x, y) = \lim_{h \to 0^+} \sup \frac{1}{h} \Big[V(t, x + hf(t, x, \lambda_k(y))) - V(t, x) \Big].$$

Moreover, if $V \in \mathscr{C}^1([t_k, t_{k+1}) \times \mathbb{R}^n; \mathbb{R}_+)$, then

$$D^+V(t, x, y) = \frac{\partial V(t, x)}{\partial t} + \nabla V(t, x) \cdot f(t, x, \lambda_k(y)).$$

12.3 Comparison Method

We develop a comparison principle for nonlinear EPCA. Then, we consider some special case of EPCA and EPCAG.

Theorem 12.1 *Assume that the following conditions hold:*

(i) for $k = 0, 1, 2, \ldots$, $V \in \mathscr{C}([t_k, t_{k+1}) \times \mathbb{R}^n; \mathbb{R}_+)$, $V(t, x)$ is locally Lipschitz in x and

$$D^+V(t, x, V_{\xi_k}) \le g(t, V(t, x), \sigma_k(V_{\xi_k}))), \qquad t \in (t_k, t_{k+1}),$$

where $V_{\xi_k} = V(\xi_k, x(\xi_k))$; and

(ii) the maximal solution $\vartheta(t; t_0, u_0)$ of the auxiliary scalar EPCA (12.4) exists on $[t_0, \infty)$.

Then, $V(t_0, x_0) \le u_0$ implies $V(t, x(t)) \le \vartheta(t, t_0, u_0)$ for $t \ge t_0$.

Proof For all $t \ge t_0$ with $t_0 \in \mathbb{R}_+$, let $x(t) = x(t; t_0, x_0)$ be the solution of (12.3). Define $m(t) = V(t, x(t))$ on $[t_0, \infty)$. Then, we have

$$D^+m(t) \le g(t, m(t), \sigma_k(m_{\xi_k})), \quad t \in (t_k, t_{k+1})$$

where $m_{\xi_k} = m(\xi_k)$.

At $t = t_0$, we have $\xi_0 = t_0$ and $m_0 = m(\xi_0) = m(t_0) = V(t_0, x_0)$ and, at $t = t_1$, we have

$$m(t_1) = V(t_1, x(t_1)) = V(t_1, x_1(t_1; t_1, x_1)) = V(t_1, x_1(t_1; t_1, x_0(t_1^-; t_0, x_0))).$$

Thus, particularly, for $t \in [t_0, t_1]$, the ongoing differential inequality implies that, with aid of the classical comparison principle [6],

$$m(t) \le \vartheta_0(t; t_0, u_0), \quad t \in [t_0, t_1]$$

where $\vartheta_0(t; t_0, u_0)$ is the maximal solution of auxiliary scalar IVP

$$\begin{aligned} \dot{u}(t) &= g(t, u(t), \sigma_0(u_{\xi_0})), \\ u(t_0) &= u_0. \end{aligned}$$

Likewise, for $t \in [t_1, t_2]$, we have

$$m(t) \le \vartheta_1(t; t_1, u_1) = \vartheta_1(t; t_1, \vartheta_0(t_1; t_0, u_0)), \quad u_1 = u(t_1) = \vartheta_0(t_1; t_0, u_0)$$

where $\vartheta_1(t; t_1, u_1)$ is the maximal solution of the IVP

$$\begin{aligned} \dot{u}(t) &= g(t, u(t), \sigma_1(u_{\xi_1})), \\ u(t_1) &= u_1. \end{aligned}$$

Generally, for $t \in [t_k, t_{k+1}]$, one gets

$$m(t) \le \vartheta_k(t; t_k, u_k),$$

where $\vartheta_k(t; t_k, u_k)$ is the maximal solution of the IVP

$$\begin{aligned} \dot{u}(t) &= g(t, u(t), \sigma_k(u_{\xi_k})), \\ u(t_k) &= u_k. \end{aligned}$$

Define $u(t)$ by

$$u(t) = \begin{cases} u_0, & t = t_0 \\ \vartheta_0(t, t_0, u_0), & t \in (t_0, t_1] \\ \vartheta_1(t, t_1, u_1), & t \in (t_1, t_2], \text{ where } u_1 = \vartheta_0(t_1; t_0, u_0) \\ \cdots \\ \vartheta_k(t, t_k, u_k), & t \in (t_k, t_{k+1}], \text{ where } u_k = \vartheta_{k-1}(t_k; t_{k-1}, u_{k-1}) \\ \cdots . \end{cases}$$

Then, for $t \geq t_0$, we get

$$m(t) \leq u(t).$$

Since $\vartheta(t; t_0, u_0)$ is the maximal solution of the scalar EPCA (12.4), then for all $t \geq t_0$, we reach

$$m(t) \leq \vartheta(t; t_0, u_0).$$

This completes the proof.

In the following corollary and example, we address some special cases of EPCA and EPCAG.

Corollary 12.1 *Suppose that the conditions in Theorem 12.1 hold. Let $k = 0, 1, 2, \ldots$ and $t \in [t_k, t_{k+1}]$. If we choose that*

(i) $g(t, u, \sigma_k(u_{\xi_k})) = \beta_k u_{\xi_k}$, *with β_k being a constant for all k, then*

(1) for $\xi_k = t_k$, we have

$$V(t, x(t)) \leq \begin{cases} \Big[1 + \beta_0(t - t_0)\Big] V(t_0, x_0), & k = 0, \quad t \in (t_0, t_1] \\ \Big[1 + \beta_k(t - t_k)\Big] \prod_{j=1}^{k} \Big[1 + \beta_{j-1}(t_j - t_{j-1})\Big] V(t_0, x_0), \\ & k \in \mathbb{N}, \quad t \in (t_k, t_{k+1}], \end{cases}$$

where $t_k < t_{k+1}$ if $\beta_k > 0$ and $t_{k+1} < t_k - \frac{1}{\beta_k}$ if $\beta_k < 0$;

(2) for $t_{k-1} < \xi_k \leq t_k$ where $k \in \mathbb{N}$ and $\xi_0 = t_0$, we have

$$V(t, x(t)) = V_0(t, x(t)) \leq \Big[1 + \beta_0(t - t_0)\Big] V_0(t_0, x_0)$$

for any $t \in [t_0, t_1)$ such that $t_1 - t_0 < -\frac{1}{\beta_0}$ and

$$V(t, x(t)) = V_k(t, x(t)) \leq V_{k-1}(t_k, x(t_k)) + \beta_k(t - t_k) V_{k-1}(\xi_k, x(\xi_k))$$

for any $t \in [t_k, t_{k+1})$ such that, for any $k \in \mathbb{N}$, $t_{k+1} - t_k < -\frac{C_k}{\beta_k C_{\xi_k}}$ where $C_k = V_{k-1}(t_{k-1}, x(t_{k-1}))$ and $C_{\xi_k} = V_{k-1}(\xi_k, x(\xi_k))$;

(ii) $g(t, u, \sigma_k(u_{\xi_k})) = \alpha u(t) + \beta_k u_{\xi_k}$, *with α and β_k being constants for any k, then*

(1) for $\xi_k = t_k$, we have

$$V(t,x(t)) \le \begin{cases} \Big[(1+\frac{\beta_0}{\alpha})e^{\alpha(t-t_0)} - \frac{\beta_0}{\alpha}\Big]V(t_0,x_0), \quad k=0, \quad t\in(t_0,t_1] \\ \Big[(1+\frac{\beta_k}{\alpha})e^{\alpha(t-t_k)} - \frac{\beta_k}{\alpha}\Big] \\ \quad \times \prod_{j=1}^{k}\Big[(1+\frac{\beta_{j-1}}{\alpha})e^{\alpha(t_j-t_{j-1})} - \frac{\beta_{j-1}}{\alpha}\Big]V(t_0,x_0), \\ \qquad k\in\mathbb{N}, \quad t\in(t_k,t_{k+1}], \end{cases}$$

provided that, for $k=0,1,2,\ldots$, $t_{k+1}>t_k$ *when* $\alpha>0$ *and* $\beta_k>0$, *or when* $\alpha<0$ *and* $\beta_k>0$ *with* $\alpha>-\beta_k>0$, *and* $t_{k+1}<t_k+\frac{1}{\alpha}\ln\Big[\frac{\beta_k}{\alpha}\Big(1+\frac{\beta_k}{\alpha}\Big)^{-1}\Big]$ *when* $\alpha>0$ *and* $\beta_k<0$ *with* $\frac{\beta_k}{\alpha}\Big(1+\frac{\beta_k}{\alpha}\Big)^{-1}>1$;

(2) *for* $t_{k-1}<\xi_k\le t_k$ *where* $k\in\mathbb{N}$ *and* $\xi_0=t_0$,

$$V(t,x(t)) = V_0(t,x(t)) \le \Big[e^{\alpha(t-t_0)} + \frac{\beta_0}{\alpha}\Big(e^{\alpha(t-t_0)}-1\Big)\Big]V_0(t_0,x_0),$$

for $t\in[t_0,t_1)$ *and*

$$V(t,x(t)) = V_k(t,x(t)) \le e^{\alpha(t-t_k)}V_{k-1}(t_k,x(t_k)) + \frac{\beta_k}{\alpha}\Big[e^{\alpha(t-t_k)}-1\Big]V_{k-1}(\xi_k,x(\xi_k)), \quad t\in[t_k,t_{k-1})$$

provided that, for $k=0,1,2,\ldots$, $t_{k+1}>t_k+\frac{1}{\alpha}\ln T_k$ *when* $\alpha>0$ *and* $\beta_k>0$, *or when* $\alpha<0$ *and* $\beta_k>0$ *with* $V_{k-1}(t_k,x(t_k))+\frac{\beta_k}{\alpha}V_{k-1}(\xi_k,x(\xi_k)) <0$, *and* $t_{k+1}<t_k+\frac{1}{\alpha}\ln T_k$ *when* $\alpha>0$ *and* $\beta_k<0$ *with* $V_{k-1}(t_k,x(t_k))+\frac{\beta_k}{\alpha}V_{k-1}(\xi_k,x(\xi_k))<0$, *where* $T_k=\frac{\beta_k}{\alpha}V_{k-1}(\xi_k,x(\xi_k))\Big(V_{k-1}(t_k,x(t_k)) +\frac{\beta_k}{\alpha}V_{k-1}(\xi_k,x(\xi_k))\Big)^{-1}>1$; *and*

(*iii*) $g(t,u,\sigma_k(u_{\xi_k}))=\alpha u(t)+h(t,u,\sigma_k(u_{\xi_k}))$ *with* $\alpha\in\mathbb{R}$ $h\in\mathscr{C}(\mathbb{R}_+\times\mathbb{R}^2;\mathbb{R}_+)$, *and* $h(t,u,v)$ *is globally Lipschitz in* u *and* v, *then*

$$V(t,x(t)) \le e^{\alpha(t-t_0)}V(x_0) + \sum_{j=1}^{k}\int_{t_{j-1}}^{t_j} e^{\alpha(t-s)}h(s,V(s,x(s)),\sigma_k(V_{\xi_{j-1}}))ds + \int_{t_k}^{t} e^{\alpha(t-s)}h(s,V(s,x(s)),\sigma_k(V_{\xi_k}))ds.$$

Proof (i)(1) For $t\in[t_k,t_{k+1}]$, since $u_{\xi_k}=u_{t_k}$, the solution of the differential equation $\dot u(t)=\beta_k u_{\xi_k}$ is given by

$$u(t)=\Big[1+\beta_k(t-t_k)\Big]u_k.$$

Particularly, for $k=0$ and $t\in[t_0,t_1]$, we have

$$u(t) = \Big[1 + \beta_0(t - t_0)\Big]u_0$$

and, for $k = 1$ and $t \in [t_1, t_2]$, we have

$$u(t) = \Big[1 + \beta_1(t - t_1)\Big]\Big[1 + \beta_0(t_1 - t_0)\Big]u_0.$$

Thus, by induction, we reach

$$u(t) = \begin{cases} \Big[1 + \beta_0(t - t_0)\Big]u_0, & k = 0, \quad t \in (t_0, t_1] \\ \Big[1 + \beta_k(t - t_k)\Big]\prod_{j=1}^{k}\Big[1 + \beta_{j-1}(t_j - t_{j-1})\Big]u_0, & \\ & k \in \mathbb{N}, \quad t \in (t_k, t_{k+1}], \end{cases}$$

To complete the proof, we use comparison result developed in Theorem 12.1.

Proof (i)(2) For any k and $t \in [t_k, t_{k+1})$, we have

$$u(t) = u(t_k) + \beta_k(t - t_k)u(\xi_k).$$

Particularly, for $k = 0$, we have $\xi_0 = t_0$ and

$$u(t) = \Big[1 + \beta_0(t - t_0)\Big]u_0 =: u_0(t),$$

where the right-hand side is positive if $t < t_0 - 1/\beta_0$; so that, for $k = 1$ and $t \in [t_1, t_2)$, we get

$$u(t) = u_0(t_1) + \beta_1(t - t_1)u_0(\xi_1) =: u_1(t).$$

Thus, by induction, we reach

$$u(t) = u_k(t) = u_{k-1}(t_k) + \beta_k(t - t_k)u_{k-1}(\xi_k), \qquad t \in [t_k, t_{k+1}), \quad k \in \mathbb{N}$$

which implies the general form given in (i)(2).

Proof (ii)(1) For $t \in (t_k, t_{k+1}]$, we have the differential equation

$$\dot{u}(t) = \alpha u(t) + \beta_k u_{\xi_k}$$

and its solution is given by

$$u(t) = \Big[e^{\alpha(t - t_k)} + \frac{\beta_k}{\alpha}\Big(e^{\alpha(t - t_k)} - 1\Big)\Big]u_k, \tag{12.5}$$

from which we obtain

$$u(t) = \begin{cases} \left[(1+\frac{\beta_0}{\alpha})e^{\alpha(t-t_0)} - \frac{\beta_0}{\alpha}\right]u_0, & k=0, \quad t \in (t_0, t_1] \\ \left[(1+\frac{\beta_k}{\alpha})e^{\alpha(t-t_k)} - \frac{\beta_k}{\alpha}\right]\prod_{j=1}^{k}\left[(1+\frac{\beta_{j-1}}{\alpha})e^{\alpha(t_j-t_{j-1})} - \frac{\beta_{j-1}}{\alpha}\right]u_0, \\ \qquad k \in \mathbb{N}, \quad t \in (t_k, t_{k+1}], \end{cases}$$

where α and β_k are defined in (ii). Applying the comparison principle leads us to the required result. The proof of (ii)(2) can be obtained in a similar way used in (i)(2); thus, it is left here as an exercise.

Proof (iii) For $k = 0, 1, 2, \ldots$ and $t \in [t_k, t_{k+1}]$, we have the differential equation

$$\dot{u}(t) = \alpha u(t) + h(t, u(t), \sigma_k(u_{\xi_k}))$$

and its solution is given by

$$u(t) = e^{\alpha(t-t_k)}u_k + \int_{t_k}^{t} e^{\alpha(t-s)}h(s, u(s), \sigma_k(u_{\xi_k}))\, ds.$$

For instance, for $t \in [t_0, t_1]$, we have

$$u(t) = e^{\alpha(t-t_0)}u_0 + \int_{t_0}^{t} e^{\alpha(t-s)}h(s, u(s), \sigma_0(u_{\xi_0}))\, ds$$

and, at $t = t_1$,

$$u_1 = e^{\alpha(t_1-t_0)}u_0 + \int_{t_0}^{t_1} e^{\alpha(t_1-s)}h(s, u(s), \sigma_0(u_{\xi_0}))\, ds.$$

For $t \in [t_1, t_2]$, we have

$$\begin{aligned} u(t) &= e^{\alpha(t-t_1)}u_1 + \int_{t_1}^{t} e^{\alpha(t-s)}h(s, u(s), \sigma_1(u_{\xi_1}))\, ds \\ &= e^{\alpha(t-t_1)}\left\{e^{\alpha(t_1-t_0)}u_0 + \int_{t_0}^{t_1} e^{\alpha(t_1-s)}h(s, u(s), \sigma_0(u_{\xi_0}))\, ds\right\} \\ &\quad + \int_{t_1}^{t} e^{\alpha(t-s)}h(s, u(s), \sigma_1(u_{\xi_1}))\, ds \\ &= e^{\alpha(t-t_0)}u_0 + \int_{t_0}^{t_1} e^{\alpha(t-s)}h(s, u(s), \sigma_0(u_{\xi_0}))\, ds + \int_{t_1}^{t} e^{\alpha(t-s)}h(s, u(s), \sigma_1(u_{\xi_1}))\, ds. \end{aligned}$$

For $t \in [t_2, t_3]$, we have

$$u(t) = e^{\alpha(t-t_0)}u_0 + \int_{t_0}^{t_1} e^{\alpha(t-s)}h(s, u(s), \sigma_0(u_{\xi_0}))\,ds + \int_{t_1}^{t_2} e^{\alpha(t-s)}h(s, u(s), \sigma_1(u_{\xi_1}))\,ds$$
$$+ \int_{t_2}^{t} e^{\alpha(t-s)}h(s, u(s), \sigma_2(u_{\xi_2}))\,ds.$$

By induction for $t \in [t_k, t_{k+1}]$

$$u(t) = e^{\alpha(t-t_0)}u_0 + \sum_{j=1}^{k}\int_{t_{j-1}}^{t_j} e^{\alpha(t-s)}h(s, u(s), \sigma_{j-1}(u_{\xi_{j-1}}))\,ds$$
$$+ \int_{t_k}^{t} e^{\alpha(t-s)}h(s, u(s), \sigma_k(u_{\xi_k}))\,ds$$

and for $t \geq t_0$, we have

$$u(t) = e^{\alpha(t-t_0)}u_0 + \sum_{j=1}^{\infty}\int_{t_{j-1}}^{t_j} e^{\alpha(t-s)}h(s, u(s), \sigma_{j-1}(u_{\xi_{j-1}}))\,ds.$$

Using the comparison result gives

$$V(t, x(t)) \leq e^{\alpha(t-t_0)}V(t_0, x_0) + \sum_{j=1}^{\infty}\int_{t_{j-1}}^{t_j} e^{\alpha(t-s)}h(s, V(s, x(s)), \sigma_{j-1}(V_{\xi_{j-1}}))\,ds.$$

The proof completes the proof.

12.4 Stability Analysis

Having established the comparison results in Theorem 12.1, we prove some stability notions for the nonlinear EPCA.

Theorem 12.2 *In addition to the conditions in Theorem 12.1, assume further that there exist class−$\mathcal{K}$ function a and b such that*

$$b(\|x\|) \leq V(t, x) \leq a(\|x\|) \tag{12.6}$$

hold. Then, the stability properties of the trivial solution $u \equiv 0$ of the auxiliary scalar system of EPCA in (12.4) imply the corresponding stability properties of the trivial solution $x \equiv 0$ of system (12.3).

Proof Let $t_0 \in \mathbb{R}_+$ and $\varepsilon > 0$ be given. Suppose that $u \equiv 0$ is stable. Then, for given $b(\varepsilon) > 0$, there is a $\delta_1 = \delta_1(t_0, \varepsilon) > 0$ such that

$$0 \le u_0 \le \delta_1 \quad \text{implies} \quad u(t; t_0, u_0) \le b(\varepsilon), \qquad \forall t \ge t_0$$

where $u(t; t_0, u_0)$ is any solution of (12.4). Choose $\delta_2 = \delta_2(\varepsilon)$ such that $a(\delta_2) < b(\varepsilon)$. Define $\delta = \min\{\delta_1, \delta_2\}$. We claim that the trivial solution $x \equiv 0$ is stable; that is, if $\|x_0\| < \delta$, then $\|x(t)\| < \varepsilon$, for $t \ge t_0$, where $x(t) = x(t; t_0, x_0)$ is any solution of (12.3). If our claim were not true, then there would exist a $t^* > t_0$ and $t_k < t^* \le t_{k+1}$ for which $\|x_0\| < \delta$ and

$$\begin{aligned} \|x(t)\| &< \varepsilon \quad \text{for} \quad t_0 \le t \le t_k \\ \|x(t)\| &\ge \varepsilon \quad \text{for} \quad t_k \le t^* \le t_{k+1}. \end{aligned} \tag{12.7}$$

From (12.7), we have $\|x(t_k)\| < \varepsilon$. Hence, we can find a $\tilde{t}$ such that $t_k < \tilde{t} \le t^*$ and at which

$$\varepsilon \le \|x(\tilde{t})\|.$$

Let $u_0 = a(\|x_0\|) < \delta_1$, and define $m(t) = V(t, x(t))$ for $t_0 < t \le \tilde{t}$. Then, by Theorem 12.1, we have

$$V(t, x(t)) \le \vartheta(t; t_0, a(\|x_0\|)), \qquad t_0 \le t \le \tilde{t}$$

where $\vartheta(t; t_0, a(\|x_0\|))$ is the maximal solution of auxiliary scalar system (12.4). Then, we obtain with the aid of the left inequality in (12.6)

$$b(\varepsilon) \le b(\|x(\tilde{t})\|) \le V(\tilde{t}, x(\tilde{t})) \le \vartheta(\tilde{t}, t_0, a(\|x_0\|)) < b(\varepsilon)$$

which is a contradiction. This shows $x \equiv 0$ is stable. If, moreover, δ is independent of t_0, then $x \equiv 0$ is uniformly stable.

To prove asymptotic stability of $x \equiv 0$, it suffices to show attractivity of this solution. Suppose that $u \equiv 0$ is asymptotically stable. Then, it implies that $x \equiv 0$ is stable; that is, for each $\varepsilon > 0$, there is a $\delta = \delta(t_0, \varepsilon)$ such that

$$\|x_0\| < \delta \quad \text{implies} \quad \|x(t)\| < \varepsilon, \quad \forall t \ge t_0.$$

Since $u \equiv 0$ is attractive, given $b(\varepsilon) > 0$ and $t_0 \in \mathbb{R}_+$, there is a $\delta_0^* = \delta_0^*(t_0) > 0$ and $T = T(t_0, \varepsilon) > 0$ such that

$$0 \le u_0 \le \delta_0^* \quad \text{implies} \quad u(t, t_0, u_0) < b(\varepsilon), \quad \forall t \ge t_0 + T.$$

Choose a $\tilde{\delta}$ such that $a(\tilde{\delta}) < \delta_0^*$. Define $\rho = \min\{\delta_0^*, \tilde{\delta}\}$, and let $\|x_0\| < \rho$. Then, as we did in proving the stability of $x \equiv 0$, we can get

$$b(\|x(t)\|) \le V(t, x(t)) \le \vartheta(t, t_0, a(\|x_0\|)) < b(\varepsilon).$$

This implies that $\|x(t)\| < \varepsilon$ for all $t \geq t_0 + T$; i.e., $x \equiv 0$ is attractive. Hence, $x \equiv 0$ is asymptotically stable. If T is independent of t_0, then $x \equiv 0$ is uniformly asymptotically stable. This completes the proof.

Corollary 12.2 *In Theorem 12.2, let $g(t, u(t), \sigma_k(u_{\xi_k})) = \beta_k u_{\xi_k}$ with β_k being a constant for all k.*

(i) In the case $\xi_k = t_k$,

(1) if $\beta_k > 0$ for any k and the infinite series

$$\sum_{j=1}^{\infty} \beta_{j-1}(t_j - t_{j-1}) \tag{12.8a}$$

converges, then $x \equiv 0$ is uniformly stable;

(2) while if $\beta_k < 0$ for any k and in addition to assumption in (i)(1), for any j, the following inequality holds

$$0 < t_j - t_{j-1} < -\frac{1}{\beta_{j-1}}, \tag{12.8b}$$

then $x \equiv 0$ is uniformly asymptotically stable.

(ii) In the case $\beta_k < 0$ and $t_{k-1} < \xi_k \leq t_k$ for any $k = 0, 1, 2, \ldots$ with $\xi_0 = t_0$, if $u_k(t) \leq L$ for some positive constant L, where $u_k(t)$ is defined in Corollary 12.1 for any k and $t \in [t_k, t_{k+1})$, then $u \equiv 0$ is uniformly stable; if, in addition, $u_k(t) \leq L_k$ for any k and $t \in [t_k, t_{k+1})$ and $\sum_{k=0}^{\infty} L_k < \infty$, then the trivial solution $u \equiv 0$ and, hence, $x \equiv 0$ is uniformly asymptotically stable. Particularly, one may define $L = \sup\{L_k \mid k = 0, 1, 2, \ldots\}$.

Proof (i)(1) The solution of the auxiliary scalar EPCA

$$\begin{aligned} \dot{u}(t) &= \beta_k u_{\xi_k}, \qquad t \in [t_k, t_{k+1}], \quad k = 0, 1, 2, \ldots \\ u(t_0) &= u_0 \end{aligned}$$

is given by

$$u(t) = \Big(1 + \beta_k(t - t_k)\Big) \prod_{j=1}^{k} \Big[1 + \beta_{j-1}(t_j - t_{j-1})\Big] u_0.$$

By (12.8a), the product $\prod_{j=1}^{\infty} \Big[1 + \beta_{j-1}(t_j - t_{j-1})\Big]$ converges. So that, defining $M = \prod_{j=1}^{\infty} \Big[1 + \beta_{j-1}(t_j - t_{j-1})\Big] < \infty$ yields

$$u(t, t_0, u_0) = M u_0 < M\sigma, \qquad \text{for some } \sigma > 0 \text{ such that } u_0 < \sigma,$$

meaning that the trivial solution $u \equiv 0$ is uniformly stable which implies, by Theorem 12.2, the uniform stability of the trivial solution $x \equiv 0$. In particular, for $k = 0, 1, 2, \ldots$, one may choose that $\beta_k = \frac{1}{2^k}$, $t_{k+1} - t_k < \delta$ for some $\delta > 0$.

Proof (i)(2) The assumption (12.8b) is equivalent to $0 < 1 + \beta_{j-1}(t_j - t_{j-1}) < 1$. So that, if we choose $1 + \beta_{j-1}(t_j - t_{j-1}) = \frac{1}{e}$, then M approaches zero; this proves the uniform asymptotic stability of $u \equiv 0$ and $x \equiv 0$.

Proof (ii) The proof is straightforward; thus, it is left here as an exercise.

Remark 12.2 It is worth noting that, for any k and $t \in [t_k, t_{k+1})$, the assumption $0 < u_k(t) \leq L_k$ is equivalent to

$$\frac{L_k - C_k}{\beta_k C_{\xi_k}} < t - t_k < \frac{-C_k}{\beta_k C_{\xi_k}},$$

where C_k and C_{ξ_k} are defined in Corollary 12.1.

Corollary 12.3 *In Theorem 12.2, let $g(t, u(t), \sigma_k(u_{\xi_k})) = \alpha u(t) + \beta_k u_{\xi_k}$, where $\alpha > 0$, $\beta_k < 0$ and $\xi_k = t_k$ for $k \in \mathbb{N}$. Then, the trivial solution $x \equiv 0$ is uniformly stable if infinite series*

$$\sum_{j=1}^{\infty} \Big[(1 + \frac{\beta_{j-1}}{\alpha})e^{\alpha(t_j - t_{j-1})} - \frac{\beta_{j-1}}{\alpha}\Big] \tag{12.9}$$

converges. Furthermore, if, in addition, the terms in corresponding infinite product are all less than unity, then $x \equiv 0$ is uniformly asymptotically stable.

Proof Since the infinite series in (12.9) converges, so does the infinite product

$$\prod_{j=1}^{\infty} \Big[(1 + \frac{\beta_{j-1}}{\alpha})e^{\alpha(t_j - t_{j-1})} - \frac{\beta_{j-1}}{\alpha}\Big].$$

So that, let

$$M = \prod_{j=1}^{\infty} \Big[(1 + \frac{\beta_{j-1}}{\alpha})e^{\alpha(t_j - t_{j-1})} - \frac{\beta_{j-1}}{\alpha}\Big].$$

Then, we have

$$u(t) < M\sigma,$$

for some positive σ for which $u_0 < \sigma$; that is, $u \equiv 0$ is uniform stability. Employing our comparison result, the uniform stability of $x \equiv 0$ will be a subsequence of this stability property. Finally, by our assumption, if, for instance, every term in the infinite product is less than or equal $1/e$, then

$$u(t) = \prod_{j=1}^{\infty}\Big[(1+\frac{\beta_{j-1}}{\alpha})e^{\alpha(t_j-t_{j-1})} - \frac{\beta_{j-1}}{\alpha}\Big]u_0 \to 0.$$

That is, $u \equiv 0$ and, hence, $x \equiv 0$ is uniformly asymptotically stable.

Remark 12.3

(i) The interesting finding of Corollary 12.3 is that the system has unstable ordinary part which is stabilized by negative piecewise constants evaluated at an individual point in each subinterval.
(ii) Assuming that the product terms equal or less than some positive constant c less than unity results in, for $\xi_k = t_k$,

$$t_{k+1} - t_k > \frac{1}{\alpha}\ln\Big[\Big(c+\frac{\beta_k}{\alpha}\Big)\Big(1+\frac{\beta_k}{\alpha}\Big)^{-1}\Big],$$

where $\left(c+\frac{\beta_k}{\alpha}\right)\left(1+\frac{\beta_k}{\alpha}\right)^{-1} > 1$ so long as $\left(1+\frac{\beta_k}{\alpha}\right) < 0$ and $c < 1$.

Corollary 12.4 *In Theorem 12.2, let $g(t, u(t), \sigma_k(u_{\xi_k})) = -\omega(u) + \beta_k u_{\xi_k}$ with $\omega \in \mathcal{K}$, $\beta_k \geq 0$ and $\xi_k = t_k$ for all k. Then, $x \equiv 0$ is uniformly asymptotically stable provided that the series $\sum_{j=1}^{\infty} \beta_j(t_j - t_{j-1})$ converges.*

Proof Since $D^+V(x, V_{\xi_k}) \leq -\omega(V(x)) + \beta_k V_{\xi_k}$ implies

$$D^+V(x, V_{\xi_k}) \leq \beta_k V_{\xi_k},$$

then it follows from Corollary 12.3 that $u \equiv 0$ of the auxiliary scalar EPCA

$$\dot{u}(t) = -\omega(u(x)) + \beta_k u_{\xi_k} \tag{12.10a}$$
$$u(t_0) = u_0 \tag{12.10b}$$

is uniformly stable. Thus, for a fixed $\rho > 0$, there is a $\sigma = \sigma(\rho) > 0$ such that

$$0 \leq u_0 \leq \sigma \quad \text{implies} \quad u(t; t_0, u_0) < \rho, \quad t \geq t_0 \tag{12.11}$$

for any solution of (12.10). Let $\varepsilon \in (0, \rho)$ be given and $\delta = \delta(\varepsilon)$. In the rest of the proof, we need to show that $u \equiv 0$ is attractive; it suffices to show that there exists a $T = T(\varepsilon) > 0$ such that

$$u(t^*; t_0, u_0) < \delta = \delta(\varepsilon), \tag{12.12}$$

for any $t^* \in [t_0, t_0 + T]$ and any solution of $u(t; t_0, u_0)$ of (12.10) that satisfies (12.11).

Since $\sum_{j=1}^{\infty} \beta_j (t_j - t_{j-1})$ converges, define

$$M = \sum_{j=1}^{\infty} \beta_j (t_j - t_{j-1}) < \infty.$$

Choose $T_1 = T_1(\varepsilon) > 0$ such that

$$T_1 > 2\rho M[\omega(\delta)]^{-1}. \tag{12.13}$$

Define

$$T = \max\left\{T_1, \frac{2(\sigma + 1)}{\omega(\delta)}\right\}. \tag{12.14}$$

We claim that (12.12) is true for T given in (12.14). If this were not true, suppose, for contradiction, that there would be a solution $u(t) = u(t; t_0, u_0)$ of (12.10) with $u_0 < \sigma$ such that

$$u(t) \geq \delta, \qquad t \in [t_0, t_0 + T]. \tag{12.15}$$

Integrating (12.10) from t_0 to $t_0 + T$ yields

$$\begin{aligned}
0 \leq u(t_0 + T) &= u_0 - \int_{t_0}^{t_0+T} \omega(u(s))\, ds \\
&\quad + \sum_{j=1}^{k} \beta_{j-1} u_{\xi_{j-1}} (t_j - t_{j-1}) + \beta_k u_{\xi_k} (t_0 + T - t_k) \\
&\leq \sigma - T\omega(\delta) + \rho M \\
&= \sigma - \frac{T\omega(\delta)}{2} - \frac{T\omega(\delta)}{2} + \rho M \\
&\leq \sigma - \frac{T\omega(\delta)}{2} < -1 < 0,
\end{aligned}$$

which is a contradiction. Thus, (12.12) must be true; that is,

$$u(t^*; t_0, u_0) < \delta,$$

for any solution of $u(t; t_0, u_0)$ of (12.10) with $u_0 < \sigma$. Hence, $u \equiv 0$ is uniformly attractive and consequently uniformly asymptotically stable which in turn implies that $x \equiv 0$ is uniformly asymptotically stable.

Corollary 12.5 *Let $g(t, u(t), \sigma_k(u_{\xi_k})) = \alpha u(t) + h(t, u(t), \sigma_k(u_{\xi_k}))$ with $\alpha < 0$. Then, $x \equiv 0$ is uniformly asymptotically stable provided that the sum*

$$\sum_{j=1}^{\infty} \int_{t_{j-1}}^{t_j} e^{\alpha(t-s)} h(s, u(s), \sigma_{j-1}(u_{\xi_{j-1}}))\, ds$$

converges. In particular, $h(t, u(t), \sigma_k(u_{\xi_k})) = 0$ when k (or t) $\to \infty$.

Proof The proof is straightforward since, from the solution

$$u(t) = e^{\alpha(t-t_0)} u_0 + \sum_{j=1}^{\infty} \int_{t_{j-1}}^{t_j} e^{\alpha(t-s)} h(s, u(s), \sigma_{j-1}(u_{\xi_{j-1}}))\, ds,$$

we get $\lim_{t\to\infty} u(t) = 0$.

12.5 Numerical Examples

To illustrate these results, we take some examples.

Example 12.1 Consider the nonlinear EPCA

$$\begin{cases} \dot{x} = 2x + 2\beta_k e^{y^2} y_{\xi_k}, & t \in [t_k, t_{k+1}], \quad k = 0, 1, 2, \ldots \\ \dot{y} = y + \beta_k (1 + x^2) x_{\xi_k}, \end{cases} \tag{12.16}$$

where $\beta_k = -3.5$ for all k. Clearly, the ordinary part is unstable. Let $V(x, y) = x + y$ for $x > 0$ and $y > 0$. Then, one may get

$$\dot{V} \leq \alpha V + \beta_k V_{\xi_k},$$

where $\alpha = 2$. The solution of the differential inequality is given in Corollary 12.1(ii) and, by Corollary 12.3, the trivial solution $x \equiv 0$ of (12.16) is uniformly asymptotically stable. If $\xi_k = t_k$, then $t_{k+1} \in (0.15, 0.34)$, where $c = 0.6$. Figure 12.1 shows the simulation result in the case $\xi_k = t_k$ for all k.

Example 12.2 Consider the nonlinear EPCA

$$\begin{cases} \dot{x} = y - x[1 + \theta(x^2 + y^2)], & t \in [t_k, t_{k+1}], \quad k = 0, 1, 2, \ldots \\ \dot{y} = -x - y[1 + \theta(x^2 + y^2)] + \frac{2y_{\xi_k}}{2^k}, \end{cases} \tag{12.17}$$

where $0 < \theta \ll 1$. Let $V(x, y) = \frac{1}{2}(x^2 + y^2)$. Then

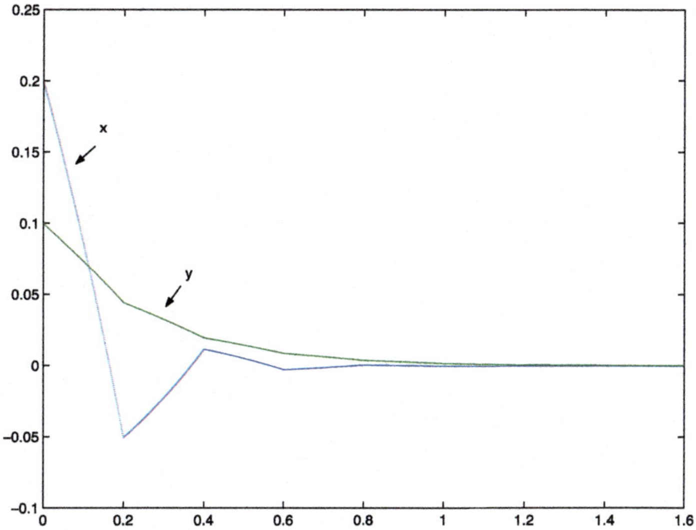

Fig. 12.1 Simulation result of Example 12.1

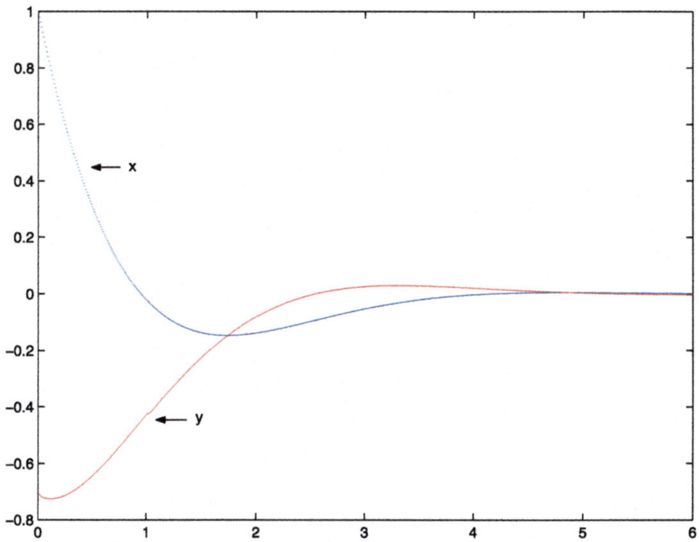

Fig. 12.2 Simulation result of Example 12.2

$$\begin{aligned}\dot{V}(x, y) &= xy - x^2[1 + \theta(x^2 + y^2)] - xy - y^2[1 + \theta(x^2 + y^2)] + \frac{2yy_{\xi_k}}{2^k} \\ &\leq -(x^2 + y^2) - \theta(x^2 + y^2)^2 + \frac{1}{2^k}(x^2 + y^2) + \frac{1}{2^k}(x_{\xi_k}^2 + y_{\xi_k}^2) \\ &= -\theta V^2(x, y) + \beta_k V(x_{\xi_k}, y_{\xi_k})\end{aligned}$$

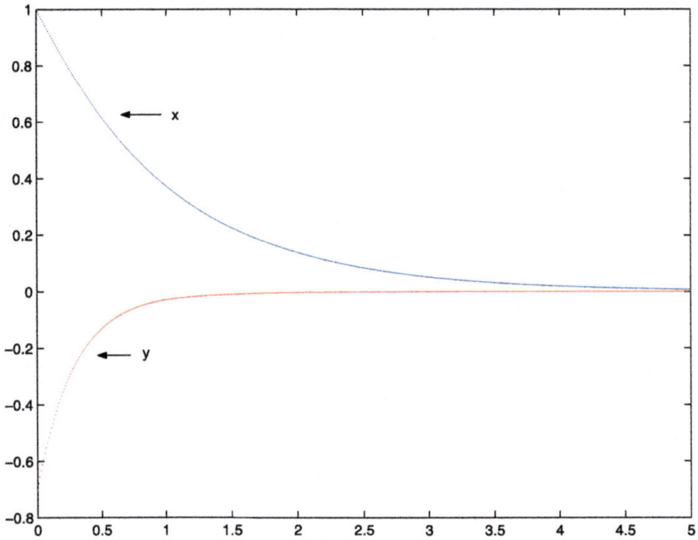

Fig. 12.3 Simulation result of Example 12.3

Let $\omega(u) = \theta u^2$. Then, by Corollary 12.4, the trivial solution of (12.17) is uniformly asymptotically stable. Simulation result is shown in Fig. 12.2, where $\theta = 0.01$, $\xi_k = t_k$ and $t_{k+1} - t_k = 1$ for all $k = 0, 1, 2, \ldots$.

Example 12.3 Consider the nonlinear EPCA

$$\begin{cases} \dot{x} = -x, t \in [t_k, t_{k+1}], \quad k = 0, 1, 2, \ldots \\ \dot{y} = -2y + \frac{x_{\xi_k} \sin y_{\xi_k}}{1+x^2} y e^{-t}. \end{cases} \tag{12.18}$$

Let $V(x, y) = \frac{1}{2}(x^2 + y^2)$. Then, one can get

$$\begin{aligned} \dot{V}(x, y) &= -x^2 - 2y^2 + \frac{x_{\xi_k} \sin y_{\xi_k}}{1 + x^2} y^2 e^{-t} \\ &\leq -(x^2 + y^2) + \frac{1}{2}(y^4 + x_{\xi_k}^2) e^{-t} \\ &= -2V(x, y) + \Big(2V^2(x, y) + V(x_{\xi_k}, y_{\xi_k})\Big) e^{-t} \\ &= \alpha V + h(t, V, V_{\xi_k}), \end{aligned}$$

where $\alpha = -2$ and $h(t, V, V_{\xi_k}) = (2V^2 + V_{\xi_k}) e^{-t}$. By Corollary 12.5, the trivial solution of (12.18) is uniformly asymptotically stable. Figure 12.3 shows the asymptotic stability of the trivial solution of (12.18).

12.6 Application

Consider the logistic population growth model undergoing density-dependence harvesting whose dynamics are given by

$$\dot{N}(t) = rN(t)\big(1 - aN(t) - bN(\gamma(t))\big), \qquad t > 0, \tag{12.19}$$

The system, here, is viewed as a switched system in which $\gamma(t)$ represents the switching signal for all $t \in [t_{k-1}, t_k)$ with $k \in \mathbb{N}$ and takes values in $\{\xi_k\}_{k=0}^{\infty}$ with $\xi_k = t_k$ for all k. Clearly, the system has two equilibria, $N_1 = 0$ and $N_2 = \frac{1}{a+b} > 0$. To analyze the stability properties of N_2, we use comparison method of this chapter.

For convenience, we transfer the desired equilibrium solution to the origin by applying the change of variable $x = b(N - N_2)$ to obtain

$$\dot{x}(t) = -r\big(x(t) + \frac{1}{1+\alpha}\big)\big(\alpha x(t) + x(\xi_k)\big),$$

where $\alpha > 1$. Define $V(x) = x$ as the Lyapunov function candidate. Then, one may get

$$\dot{V}\big(x, x(\xi_k)\big) \leq -\lambda V(x) - \mu V(x(\xi_k)),$$

where $\lambda = \frac{r\alpha}{1+\alpha}$ and $\mu = \frac{r}{1+\alpha}$. Consider the comparison system

$$\begin{aligned} &\dot{u}(t) = -\lambda u(t) - \mu u(\xi_k), \qquad t \in [t_{k-1}, t_k), \qquad k \in \mathbb{N}, \\ &u(t_0) = u_0 > 0. \end{aligned}$$

One can easily show that

$$u(t) = \Big[\big(1 - \frac{\mu}{\lambda}\big)e^{-\lambda(t-t_k)} - \frac{\mu}{\lambda}\Big]\prod_{j=1}^{k}\Big[\big(1 - \frac{\mu}{\lambda}\big)e^{-\lambda(t_j - t_{j-1})} - \frac{\mu}{\lambda}\Big]u_0.$$

As illustrated in Corollary 12.3, assume that infinite series

$$\sum_{j=1}^{\infty}\Big[\big(1 - \frac{\mu}{\lambda}\big)e^{-\lambda(t_j - t_{j-1})} - \frac{\mu}{\lambda}\Big]$$

converges. Furthermore, if, in addition, the terms in corresponding infinite product are all less than unity, say $c < 1$, then $x \equiv 0$ is uniformly asymptotically stable. This implies that, for $\xi_k = t_k$,

$$t_{k+1} - t_k > \frac{1}{-\lambda}\ln\Big(\frac{c\lambda + \mu}{\lambda - \mu}\Big),$$

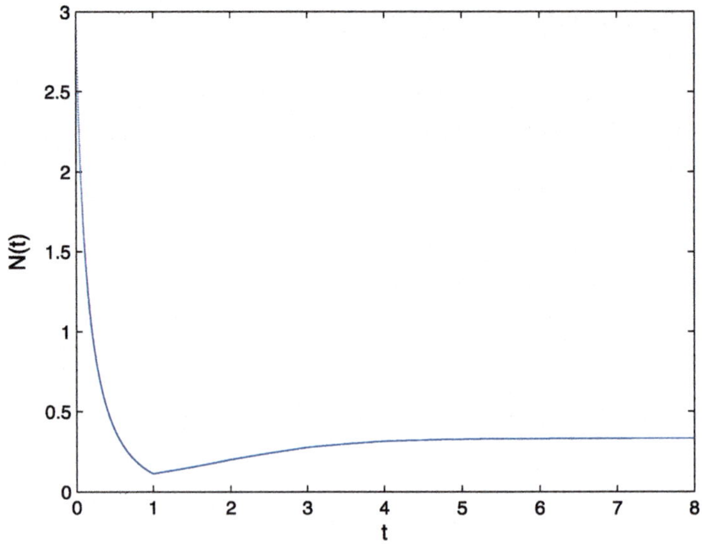

Fig. 12.4 Simulation result of the logistic growth model

which represent the dwell times, where $\frac{c\lambda+\mu}{\lambda-\mu} < 1$ so long as $\lambda > \mu$ and $c < 1$. For the simulation purposes, we take $r = 1$, $a = 2$ and $b = 1$ to get from the last inequality $t_k - t_{k-1} > 0.6$ for any $k \in \mathbb{N}$. Figure 12.4 shows the asymptotic stability of the positive equilibrium point $N_2 = \frac{1}{a+b} = 0.33$, where $t_{k+1} - t_k = 1$ for all $k = 0, 1, \ldots$.

12.7 Notes and Comments

In this chapter, systems of nonlinear EPCA have been viewed as a switched system. We also presented a comparison method (Theorem 12.1) for the systems which has been later used to prove some stability properties by the classical Lyapunov method. We have also shown that piecewise constant arguments do contribute to stabilize unstable systems of ordinary differential equations (Corollary 12.3). Finally, the stability result has been applied to address the asymptotic stability of a population growth model. The material of this chapter is taken from [7].

Initially, the theory of EPCA was developed in [1]. Later, it was well discussed in the survey paper [2] and book [3]. A general type of EPCA (EPCAG) in which the piecewise constant real function γ takes values over discrete subintervals instead of at the most left endpoint of each subinterval, has appeared in some works [8, 9]. In those works, the solutions of linear and quasi-linear EPCAG are determined by a unique initial datum at an initial moment t_0, rather than by a countable set of initial data defining at discrete moments n for nonnegative integers n or, as in the case of functional differential equations, by an initial function defining on some interval from the past history. In either case, EPCA or EPCAG, functional differential equations reduce to ordinary ones. Consequently, one can use the theory of ordinary differential equations.

The logistic population growth model had been studied in [10], where $\gamma(t) = [t]$ for all $t \in [n, n+1)$ with $n = 0, 1, 2, \ldots$. It was shown that the positive equilibrium solution is globally asymptotically stable if $\alpha = a/b \geq 1$ and, whenever $N(n) > 0$, $N(t) > 0$ for all $t \in [n, n+1)$, $n = 0, 1, 2, \ldots$. Later, differential equation (12.19) in which $\gamma(t) = [t]$ for all $t > 0$ was considered in [11], where stability results were established by using Lyapunov–Razumikhin method.

References

1. Cooke KL, Wiener J (1984) Retarded differential equations with piecewise constant delays. J Math Anal Appl 99:265–297
2. Cooke KL, Wiener J (1991) A survey of differential equations with piecewise constants arguments. In: Busenberg S, Martelli M (eds) Delay differential equations and dynamical systems, vol 1475. Lecture notes in mathematics. Springer, Heidelberg-New York, pp 1–15
3. Wiener J (1993) Generalized solutions of functional differential equations. World Scientific, Singapore
4. Busenberg S, Cooke KL (1982) Models of vertically transmitted with sequential-continuous dynamics. In: Lakshmikantham V (ed) Nonlinear phenomena in mathematical sciences. Academic Press, New York, pp 179–187
5. Lakshmikantham V, Liu XZ (1998) Impulsive hybrid systems and stability theory. Dyn Syst Appl 7:1–10
6. Lakshmikantham V, Leela S (1969) Differential and integrals inequalities. Academic Press, New York
7. Alwan MS, Liu XZ, Xie W-C (2013) Comparison principle and stability results for nonlinear differential equations with piecewise constant arguments. J Frankl Inst 350:211–230
8. Akhmet MU (2008) Stability of differential equations with piecewise constant arguments of generalized type. Nonlinear Anal 68:794–803
9. Akhmet MU, Aruğtaslan D, Liu XZ (2008) Permanence of nonautonomous ratio-dependent predator-prey systems with piecewise constant argument of generalized type. Dyn Contin Discret Impuls Syst Ser A Math Anal 15:37–51
10. Gopalsamy K, Liu P (1998) Persistence and global stability in a population model. J Math Anal Appl 224:59–80
11. Akhmet MU, Aruğaslan D (2009) Lyapunov-Razumikhin method for differential equations with piecewise constant argument. Discret Contin Dyn Syst 25(2):457–466

Chapter 13
Existence, Uniqueness and Stability of Stochastic EPCA

In this chapter, we consider systems of stochastic EPCA (or SEPCA). We start with the problem of existence and uniqueness of solutions. Then, we address the comparison method and the stability notion of the solution.

Consider the nonlinear systems with SEPCA of the form

$$dx(t) = f(t, x(t), \lambda_{\varrho(t)}(x(\gamma(t))))dt + g(t, x(t), \lambda_{\varrho(t)}(x(\gamma(t))))dW(t), \quad (13.1a)$$

$$x(t_0) = x_0, \quad (13.1b)$$

where $x \in \mathbb{R}^n$ is the system state and, for all $t \geq t_0$ with $t_0 \in \mathbb{R}_+$, $\varrho(t)$ and $\gamma(t)$ are piecewise constant functions taking values in the sets $K = \{k\}_{k=0}^{\infty}$ and $\Xi = \{\xi_k\}_{k=0}^{\infty}$, respectively, where $t_k \leq \xi_k < t_{k-1}$ for any $k = 0, 1, 2, \ldots$. As stated in the previous chapter, these functions represent the switching signals of the system switching between the piecewise constant argument λ_k and the values of its state argument x.

Accordingly, one may define system (13.1) as follows: for all $t \in [t_k, t_{k+1})$,

$$dx(t) = f(t, x(t), \lambda_k(x(\xi_k)))dt + g(t, x(t), \lambda_k(x(\xi_k)))dW(t), \quad (13.2a)$$

$$x(t_0) = x_0 \quad (13.2b)$$

or, equivalently,

$$x(t) = x_0 + \int_{t_0}^{t} f(s, x(s), \lambda_k(x(\xi_k)))ds + \int_{t_0}^{t} g(s, x(s), \lambda_k(x(\xi_k)))dW(s). \quad (13.3)$$

The following definitions will be needed in this chapter.

Definition 13.1 For any α, $\beta \in \mathbb{R}$, an $\mathbb{R}^n$−valued stochastic process $x : (\alpha, \beta) \to \mathbb{R}$ is said to be a *solution* of (13.1) if the following hold:

M. S. Alwan and X. Liu, *Theory of Hybrid Systems: Deterministic and Stochastic*, Nonlinear Physical Science, https://doi.org/10.1007/978-981-10-8046-3_13

(i) $x(t)$ is continuous and $\mathscr{F}_t$−adapted for all $t \in (\alpha, \beta)$;
(ii) $f(t, x(t), \lambda_k(x(\xi_k))) \in \mathscr{L}_{ad}(\Omega, L^1(\alpha, \beta))$ and
$g(t, x(t), \lambda_k(x(\xi_k))) \in \mathscr{L}_{ad}(\Omega, L^2(\alpha, \beta))$; and
(iii) the stochastic integral equation (13.3) holds w.p.1.

Definition 13.2 For all $t \in [a, b]$, an $\mathbb{R}^n$−valued $\mathscr{F}_t$−adapted process $f(t)$ with $\int_a^b \|f(t)\|^p dt < \infty$ (a.s.) (i.e., $f \in \mathscr{L}_{ad}(\Omega; L^p[a, b])$) is said to be in $\mathscr{M}^p([a, b]; \mathbb{R}^n)$ if $\mathbb{E}\big[\int_a^b \|f(t)\|^p dt\big] < \infty$.

Definition 13.3 An $\mathbb{R}^n$−valued $\mathscr{F}_t$−adapted integrable process $X(t)$ is said to be a *martingale with respect to the filtration* $\{\mathscr{F}_t\}_{t\geq 0}$ if

$$\mathbb{E}[X(t)|\mathscr{F}_s] = X(s), \quad \text{(a.s.), for all } 0 \leq s < t < \infty,$$

where $\mathbb{E}[X(t)|\mathscr{F}_s]$ stands for the conditional expectation of process $X(t)$ with respect to the filtration $\mathscr{F}_s$.

Doob's martingale inequality. For all $t \geq 0$, let $X(t)$ be an $\mathbb{R}^n$-valued martingale and $[a, b]$ be a bounded interval of $\mathbb{R}$. If $p > 1$ and $X(t) \in L^p(\Omega; \mathbb{R}^n)$, then

$$\mathbb{E}\Big[\sup_{a\leq t\leq b} \|X(t)\|^p\Big] \leq \Big(\frac{p}{p-1}\Big)^p \mathbb{E}[\|X(b)\|^p].$$

Borel–Cantelli's lemma. If $\{A_k\}_{k=1}^{\infty} \subset \mathscr{F}$ and $\sum_{k=1}^{\infty} \mathbb{P}(A_k) < \infty$, then

$$\mathbb{P}(\limsup_{k\to\infty} A_k) = 0.$$

13.1 Existence and Uniqueness of Solutions

In this section, we address the problem of existence of a unique solution of SEPCA given in (13.1) or (13.2). As will be seen, the technique followed here is to generate a convergent Cauchy sequence of solutions. For this purpose, we assume that the system vector fields are bounded by a linear growth estimate and satisfy the Lipschitz condition. The first condition is to avoid a finite escape time that a solution may have when time evolves. The second condition is made to be used in proving the convergence of the generated sequence of the solution and to guarantee the uniqueness of the solution.

Before we prove this theorem, the following lemma is needed.

Lemma 13.1 *For any* $k = 0, 1, 2, \ldots$, *assume that the linear growth condition holds. Then, solution* x *cannot grow faster than the following exponential estimate*

$$\mathbb{E}\Bigg(\sup_{t_k\leq t\leq t_{k+1}} \|x(t)\|^2\Bigg) \leq (1 + c_k)e^{3L_1(t_{k+1}-t_k+4)(t_{k+1}-t_k)},$$

where $c_k = 3\mathbb{E}[\|x_0\|^2] + 3L_1(t_{k+1} - t_k + 4)(t_{k+1} - t_k)\mathbb{E}[\|\lambda_k(x_{\xi_k})\|^2] < \infty$. *In other words,* $x \in \mathscr{M}^2([t_k, t_{k+1}); \mathbb{R}^n)$ *with* $0 < t_{k+1} - t_k \leq \theta < \infty$ *for any* k.

Proof Choose k arbitrarily and, for any $l \geq 1$, define a sequence of stopping times

$$\tau_l = t_{k+1} \wedge \inf\{t \in [t_k, t_{k+1}) \mid \|x(t)\| \geq l\},$$

where $\lim_{l\to\infty} \tau_l = t_{k+1}$ (a.s.). For simplicity of notation, we set $x_l(t) = x(t \wedge \tau_l)$ for all $t \in [t_k, t_{k+1})$. Then, from system (13.1), we get

$$x_l(t) = x_k + \int_{t_k}^{t} f(s, x_l(s), \lambda_k(x_{\xi_k}))1_{[t_k,\tau_l]}ds + \int_{t_k}^{t} g(s, x_l(s), \lambda_k(x_{\xi_k}))1_{[t_k,\tau_l]}dW(s),$$

where 1_A is the indicator function of a set A. In virtue of (i) and using Doob's martingale inequality to the stochastic Itô integral, one may get

$$\begin{aligned}
&\mathbb{E}\left(\sup_{t_k\leq t\leq t} \|x_l(t)\|^2\right)\\
&\leq 3\mathbb{E}[\|x_k\|^2] + 3L_1(t_{k+1} - t_k)\int_{t_k}^{t}(1 + \mathbb{E}[\|x_l(s)\|^2] + \mathbb{E}[\|\lambda_k(x_{\xi_k})\|^2])ds\\
&\quad + 12L_1\int_{t_k}^{t}(1 + \mathbb{E}[\|x_l(s)\|^2] + \mathbb{E}[\|\lambda_k(x_{\xi_k})\|^2])ds\\
&\leq 3\mathbb{E}[\|x_k\|^2] + 3L_1(t_{k+1} - t_k + 4)\int_{t_k}^{t}(1 + \mathbb{E}[\|x_l(s)\|^2])ds\\
&\quad + 3L_1(t_{k+1} - t_k + 4)(t_{k+1} - t_k)\mathbb{E}[\|\lambda_k(x_{\xi_k})\|^2],
\end{aligned}$$

which implies that

$$\begin{aligned}
&1 + \mathbb{E}\left(\sup_{t_k\leq t\leq t} \|x_l(t)\|^2\right)\\
&\leq 1 + c_k + 3L_1(t_{k+1} - t_k + 4)\int_{t_k}^{t}(1 + \mathbb{E}[\|x_l(s)\|^2])ds\\
&\leq 1 + c_k + 3L_1(t_{k+1} - t_k + 4)\int_{t_k}^{t}(1 + \mathbb{E}[\sup_{t_k\leq t\leq \tau_l} \|x_l(s)\|^2])ds.
\end{aligned}$$

By the Gronwall inequality, we get

$$\mathbb{E}\left(\sup_{t_k\leq t\leq t} \|x_l(t)\|^2\right) \leq (1 + c_k)e^{3L_1(t_{k+1}-t_k+4)(t_{k+1}-t_k)}.$$

The desired result is implied by letting $l \to \infty$. This completes the proof.

Theorem 13.1 *Assume the following assumptions hold:*

(i) the vector fields functions f and g satisfy the linear growth condition; i.e., there exists a positive L_1 such that

$$\|f(t,x,y)\|^2 + \|g(t,x,y)\|^2 \le L_1(1+\|x\|^2+\|y\|^2), \quad (a.s.),$$

for all $(t,x,y) \in [t_k, t_{k+1}) \times \mathbb{R}^n \times \mathbb{R}^n$;

(ii) f and g satisfy a global Lipschitz condition; i.e., there exists a positive constant L_2 such that

$$\|f(t,x_1,y_1 - f(t,x_2,y_2))\|^2 + \|g(t,x_1,y_1) - g(t,x_2,y_2))\|^2 \\ \le L_2\|x_1-x_2\|^2 + \|y_1-y_2\|^2, \quad (a.s.),$$

for all $(t,x,y) \in [t_k, t_{k+1}) \times \mathbb{R}^n \times \mathbb{R}^n$.

Then, system (13.1) or (13.2) has a unique solution, $x(t)$, defined for all $t \ge t_0$.

Proof The proof is given here for all $t \in [t_0, t_1)$ since the rest will be similar. Define the sequence x_n with the initial state, x_0, by the following iteration

$$x_n(t) = x_0 + \int_{t_0}^{t} f(s, x_{n-1}(s), \lambda_k(x_{n-1_{\xi_0}}))ds + \int_{t_0}^{t} g(s, x_{n-1}(s), \lambda_k(x_{n-1_{\xi_0}}))dW(s), \tag{13.4}$$

where $x_{j_{\xi_0}} = x_j(\xi_0) = x_j(t_0)$. By Lemma 13.1, $x_0 \in \mathscr{M}^2([t_k, t_{k+1}); \mathbb{R}^n)$, and by the mathematical induction, we can see that $x_n(t) \in \mathscr{M}^2([t_k, t_{k+1}); \mathbb{R}^n)$ as follows:

$$\mathbb{E}[\|x_n(t)\|^2] \le C_1 + 3L_1(t+t_1)\int_{t_0}^{t} \mathbb{E}[\|x_{n-1}(s)\|^2]ds,$$

where $C_1 = 3\mathbb{E}[\|x_0\|^2] + 3L_1t_1(1+t_1)\left(1+\mathbb{E}[\|\lambda_k(x_{n-1_{\xi_0}})\|^2]\right) < \infty$, where we used the fact $t_1 - t_0 < t_1$. This also implies that, for an arbitrary j,

$$\begin{aligned}\max_{1\le n\le j} \mathbb{E}[\|x_n(t)\|^2] &\le C_1 + 3L_1(t+t_1)\int_{t_0}^{t} \max_{1\le n\le j} \mathbb{E}[\|x_{n-1}(s)\|^2]ds \\ &\le C_1 + 3L_1(t+t_1)\int_{t_0}^{t}\left(\mathbb{E}[\|x_0\|^2] + \max_{1\le n\le j}\mathbb{E}[\|x_n(s)\|^2]\right)ds \\ &= C_2 + 3L_1(t+t_1)\int_{t_0}^{t} \max_{1\le n\le j}\mathbb{E}[\|x_n(s)\|^2]ds,\end{aligned}$$

where $C_2 = C_1 + 3L_1t_1(1+t_1)\mathbb{E}[\|x_0\|^2]$. By the Gronwall inequality

$$\max_{1\le n\le j}\mathbb{E}[\|x_n(t)\|^2] \le C_2e^{3L_1t_1(1+t_1)}.$$

Since j is arbitrary, we get

$$\mathbb{E}[\|x_n(t)\|^2] \le C_2 e^{3L_1 t_1(1+t_1)}, \tag{13.5}$$

i.e., for all n, $x_n \in \mathscr{M}^2([t_k, t_{k+1}); \mathbb{R}^n)$; that is, $x_n(t)$ is bounded over $[t_0, t_1)$.
Now, we want to prove that this sequence is convergent. Note that

$$\begin{aligned}\|x_1(t) - x_0(t)\|^2 &= \|x_1(t) - x_0\|^2 \\ &\le 2\left\| \int_{t_0}^{t} f(s, x_0, \lambda_k(x_{0_{\xi_0}}))ds \right\|^2 + 2\left\| \int_{t_0}^{t} g(s, x_0, \lambda_k(x_{0_{\xi_0}}))dW(s) \right\|^2,\end{aligned}$$

which implies, after taking the mathematical expectation,

$$\begin{aligned}&\mathbb{E}[\|x_1(t) - x_0(t)\|^2] \\ &\quad \le 2L_1 \left[(t_1 - t_0)(1 + (t_1 - t_0))\right] \left(1 + \mathbb{E}[\|x_0\|^2] + \mathbb{E}[\|\lambda_k(x_{\xi_0})\|^2]\right) = C,\end{aligned}$$

i.e., $\mathbb{E}[\|x_1(t) - x_0(t)\|^2] \le C$, where

$$C = 2L_1 \left[(t_1 - t_0)(1 + (t_1 - t_0))\right] \left(1 + \mathbb{E}[\|x_0(t)\|^2] + \mathbb{E}[\|\lambda_k(x_{\xi_0})\|^2]\right).$$

We will show by mathematical induction that, for any $n \ge 0$ and $t \in [t_0, t_1)$,

$$\mathbb{E}[\|x_{n+1}(t) - x_n(t)\|^2] \le \frac{C[M(t - t_0)]^n}{n!} \tag{13.6}$$

with $M = 2L_2(t_1 - t_0 + 1)$. Obviously, the relation is true for $n = 0, 1$. Assume that it is also true for some $n \ge 0$. As for the case of $n + 1$, we have

$$\begin{aligned}&\|x_{n+2}(t) - x_{n+1}(t)\|^2 \\ &\quad \le 2\left\| \int_{t_0}^{t} \Big(f(s, x_{n+1}(s), \lambda_k(x_{n+1_{\xi_0}})) - f(s, x_n(s), \lambda_k(x_{n_{\xi_0}}))\Big)ds \right\|^2 \\ &\qquad + 2\left\| \int_{t_0}^{t} \Big(g(s, x_{n+1}(s), \lambda_k(x_{n+1_{\xi_0}})) - g(s, x_n(s), \lambda_k(x_{n_{\xi_0}}))\Big)dW(s) \right\|^2.\end{aligned}$$

Taking the mathematical expectation and using the Lipschitz condition give

$$\begin{aligned}\mathbb{E}[\|x_{n+2}(t) - x_{n+1}(t)\|^2] &\le 2L_2(t - t_0 + 1)\mathbb{E} \int_{t_0}^{t} \Big(\|x_{n+1}(s) - x_n(s)\|^2 \\ &\quad + \|\lambda_k(x_{n+1_{\xi_0}}) - \lambda_k(x_{n_{\xi_0}})\|^2\Big)ds \\ &= M \int_{t_0}^{t} \mathbb{E}[\|x_{n+1}(s) - x_n(s)\|^2]ds \\ &\le M \int_{t_0}^{t} \frac{C[M(s - t_0)]^n}{n!} ds = \frac{C[M(t - t_0)]^{n+1}}{(n + 1)!}\end{aligned}$$

because $\lambda_k(x_{n+1_{\xi_0}}) - \lambda_k(x_{n_{\xi_0}}) = 0$ for any $n \geq 0$; for instance, for $n = 0$, we have

$$\lambda_k(x_{1_{\xi_0}}) - \lambda_k(x_{0_{\xi_0}}) = \lambda_k(x_1(t_0)) - \lambda_k(x_0(t_0)) = 0 \qquad \text{(a.s.)}.$$

This is because $x_0(t) = x_0$ for all t, and by the solution sequence (13.4), we have $x_1(t_0) = x_0(t_0) = x_0$. Thus, the relation is true for $n + 1$.

To prove that x_n is a Cauchy sequence, replace n by $n - 1$ and consider

$$\begin{aligned}
&\sup_{t_0 \leq t \leq t_1} \|x_{n+1}(t) - x_n(t)\|^2 \\
&\quad \leq 2 \sup_{t_0 \leq t \leq t_1} \left\| \int_{t_0}^{t} [f(s, x_n(s), \lambda_k(x_{n_{\xi_0}})) - f(s, x_{n-1}(s), \lambda_k(x_{n-1_{\xi_0}}))]ds \right\|^2 \\
&\quad + 2 \sup_{t_0 \leq t \leq t_1} \left\| \int_{t_0}^{t} [g(s, x_n(s), \lambda_k(x_{n_{\xi_0}})) - g(s, x_{n-1}(s), \lambda_k(x_{n-1_{\xi_0}}))]dW(s) \right\|^2,
\end{aligned}$$

which implies, after taking the mathematical expectations and using the Doob's martingale inequality

$$\begin{aligned}
&\mathbb{E}\left(\sup_{t_0 \leq t \leq t_1} \|x_{n+1}(t) - x_n(t)\|^2 \right) \\
&\leq 2L_2(t_1 - t_0 + 4) \int_{t_0}^{t_1} \mathbb{E}\left[\|x_n(s) - x_{n-1}(s)\|^2 + \|\lambda_k(x_{n_{\xi_0}}) - \lambda_k(x_{n-1_{\xi_0}})\|^2 \right] ds \\
&= 2L_2(t_1 - t_0 + 4) \int_{t_0}^{t_1} \mathbb{E}\left[\|x_n(s) - x_{n-1}(s)\|^2 \right] ds
\end{aligned} \tag{13.7}$$

because $\lambda_k(x_{n_{\xi_0}}) - \lambda_k(x_{n-1_{\xi_0}}) = 0$ for any $n \geq 1$. For instance, for $n = 1$, we have

$$\lambda_k(x_{1_{\xi_0}}) - \lambda_k(x_{0_{\xi_0}}) = \lambda_k(x_1(t_0)) - \lambda_k(x_0(t_0)) = 0 \qquad \text{(a.s.)}.$$

This is because $x_0(t) = x_0$ for all t, and by the solution sequence (13.4), we have $x_1(t_0) = x_0(t_0) = x_0$. Therefore, from (13.7), it follows that

$$\begin{aligned}
\mathbb{E}\Big(\sup_{t_0 \leq t \leq t_1} \|x_{n+1}(t) - x_n(t)\|^2 \Big) &\leq 4M \int_{t_0}^{t_1} \frac{4C[M(s - t_0)]^{n-1}}{(n-1)!} ds \\
&= \frac{4C[M(t_1 - t_0)]^n}{n!},
\end{aligned}$$

from which, we get

$$\mathbb{P}\Big\{ \sup_{t_0 \leq t \leq t_1} \|x_{n+1}(t) - x_n(t)\|^2 > \frac{1}{2^n} \Big\} \leq \frac{4C[M(t_1 - t_0)]^n}{n!}.$$

Since series $\sum_{n=0}^{\infty} \frac{4C[M(t_1-t_0)]^n}{n!}$ is convergent, by the Borel–Cantelli's lemma, we have

$$\sup_{t_0 \le t \le t_1} \|x_{n+1}(t) - x_n(t)\|^2 \le \frac{1}{2^n}.$$

It follows that, w.p.1, the partial sums

$$x_n(t) = x_0(t) + \sum_{j=0}^{n-1} (x_{j+1}(t) - x_j(t))$$

are convergent over $[t_0, t_1]$. Therefore, we conclude that sequence x_n is Cauchy; i.e., there exists a limit point x such that $\lim_{n\to\infty} x_n(t) = x(t)$, which implies that, for all $t \in [t_0, t_1)$,

$$x(t) = x_0 + \int_{t_0}^{t} f(s, x(s), \lambda_k(x_{\xi_0}))ds + \int_{t_0}^{t} g(s, x(s), \lambda_k(x_{\xi_0}))dW(s). \quad (13.8)$$

Similarly, one can show this relation holds for any $t \in [t_k, t_{k+1})$. We should mention that the inequality in (13.8) is still true for any k because, by defining the general form of the solution sequence for any $t \in [t_k, t_{k+1})$, we have

$$\begin{aligned} x_n(t) = x_0(t_k) &+ \int_{t_k}^{t} f(s, x_{n-1}(s), \lambda_k(x_{n-1_{\xi_k}}))ds \\ &+ \int_{t_k}^{t} g(s, x_{n-1}(s), \lambda_k(x_{n-1_{\xi_k}}))dW(s), \end{aligned} \quad (13.9)$$

where $x_{j_{\xi_k}} = x_j(\xi_k) = x_j(t_k)$; for instance, if $n = 2$, we obtain

$$\begin{aligned} \lambda_k(x_{2_{\xi_k}})) - \lambda_k(x_{1_{\xi_k}})) &= \lambda_k(x_2(t_k)) - \lambda_k(x_1(t_k)) \\ &= \lambda_k(x_0(t_k)) - \lambda_k(x_0(t_k)) \\ &= 0, \end{aligned}$$

w.p.1. Due to the continuity of solution x, $\lim_{t\to t_{k+1}^-} x(t) = x(t_{k+1})$. Thus, the constructed solution is continuous and $\mathscr{F}_t$–adapted for all $t \ge t_0$. Furthermore, from (13.6), for all $t \ge t_0$, sequence $x_n(t)$ is Cauchy in L^2, which implies that $\lim_{n\to\infty} x_n(t) = x(t)$ in L^2. It follows that, by letting $n \to \infty$ in (13.5),

$$\mathbb{E}[\|x(t)\|^2] \le C_2 e^{3L_1 t_1(1+t_1)}, \quad \text{for all } t \ge t_0,$$

i.e., $x \in \mathscr{M}^2(\mathbb{R}_+; \mathbb{R}^n)$. Next, we will show that x satisfies the stochastic integral equation in (13.3), for all $t \in [t_k, t_{k+1}]$ and every k, as follows:

$$\mathbb{E}\left\|\int_{t_0}^{t} f(s, x_n(s), \lambda_k(x_{n_{\xi_k}}))ds - f(s, x(s), \lambda_k(x_{\xi_k}))ds\right\|^2$$
$$+ \mathbb{E}\left\|\int_{t_0}^{t} g(s, x_n(s), \lambda_k(x_{n_{\xi_k}}))dW(s) - g(s, x(s), \lambda_k(x_{\xi_k}))dW(s)\right\|^2$$
$$\leq L_2(t_{k+1} - t_0 + 1)\int_{t_0}^{t_{k+1}} \mathbb{E}\|x_n(s) - x(s)\|^2 ds \to 0, \quad \text{as} \quad n \to \infty.$$

Therefore, by letting $n \to \infty$ in (13.4), we get the required result. Finally, to prove the uniqueness, assume that there is another solution, say $y(t)$. Then,

$$x(t) - y(t) = \int_{t_0}^{t} \big(f(s, x(s), \lambda_k(x_{\xi_k})) - f(s, y(s), \lambda_k(y_{\xi_k}))\big)ds$$
$$+ \int_{t_0}^{t} \big(f(s, x(s), \lambda_k(x_{\xi_k})) - f(s, y(s), \lambda_k(y_{\xi_k}))\big)dW(s),$$

which implies that, after applying Hölder's inequality, Doob's martingale inequality and Lipschitz condition,

$$\mathbb{E}\Big[\sup_{t_0\leq s\leq t} \|x(s) - y(s)\|^2\Big] \leq 2L_2(t_{k+1} + 4)\int_{t_0}^{t} \mathbb{E}[\sup_{t_0\leq u\leq s} \|x(u) - y(u)\|^2]ds.$$

By the Gronwall inequality, we obtain

$$\mathbb{E}\Big[\sup_{t_0\leq s\leq t} \|x(s) - y(s)\|^2\Big] = 0.$$

Thus, processes x and y are indistinguishable for all t. Hence, system (13.1) has a unique solution $x(t)$ for all $t \geq t_0$. This completes the proof.

13.2 Comparison Method

Having established the existence of a unique solution, in this section we deal with the comparison method and stability properties of the trivial solution of system (13.1).

Theorem 13.2 *Assume that the following assumptions hold:*

(i) for any $k = 0, 1, 2 \ldots$, $V \in \mathscr{C}^{1,2}([t_k, t_{k+1}) \times \mathbb{R}^n; \mathbb{R}_+)$, V is bounded below and satisfies

$$\mathscr{L}V(t, x, y) \leq h(t, x, \sigma_k(y)), \quad \textit{(a.s.)}, \qquad t \in [t_k, t_{k+1}),$$

where the function h is concave and nondecreasing in x and σ_k with σ_k being a concave function; and

(ii) the auxiliary scalar comparison system

$$\begin{aligned}\dot{u}(t) &= h(t, u(t), \sigma_k(u_{\xi_k})), \qquad t \in [t_k, t_{k+1}), \\ u(t_0) &= u_0\end{aligned} \tag{13.10}$$

has a maximal solution $\nu(t; t_0, u_0)$ for all $t \geq t_0$.

Then, $\mathbb{E}[V(t_0, x_0)] \leq u_0$ implies $\mathbb{E}[V(t, x)] \leq \nu(t; t_0, u_0)$ for any solution x of (13.1) for all $t \geq t_0$.

Proof For any $k = 0, 1, 2 \ldots$ and all $t \in [t_k, t_{k+1})$, let $x(t)$ be the solution of system (13.1) that is guaranteed by Theorem 13.1. Let τ_{k_l} or, for simplicity τ_l (for $l \geq 1$), be the first exit time of the process from the ball

$$B_l(x) = \{x \in \mathbb{R}^n \mid \|x\| \leq l\},$$

i.e., $\tau_l = \inf\{t \in [t_k, t_{k+1}) \mid \|x(t)\| > l\}$.

Define $\tau_l(t) = \min\{\tau_l, t\}$. Then, by the Itô formula, we have, for all $t \in [t_k, \tau_l(t)]$,

$$\begin{aligned}\mathbb{E}[V(\tau_l(t), x(\tau_l(t)))] &= \mathbb{E}[V(t_k, x(t_k))] + \mathbb{E}\int_{t_k}^{\tau_l(t)} \mathscr{L}V(s, x(s), \sigma_k(V_{\xi_k}))ds \\ &\leq \mathbb{E}[V(t_k, x(t_k))] + \mathbb{E}\int_{t_k}^{\tau_l(t)} h(s, V(s, x(s)), \sigma_k(V_{\xi_k}))ds,\end{aligned}$$

where $V_{\xi_k} = V(\xi_k, x(\xi_k))$. Define $m(t) = \mathbb{E}[V(s, x(s))]$ for all $t_k \leq s \leq \tau_l(t)$. Thus, by the properties of h and σ_k, the last inequality becomes

$$m(t) \leq m(t_k) + \int_{t_k}^{s} h(r, m(r), \sigma_k(m_{\xi_k}))dr, \qquad t_k \leq r \leq s \leq \tau_l(t),$$

where $m_{\xi_k} = m(\xi_k) = \mathbb{E}[V(\xi_k, x(\xi_k))]$.

By Theorem 12.1, we obtain

$$m(t) \leq \nu_k(t; t_k, m_{\xi_k}), \qquad t \in [t_k, \tau_l(t)]$$

and by letting $l \to \infty$, we obtain, for all $t \in [t_k, t_{k+1})$, $m(t) \leq \nu_k(t; t_k, m_{\xi_k})$. Particularly, for $t \in [t_0, t_1)$, we have

$$m(t) \leq \nu_0(t; t_0, m_{\xi_0}) = \nu_0(t; t_0, m(t_0)) \leq \nu_0(t; t_0, u_0) =: \nu(t; t_0, u_0),$$

where $\nu_0(t; t_0, u_0)$ is the maximal solution of the auxiliary comparison system (13.10) for $t \in [t_0, t_1)$ with $m(t_0) = \mathbb{E}[V(t_0, x(t_0))] \leq u_0$, as given initially.

For $t \in [t_1, t_2)$, we have

$$m(t) \le \nu_1(t; t_1, m_{\xi_1}) = \nu_1(t; t_1, m(t_1)) = \nu_1(t; t_1, \nu_1(t_1; t_0, u_0)) \\ =: \nu(t; t_0, u_0),$$

or

$$m(t) \le \nu(t; t_0, u_0), \qquad t \in [t_0, t_2).$$

In general, one obtains

$$m(t) = \mathbb{E}[V(t, x(t))] \le \nu(t; t_0, u_0), \qquad t \ge t_0,$$

where $\nu(t; t_0, u_0)$ is the maximal solution of the comparison system (13.10) for all $t \ge t_0$. This completes the proof.

13.3 Stability Analysis

In the following theorem, we prove some stability properties of the trivial solution of (13.1).

Theorem 13.3 *Assume that the conditions of Theorem 13.2 hold. Suppose also that there exist two functions $b \in \mathscr{K}_1$ and $a \in \mathscr{K}_2$ such that*

$$b(\|x\|^2) \le V(t, x) \le a(\|x\|^2), \qquad (a.s.). \tag{13.11}$$

Then, the stability properties of the trivial solution $u \equiv 0$ of system (13.10) imply the stability properties (in the m.s.) of the trivial solution $x \equiv 0$ of system (13.1).

Proof Assume that the trivial solution $u \equiv 0$ of comparison system (13.10) is stable. Then, for every $\varepsilon > 0$, there exists $\delta = \delta(t_0, \varepsilon) > 0$ for which

$$\nu(t, t_0, u_0) < b(\varepsilon), \qquad \text{whenever} \qquad u_0 \le \delta, \qquad \forall t \ge t_0 \ge 0, \tag{13.12}$$

where $\nu(t, t_0, u_0)$ is the maximal solution of comparison system (13.10).

To investigate the stability at t_0, we choose $\delta = \delta(t_0, \varepsilon) \le \delta_1$ (for the same ε) with $a(\delta_1) < b(\varepsilon)$ and let $u_0 = a(\mathbb{E}[\|x_0\|^2]) \le \delta_1$. Now, let $\mathbb{E}[\|x_0\|^2] \le \delta$. Then, from (13.11), we obtain

$$b(\mathbb{E}[\|x(t_0)\|^2]) \le \mathbb{E}[V(t_0, x_0)] \le a(\mathbb{E}[\|x_0\|^2]) \le a(\delta) \le b(\varepsilon),$$

i.e., $\mathbb{E}[\|x_0\|^2] \le \varepsilon$, whenever $\mathbb{E}[\|x_0\|^2] \le \delta$.

Under the given assumptions, we claim that the trivial solution $x \equiv 0$ of SEPCA (13.1) is stable in the m.s. for all $t > t_0$; i.e., for the assigned ε and δ, the following statement

$$\mathbb{E}[\|x_0\|^2] \leq \delta \quad \text{implies} \quad \mathbb{E}[\|x(t)\|^2] < \varepsilon, \quad \forall t > t_0$$

holds. If our claim were not true, there would be a $t^* > t_k > t_0$, specifically $t_k < t^* \leq t_{k+1}$, such that $\mathbb{E}[\|x_0\|^2] \leq \delta$ and

$$\mathbb{E}[\|x(t)\|^2] < \varepsilon, \quad t_k \leq t < t^*, \tag{13.13}$$

$$\mathbb{E}[\|x(t^*)\|^2] = \varepsilon. \tag{13.14}$$

Recall that, by Theorem 13.2, we have shown $\mathbb{E}[V(t, x(t))] \leq \nu(t; t_0, u_0)$ for all $t \geq t_0$. This, together with (13.12), implies

$$\mathbb{E}[V(t^*, x(t^*))] \leq \nu(t^*; t_0, u_0) = \nu(t^*; t_0, a(\mathbb{E}[\|x_0\|^p])) < b(\varepsilon).$$

We also have, by (13.11) and (13.14),

$$b(\varepsilon) = b(\mathbb{E}[\|x(t^*)\|^2]) \leq \mathbb{E}[V(t^*, x(t^*))].$$

Combining the last two inequality results in a contradiction. Therefore, our claim must be true; i.e., the trivial solution $x \equiv 0$ is stable in the m.s. for all $t \geq t_0$. As for the uniformity property, it suffices to choose δ independently of t_0.

To prove the m.s. asymptotic stability property of $x \equiv 0$, we need only to establish attractivity of this solution. Assume that $u \equiv 0$ is asymptotic stable, which implies the existence of $\delta_2 = \delta(t_0)$ and $T = T(t_0, \varepsilon) > 0$, for any given ε, such that

$$u_0 \leq \delta_2 \quad \text{implies} \quad \nu(t, t_0, u_0) < b(\varepsilon), \quad \forall t \geq t_0 + T.$$

Following the same argument of the first part, we choose $u_0 = a(\mathbb{E}[\|x_0\|^2]) \leq \delta_2$ and $\delta_3 < \delta_2$ such that $\mathbb{E}[\|x_0\|^2] \leq \delta_3$. Then,

$$b(\mathbb{E}[\|x(t)\|^2]) \leq \mathbb{E}[V(t, x(t))] \leq \nu(t, t_0, a(\mathbb{E}[\|x_0\|^2])) \leq b(\varepsilon),$$

i.e., $\mathbb{E}[\|x(t)\|^2] \leq \varepsilon$ for all $t \geq t_0 + T$. We have proved that $x \equiv 0$ is asymptotic stability in the m.s. Furthermore, choosing $T = T(\varepsilon)$ leads to the uniformity property. In the following, we illustrate our theoretical result through a numerical example with simulation.

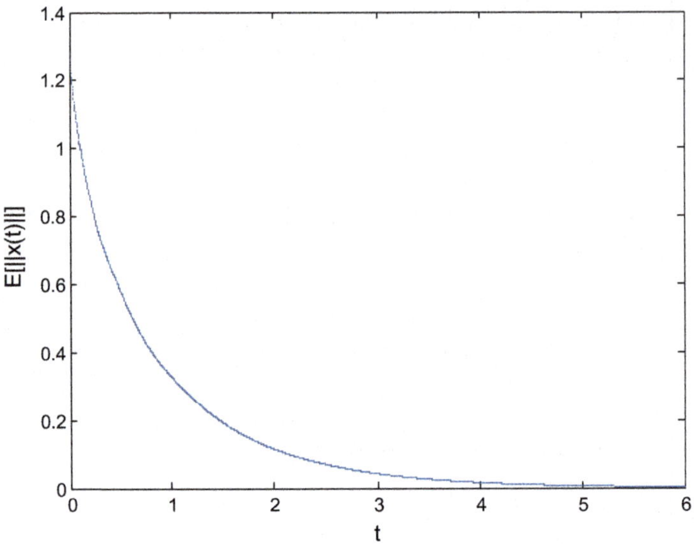

Fig. 13.1 First moment asymptotic stability of $(x\ y)^T = (0\ 0)$

Example 13.1 Consider the following SEPCA

$$\begin{aligned} dx &= \big(-x[\lambda + \theta(x^2 + y^2) + \beta_k x_{\xi_k}]\big)dt + axdW_1, \\ dy &= bydt - x^2 dW_1 + \gamma_{\xi_k} y_{\xi_k} e^{-x^2} dW_2. \end{aligned} \tag{13.15}$$

Taking $V(x, y) = \frac{1}{2}(x^2 + y^2)$ as a Lyapunov function candidate implies

$$\begin{aligned} \mathscr{L}V &\le -(\lambda + \frac{\beta_k^2}{2} + \frac{a^2}{2})x^2 + by^2 + \frac{\beta_k^2}{2}x_{\xi_k}^2 + \frac{\gamma_k^2}{2}y_{\xi_k}^2 \\ &\le \frac{\theta^*}{2}(x^2 + y^2) + \frac{1}{2}\xi_k(x_{\zeta_k}^2 + y_{\xi_k}^2) \\ &= \theta^* V(x, y) + \zeta_k V_{\xi_k}, \end{aligned}$$

where $\theta^* = 2\min\{-(\lambda + \frac{\beta_k^2}{2} + \frac{a^2}{2}), b\} < 0$ and $\zeta_k = \max\{\beta_k^2, \gamma_k^2\} > 0$. Choose $\lambda = 2$, $\theta = 1$, $a = 1$, $b = -1$, $\beta_k = \gamma_k = 1/2^k$ and $a = b = V = \frac{1}{2}\|(x, y)\|^2$. Clearly, the trivial solution of the comparison system is asymptotically stable. This conclusion can be checked with Corollary 12.5, where $w(s) = s > 0$, $\beta_k = \zeta_k$ and $t_k - t_{k-1} = 1$ for any k. We deduce that $(x, y)^T = (0, 0)$ is asymptotically stable in the m.s. Figures 13.1 and 13.2 show the simulation results of the mean and m.s. of the solution.

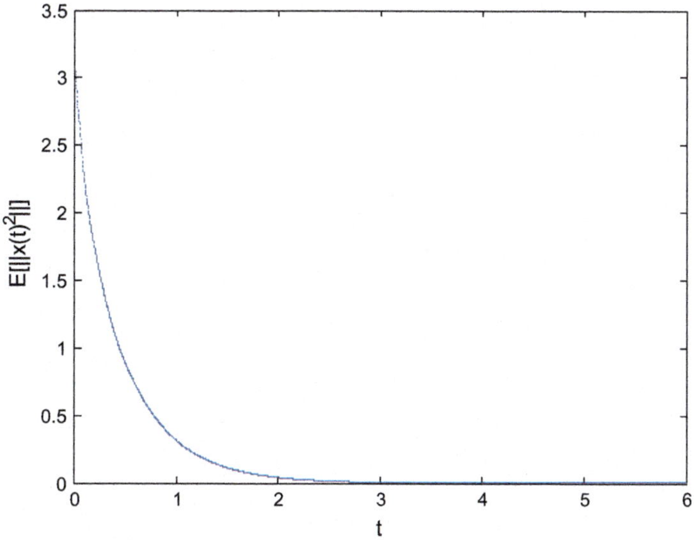

Fig. 13.2 Mean square asymptotic stability of $(x\ y)^T = (0\ 0)$

13.4 Notes and Comments

In this chapter, we have considered systems with SEPCA, which have been treated as switched systems. The material of this chapter is adapted from [1]. Particularly, we have addressed the problems of existence and uniqueness of solutions. Then, we demonstrated some stability properties of the system. As for the existence result, the vector fields have been assumed to be bounded above by some linear growth estimation. Therefore, one can extend this result by considering a nonlinear growth bound. The second part of this chapter has dealt with developing stability results, where we have used the comparison method and Lyapunov function criteria. We should mention that Definitions 13.2 and 13.2, and Borel–Cantelli's lemma and Doob's martingale inequality are taken from [2].

References

1. Alwan MS, Liu XZ, Xie W-C Existence, uniqueness, and stability results for stochastic differential equations with piecewise constant arguments (Submitted)
2. Mao X, Yuan C (2006) Stochastic differential equations with Markovian switching. Imperial College Press

Printed by Printforce, the Netherlands